PRODU
TABLE OLIVES

Stan Kailis and David Harris

LAND
LINKS

© Stanley George Kailis 2007.
All photographs are courtesy of Stanley George Kailis, unless otherwise noted.
All rights reserved. Except under the conditions described in the *Australian Copyright Act* 1968 and subsequent amendments, no part of this publication may be reproduced, stored in a retrieval system or transmitted in any form or by any means, electronic, mechanical, photocopying, recording, duplicating or otherwise, without the prior permission of the copyright owner. Contact Landlinks Press for all permission requests.

National Library of Australia Cataloguing-in-Publication entry
  Kailis, Stanley G.
  Producing table olives.

  Bibliography.
  Includes index.
  ISBN 978 0 643092 03 7.

  1. Olive industry and trade – Australia. 2. Olive – Australia. I. Harris, David John, 1947– . II. Title.

  634.630994

Published by and available from
Landlinks Press
150 Oxford Street (PO Box 1139)
Collingwood VIC 3066
Australia

Telephone: +61 3 9662 7666
Local call: 1300 788 000 (Australia only)
Fax: +61 3 9662 7555
Email: publishing.sales@csiro.au
Web site: www.landlinks.com

Landlinks Press is an imprint of **CSIRO** PUBLISHING

Front cover: Photograph by istockphoto.
Back cover (clockwise from top left): Leccino olives; Olives for sale at a market in Melbourne; *Arbequina* olives; Provencale-style black *Leccino* olives. Photographs by Stanley George Kailis.

Set in 10.5 Minion
Cover and text design by James Kelly
Typeset by J & M Typesetting
Index by Indexicana
Printed in Australia by BPA Print Group

Disclaimer: This manual has been prepared in good faith on a non-liability basis by the authors and should only be used as a guide for the production of table olives. The authors take no responsibility with respect to health and safety problems or commercial loss associated with the use of the manual. Commercial processors are advised to have their processing protocols prepared and checked by a qualified consultant and final products cleared by a food technologist before sale.

Foreword

Table olives available in Australia are either produced domestically or imported from overseas. The Australian and international food industries have recognised the importance of quality assurance and in particular the implementation of internationally recognised quality systems in processing foods. With table olives, quality and safety begins in the orchard and continues to the consumer. As table olives are generally produced either by fermentation, chemical or drying methods, it is essential that those growing or producing table olives be provided with scientifically based technologies with the necessary quality and safety information.

The Australian olive industry is expanding in the production of olive oil and table olives. The vibrancy of the industry is reflected in the large number of plantings that are taking place or are planned for the future. The production of olive oil has been well established in Australia, but the table olive industry is in its infancy. It is estimated that within five to 10 years, table olive production in Australia could replace imported table olives and the country will become an exporter of table olives.

Producing Table Olives, which provides predominantly an Australian perspective on table olives, will have universal application for all table olive producers including the home processor. This reference covers all theoretical and practical aspects essential for the production of safe, nutritious and marketable table olives.

Australia has ideal conditions for growing and processing table olives. The industry is in its infancy and most producers are at the boutique/small-scale level. In a climate where the majority of table olives eaten by Australians are imported, real opportunities exist for a domestic table olive industry. Attention to quality and safety will ensure that Australian table olive producers are in a position to tackle and make inroads into the international export market.

The development of *Producing Table Olives* would not have been possible without support from the Rural Industries Research and Development Corporation (RIRDC) and olive growers in Western Australia, the resources of the Chemistry Centre (WA) and The University of Western Australia. Much of the focus of *Producing Table Olives* has come from table olive workshops conducted around Australia by the authors where substantial interaction with olive growers, processors and consumers has occurred.

We warmly thank Dr Antonio Garrido Fernández (Instituto de la Grasa, Seville, Spain), Dr Vincenzo Marsilio (Istituto Sperimentale per la Elaiotecnica, Pescara, Italy), Dr Louise Ferguson (University of California, Davis, USA) and Mr Jim Smyth (Olive Skills Pty Ltd, Loxton, South Australia) for reading the primary manuscript and providing their valuable comments.

Stanley George Kailis and David Harris

Contents

Foreword..iii

Preface..xi

Acknowledgments..xiii

About the authors..xv

Chapter 1. Table olive perspectives 1
   Introduction 1
   Historical aspects of table olives 2
   Historical aspects of Australian table olives 4
   Cultural perspectives regarding table olives.............................7
   International table olive trade..7
   Australian table olive trade activities 8
   Australian table olive perspectives 9
   Table olive activities in Australia.....................................10
      Australian table olive trade......................................10
      Activities undertaken by Australian table olive growers/processors 11
   Australian table olive production 12
   Australian table olive imports..12
   Australian table olive exports 13
   Australian table olive consumption 13
   Olive growing in Australia..14
   Conceptual approach to the Australian table olive industry 15

Chapter 2. The olive tree *Olea europaea* 17
   The olive tree..17
   Natural history of the olive tree 18
      Natural history of olive production 20
   Principal components of the olive tree..................................21
      Olive leaves 21
      Olive roots...24
      Olive trunk, branches and stems...................................24
      Olive suckers...25
   Annual olive events...25
      Critical water requirements.......................................26

Vegetative growth in the olive 26
Reproductive development in the olive 28
Olive flower fertilisation and fruit set... 31
Olive fruit development and growth 33
   Stages of olive fruit growth 34
Biennial bearing (alternate bearing) and the olive tree......................... 36
Olive fruit 37
Description of olive fruit.. 38
Olive fruit shape and size... 39
General physical features of olive fruit... 40
Olive exocarp (epidermis and cuticle) 40
Olive fruit flesh (mesocarp)... 41
Olive endocarp and embryo 42
Moisture content of raw olives 42
Flesh to stone ratio of olives 43
Oil content of raw olive flesh.. 43
Carbohydrate levels in raw olive flesh... 45
Phenolic substances in raw olive flesh 46
Protein levels in raw olive flesh 48
Pigments in raw olive flesh 48
Mineral content of raw olive flesh... 49
Further components in raw olive flesh... 50
Natural microbiological flora on olive fruit...................................... 50
Changes in olive flesh resistance during maturation............................ 50
Internationally important table olive varieties 51
Important table olive varieties... 52
Selected characteristics for table olive varieties 53
Additional table olive varieties grown in Australia 64

Chapter 3.   Producing quality raw olives 67
Introduction 67
Basic requirements of raw olives... 68
Planning for table olive production 69
Starting to grow table olives 70
   Planning the table olive orchard... 70
   Local government and agency requirements for growing olives 70
   Climatic considerations.. 70
   Chilling hours.. 72
   Heat degree-days and radiation 72
   Frost prone areas and extremely low temperatures......................... 72
   Hot dry areas 73
   Water availability 73
   Site requirements 74
   Other considerations 75
Varietal considerations for table olives.. 76

    Olive pollinating varieties 78
Olive varieties and table olive processing. 78
New varietal considerations for table olives in Australia 79
Establishment of the table olive orchard 81
    Olive orchard design 81
    Soil requirements 82
Olive tree maintenance for quality olive production 90
    Soil management 90
    Fertilisers and manures 92
    Organic manures 93
    Nitrogen fertilisers 95
    Potassium fertilisers 96
    Phosphate fertilisers. 96
    Olive orchard nutrition specifics 97
    Trace elements for olive trees 102
    Fertiliser requirements and monitoring nutritional status 104
    Irrigation requirements for olive production 106
    Olive productivity 109
    Maintenance of the olive tree 110
    Pruning tips for olive trees 112
Stress factors and diseases that can affect olive trees. 112
Pests and disease management 113
Integrated pest management 120
Maturation states for table olive production 122
Harvesting for table olive production. 123
Post-harvest handling of raw olives 126
Undesirable qualities of raw olives for table olives 127
Concluding remarks. 129

Chapter 4.   Table olive processing: general aspects **131**
Introduction 131
Table olive processing methods and varieties 132
Good Manufacturing Practice (GMP) and table olives 133
Water requirements for table olive production 134
Planning table olive processing facilities 135
Functional table olive processing facility 136
General facilities and equipment required. 153
    Protocol for cleaning and sanitising equipment 155
    Protocol for cleaning and sanitising food contact surfaces 155
Table olive processing. 157
Table olive methods and styles 160
Raw olives used for table olive production. 161
Common trade preparations of table olives 161
Specialty table olive products 164
Table olive styles 165

Common table olive methods and products 167
Primary processing specifications for table olives 167
Secondary table olive processing specifications 168
Finished table olive product specifications 169
Generic processing protocol for table olives 169
Acceptance of raw olives by processors. 170
Correct storage of raw olives at the processing facility 171
Raw olives enter the table olive processing line 172
Placement of table olives into processing tanks. 174
General methods for processing table olives 175
Microorganisms relevant in table olive processing 177
Manipulation of microbial activity. 179
Fermentation and table olives 182
Anaerobic fermentation. 183
Environmental considerations with table olive processing. 185
Common methods of preservation for table olives. 186
Packaging of table olive products 187
Pasteurisation of table olives 188
    Alternative procedures. 189

Chapter 5.   Specific table olive processing methods. 191
Processed table olives available in Australia 191
Wholesale table olive trade in Australia 192
Sale of loose table olives in Australia 193
Packaged table olive products in Australia. 194
Processing olives with water (water-cured) 196
    Step-by-step procedure for water-cured olives 197
Processing olives with brine (brine-cured) 198
    Naturally black-ripe olives processed in brine. 198
    Features of anaerobic conditions 199
    Green-ripe olives processed in brine 203
    Turning colour olives processed in brine. 203
    Step-by-step method for processing untreated olives in brine under
    anaerobic conditions 203
    Bruised/cracked olives processed in brine 205
Processing Kalamata-style table olives 206
    Traditional short method for processing Kalamata-style olives. 207
    Long method for processing Kalamata-style olives. 207
Processing olives with lye (sodium hydroxide) 209
    Processing tips for treating green-ripe olives with lye 213
    Step-by-step method for processing green-ripe olives with lye 213
    Potential problems: stuck fermentations and corrective actions 214
    Stuck fermentations and their management 214
Lye treated green-ripe olives without fermentation 217
    Step-by-step procedure for lye treated green-ripe olives without fermentation . . 217

    Step-by-step method for *Olive de Nimes* 218
    Californian/Spanish-style black olives 219
        Step-by-step method for preparing Californian/Spanish-style black olives 220
        Lye treatment 1 220
        Lye treatment 2...220
        Lye treatment 3...220
        Washing step...221
        Colour enhancing step...221
        Brining step..221
        Further steps...221
    Processing dried table olives (shrivelled olives).............................223
    Salt-dried olives 223
        Step-by-step method for preparing salt-dried black olives 223
    Heat-dried table olives..225
        Step-by-step method for preparing sun/heat-dried olives 225
    Ferrandina-style table olives 226
        Step-by-step method for processing *Ferrandina* olives.......................226
    Post-processing table olive operations (secondary processing) 227
    Packing solutions for table olives 227
    Secondary table olive processing..228
        Adding vinegar and olive oil 228
        Recipe for packing solutions with brine, vinegar and olive oil 228
        Adding herbs, spices and marinades to table olives 229
        Recipes..232
        Antipasto..234
        Oil cured olives...234
        Pitting and stuffing table olives.....................................234
        Step-by-step method for stuffing table olives 235
        Olive pastes..235
        Tapenade..237
        Step-by-step method for making tapenade 238
    Final table olive products..240

Chapter 6.   Quality and safety..243
    Introduction 243
    General qualities of packed table olives 244
    Quality and safety evaluation of processed olives 245
    Physical testing/evaluation of processed table olives.........................245
    Physical quality criteria for table olives 248
    Description of physical defects for processed table olives......................249
    Chemical quality criteria for table olives 250
        Brine analysis 250
        Clinistix™ and Diastix™ test strips 251
        Clinitest™ tablets..251
        High pressure liquid chromatography................................251

    Salt refractometer...252
    Salometer 252
    Volhard titration method for sodium chloride estimation 253
    Accurate pH measurements using a pH meter 254
    Titration method with colour indicator (phenolphthalein) 254
    Titration method with pH meter 254
    Evaluation of table olive brines 255
    Table olive flesh analysis 256
Carbohydrate and sugar levels in processed olive flesh........................258
Mineral content of processed olive flesh 259
Microscopic examination of fermentation brines............................260
Microbiological evaluation of table olives 260
Microbiological analysis of processed olive flesh and brines 262
Organoleptic evaluation of table olive products 263
Proposed IOOC organoleptic assessment of table olives 270
Table olive spoilage and deterioration 273
Malodorous deterioration of processed olives...............................277
Packaging and labelling olive products.....................................279
Food-borne disease and table olives.......................................286
    Physical contaminants: foreign matter................................288
    Chemical contaminants...288
    Microbial contaminants 289
Final table olive products 294

Bibliography and additional reading.................................297

Index 322

Preface

The origin of the cultivated olive tree lies rooted to legend and tradition. It started about 5000–6000 years ago within a wide strip of land by the eastern Mediterranean Sea. The oldest reference to the olive tree in the Bible occurs in its first book, Genesis, where the flight of the dove with the olive branch announcing the end of the flood is described. Numerous Mediterranean civilisations – Phoenicians, Greeks, Hebrews and Romans – contributed to expand olive tree cultivation all around the Mediterranean Sea.

Myths, legends, customs, history, literature, cuisine, medicine and economy have always been related to the sacred tree of the Mediterranean. Greek and Spanish cultures were especially linked to it and many published references to the olive tree well illustrate these ties. The National Archaeological Museum in Madrid has a Greek vase dated circa 350 BC, on which is depicted the dispute between Pallas Athene and Poseidon over the name that should be given to what was to become Athens. The goddess made an olive tree sprout from the earth, winning from Poseidon the right to give her name to the recently founded colony of Attica. The olive tree was cultivated in the Guadalquivir Valley by pre-Roman people. Strabo writes that the olive plantations of the Baetica (the actual Andalusia) were admirably tended. The oldest written reference on how to prepare table olives is from the first century text by Lucious Junius Moderatus Columela (42 BC), *De re rustica*. The book is a detailed report on agricultural practices and food processing, and describes various methods of preparing olives for eating. His instructions give an idea of just how old the tradition is in Andalusia: 'Throw into a modius of olives a sextary of mature aniseed and mastic-tree oil and three cyati of fennel seed; in default of this last, the amount of chopped fennel that seems sufficient. Then mix with each modious of olives three heminae of ungrounded toasted salt, place the olives in amphoras and cover them with fennel.' The tradition has continued for centuries and Seville is the largest table olive producer in the world.

During the 15th and 16th centuries, Spanish colonisers spread olive plantings to the New World (South and North America). A book, dated 1530, tells that all ships bound for the Indies should transport at least a few olive samplings: 'Henceforth all ships' captains travelling to the Indies shall take in their vessels the amount of wine and olive plants they see fit, but none shall leave without at least some on board.' Spain was thus contributing to the universal fate of olive culture. The olive tree was introduced to Australia at the beginning of the 19th century.

Myths and pagan or religious customs are also tied to the olive tree and to the oil that oozes from its fruit. The olive was the traditional food of Mediterraneans, and, according to Ovid in his *Metamorphoses*, was also the food of gods. Baucis prepared a meal based on the fruit of the olive for Jupiter and Mercury. The Roman breakfast, or *ientaculum*, consisted of bread with oil and garlic, an ancestral ritual repeated today in so many parts around the Mediterranean Sea.

Olive branches have always been a sign of peace, reconciliation and pacification. In human history, the olive is associated with the noblest acts of society and with tender images; simple yet imposing. They have been used to honour heroes, exalt the virtues of great men, reward excellence in fine arts and symbolise peace. The olive tree has thus been related to the most noble and honourable feelings of the human being. Its fascination still continues all over the world and its current expansion through Argentina, Morocco and Australia is only a new step towards the conquest of new regions, towards a better understanding of civilisations.

In their book *Producing Table Olives*, Professor Stanley Kailis and Dr David Harris are symbolising the combination of tradition, modernity and future development. It represents their particular adaptation of the Mediterranean olive culture from Greece, Turkey, Italy or Spain to new lands as an example of the contribution of classical processes, not only to philosophy or theatre but also to science and technology, that have lasted for centuries. The book may be, in fact, a tribute to ancestors through Professor Kailis' grandmother Konstantinia, who introduced him to the table olive tradition.

Producing Table Olives is a very complete manual for the elaboration of table olive products. Everything is thoroughly discussed and commented on. Step-by-step protocols are given for every subject covered. It introduces everybody into the olive business and guides them through planning the production of olives, starting to grow the olives, selecting olive varieties, establishing and maintaining the orchard, designing the processing plant for any kind of method, style or commercial presentation, and packing the product and controlling product quality and safety. From this point of view the book is useful to all olive growers and processors.

This book represents a very important tool for table olive development in Australia but it also may be useful for producers from other countries or regions because at the end, cultivars and processes are not so different from one place to another.

I am sure that *Producing Table Olives* represents a valuable contribution to the spreading of the olive-tree growing land and table olive production. I hope that time will confirm this impression. In short, some areas of Australia may be, like the actual Andalusia, 'immense silver-green woods of olive trees, a land of light and gaiety, but also of grief and shadow, a land where the olive tree forms an authentic natural park'.

Antonio Garrido Fernández
Instituto de la Grasa
Seville, Spain

Acknowledgments

The authors wish to thank the Australian Rural Industries Development Corporation (RIRDC) and the following people, associations and groups for their valuable assistance in developing this publication.

Olive industry

Australian Olive Association, Pendle Hill, New South Wales, Australia.
Australian Olive Growers and Processors in Australia, Greece, Italy and South Africa.
Australian Regional Olive Associations.
Buffet Olives, Paarl, South Africa.
Costa and Son Olives, Paarl, South Africa.
Luigi and Lena Bazzani and Family, Olea Nurseries, West Manjimup, Western Australia.
David Carr, Frankland R. Olive Company, Fremantle, Western Australia.
Newton and Merryl Ellaby, Northern Rivers Olives, Casino, New South Wales, Australia.
Colin Helliar, Stellar Ridge, Cowaramup, Western Australia.
Mark Kailis, Baldivis Estate, Serpentine, Western Australia.
Graeme and Felicity Legge, 'Kingally', Fosterton, New South Wales, Australia.
Guilio Lombardi, Alta Vista Olives, Harvey, Western Australia.
Karen and Gibb Macdonald, Grace Hill Olives, Cowaramup, Western Australia.
Robert Moltoni, Koorian Olives, Belmont, Western Australia.
Ian Rowe, Fulcrum Olives, Karrinyup, Western Australia.
Jim Smyth, Olive Skills Pty Ltd, Loxton, South Australia.
Susan Sweeney, Horticultural Consultant – Olives, Rural Solutions SA, Adelaide, South Australia.
Dick Taylor, Agriculture (WA), South Perth, Western Australia.
Mark Troy, Inglewood Olive Processing Limited, Woolloongabba, Queensland, Australia.
Sam Vescio, Olive Grower, Chidlow, Western Australia.
Mark and Veronica Westlake, Don Vica Pty Ltd, Fremantle, Western Australia.
Philip Wheatley, Glen Eagle Olives, Wandering, Western Australia.

Scientific collaborators

Daniel Bride, Chemistry Centre (WA), East Perth, Western Australia.
Prof. John Considine, School of Plant Biology, Faculty of Natural and Agricultural Sciences, University of Western Australia, Crawley, Western Australia.

Dr Carlo Costa, Infruitech, Stellenbosch, South Africa.
Dr Luciano di Giovacchino, Istituto Sperimentale per la Elaiotecnica, Pescara, Italy.
Ken Dods, Chemistry Centre (WA), East Perth, Western Australia.
Tom Ganz, School of Plant Biology, Faculty of Natural and Agricultural Sciences, University of Western Australia, Crawley, Western Australia.
Prof. Apostolos Kiritsakis, Technological Educational Institution of Thessaloniki, Sindos, Greece.
Dr Rod Mailer, Wagga Wagga Agricultural Institute, Wagga Wagga, New South Wales, Australia.
Dr Vincenzo Marsilio, Istituto Sperimentale per la Elaiotecnica, Pescara, Italy.
Tom Naumovski, Chemistry Centre (WA), East Perth, Western Australia.
Dr Anh Van Pham, Lecturer Mathematics Department, University of Western Australia, Crawley, Western Australia.
Bruce Youngberg, Chemistry Centre (WA), East Perth, Western Australia.

Food industry

Dr Rex Fletcher, Fletcher Foods, Victoria, Australia.
Elizabeth Frankish, Microserve Laboratories, Claisebrook, Western Australia.
Jan Oldham, Home Economist and Food Journalist, Swanbourne, Western Australia.
Gregory Swan, Peters and Brownes Group, Balcatta, Western Australia.

Document preparation

Blanche Diane Kailis and George Efstathios Kailis, Perth, Western Australia.

About the authors

Professor Stanley George Kailis

Professor Stanley Kailis is Professorial Fellow at the School of Plant Biology, University of Western Australia, and a Fellow of Curtin University of Technology, WA. He holds qualifications in science, pharmacy and teaching and holds a doctorate in science. His antecedents came from the Greek island Megisti, and he was introduced to table olives by his grandmother Kostantinia. His interests focus on the quality aspects of olives. Stan has made presentations on olive growing, olive oil and table olives at national and international forums and to industry groups. He has published numerous research papers in national and international journals. He has conducted many courses and workshops in Australia on olive growing, olive oil and table olive production, organoleptic evaluation of olive products and olive propagation.

Dr David Harris

Dr David Harris is Principal Chemist at the Chemistry Centre (WA) and is section leader of the Food and Agricultural Chemistry Section. He gained a doctorate degree in chemistry, specialising in organic chemistry, in 1976 in Canada. His main interests are research into the organic compounds present in legumes and pulses as well as pasture legumes. Over the last five years he has become very interested in food safety and quality in Western Australia. Working with Professor Kailis for the last five years has aroused a keen interest in table olives and olive oil with regard to the chemistry associated with their production. David has presented papers at a large number of international forums and has published numerous papers in national and international journals.

1

Table olive perspectives

This chapter reviews some of the history of olive growing, with particular reference to table olives and the current features of the international olive trade. Information is provided for olive growers and processors on how the Australian table olive industry is placed within the international table olive market, particularly with respect to export and the local table olive markets. Australia is a significant importer of table olives and there are great opportunities for expanding the Australian table olive industry. This section also examines the regions of Australia in which olives can be successfully grown. Olives are grown commercially in all Australian states and possibly in the Northern Territory. A conceptual model of the Australian table olive industry is given in a schematic form.

Introduction

Table olives are either produced at home or on commercial premises. Home processing of table olives is very popular among Australians, particularly those with Mediterranean or Middle Eastern origins.

Table olives are prepared from the raw fruit of the European or domesticated olive *Olea europaea* L. Hoff. Raw olives are picked when green-ripe, turning colour or black-ripe, depending on the processing style to be used. Raw olives are inedible due to the presence of the extremely bitter glucoside, oleuropein. Processing raw olives to reduce bitterness and make them edible can be undertaken by soaking them in water, brine or dilute alkali, or by drying, salting or heating.

Today the Australian olive industry, which involves both growing olives and processing them into foodstuffs, is vibrant and dynamic. The quantities of olive oil and table olives produced are expected to increase significantly over the next few years. Substantial investment has been made into establishing olive orchards and processing facilities in the major mainland states of Australia using the latest international technology. Many parts of Australia have suitable growing conditions for olives and the olive products produced have gained interest and recognition by international competitors. Two sectors are evolving – table olive production and olive oil production. Drawing upon the rich history of the olive, spanning thousands of years, and current international research, those in the industry have the common objective of producing high quality olive products using the latest technologies. The development of the Australian table olive industry must be considered in a national and international context if it is to reach its economic potential. The success of the Australian table olive industry will depend on capturing a significant proportion of the domestic market, mostly now served by imported products, and the development of international markets.

Currently most Australian table olive enterprises are at the boutique to small-scale levels. A small number of these are processing or planning to process 100 t/year or more with one major processor having a processing capacity of 500 t/year. Capturing a significant segment of national supermarket table olive sales has so far proved difficult. So, many Australian table olive products are sold as specialty lines through food shops, wineries, gourmet centres and delicatessens. Often Australian grown and processed table olives are purchased within wine producing regions, for example McClaren Vale (South Australia), Hunter Valley (New South Wales), Margaret River (Western Australia), Rutherglen (Victoria), Kingaroy (Queensland) and Launceston (Tasmania). Table olives from a small number of processors have penetrated state and national markets. At the international level most olive growing countries have established table olive processing facilities either for domestic production, international consumption or both. Spain and Greece, which have substantial table olive industries, are major exporters of table olives.

Table olives are popular with Australians; however, most table olives that Australians eat are imported from Spain and Greece. These include: Spanish-style green, black Californian/Spanish-style, Greek/Sicilian-style green and black olives, and Kalamata-style black olives. Table olive consumption by Australians is approaching 0.9 kg/person/year and increasing. Those Australians with Mediterranean or Middle Eastern links eat substantially more. Table olives whole, cracked, stuffed, marinated or incorporated into pastes are eaten with bread and cheese, with salads and cold collations and cooked foods. Olives are commonly presented with pickled vegetables, starters, antipasti, hors d'oeuvres or mezedes.

Historical aspects of table olives

The wild olive is native to the Mediterranean area, sub-tropical and central Asia, and parts of Africa. The domesticated olive tree, one of civilised man's first achievements, is of very ancient origin, probably arising at the dawn of agriculture. A strongly held view is that the domesticated olive evolved from the wild olive, *O. oleaster*, which can still be found growing in the Mediterranean basin, especially in Greece and Italy. Wild olives have short branches, small, thick, fleshy, round to oval leaves and small round fruit with a large stone and little flesh. Heavily grazed domesticated olive trees revert to the juvenile vegetative state, having an appearance similar to *O. oleaster* except they do not fruit. These are often referred to as *O. olevaster*. The term 'wild olive' is often loosely used to denote *O. oleaster* and cultural escapes (feral olives).

The domesticated olive is thought to have originated in the Middle East, possibly near Iran, Mesopotamia and Syria, spreading south and west through Palestine and Anatolia (Fig. 1.1) to the rest of the Mediterranean basin initially through the movement and trading activities of the Phoenicians and ancient Greeks. An alternative view is that the olive was also domesticated independently of Syria in Crete at around 2500 BC. Historically, and to the present time, olives have been culturally and economically significant in the Mediterranean and Middle Eastern regions (Fig. 1.2). Although the olive is not indigenous outside these regions, it has been introduced in countries with suitable growing environments: Mediterranean-like climates. The olive was taken to countries such as South America (Chile and Argentina) and North America (Mexico and California) by Spanish missionaries and later by immigrants from the Mediterranean region. Olive trees are now also growing in Australia, South Africa, New Zealand, China, India and Japan. In parts of Asia and north-eastern Australia, with monsoonal-type climates, the boundaries for commercial olive growing are being tested.

Figure 1.1 Ancient olive tree (9–10 metres tall) in the ruins of ancient Acropolis, on the island of Megisti, Greece, opposite Anatolia, Turkey.

Figure 1.2 Table olives at a local market in Turkey.

For thousands of years olives have been an important foodstuff, possibly essential, for the people living around the Mediterranean basin and in the Middle East. Processing methods used by these groups, for example debittering, probably evolved by trial and error. It is unclear when the first olive was eaten or processed. Very ripe fruit of some olive varieties, although still bitter, were probably eaten directly off the tree or off the ground. The process of debittering olives could have developed by drying the fruit in the sun or by soaking in water. Soaking in water as an operation to make food more palatable was well understood by hunter and gatherer communities. As most olives in the Old World grew close to the sea, especially the Mediterranean, the possible use of seawater to debitter olives is understandable. Prolonged storage of raw olives in salt water would have triggered natural fermentation that would have facilitated their debittering and improved palatability. The olives were most likely flavoured with herbs gathered from nearby heaths and hillsides. The early uses of alkaline wood ash to debitter olives and later sodium hydroxide (in the mid 19th century) were the precursors of lye treatments used today.

Through trial and error, and with the development of agrarian communities, table olive processing became well established at the domestic level using traditional recipes that were handed from one family to another and down the generations with or without modifications. Quality was unpredictable due to poor understanding of processing methods, lack of hygienic practice and the absence of control measures, resulting in poor quality products of low economic value. In the last 100 years table olive processing has moved from the village to well-managed, large-scale production centres particularly in southern Europe, northern Africa, Middle Eastern countries and the Americas, especially the USA and Argentina. Countries that are scaling up their table olive activities with substantial international markets in view are: Morocco, Turkey, Argentina and Australia.

Historical aspects of Australian table olives

Australia is at the cusp of being a significant table olive producing country. Olives are not indigenous to Australia but they have been growing in this country since the settlement of Europeans. Here we can only provide a few historic highlights. The Australian olive industry has its roots in the early days of European colonisation when settlers planted the first olive trees some 200 years ago. Australian olive growing had its beginnings in New South Wales where George Suttor, a market gardener, planted the first olive tree that was brought in as part of a shipment of plants consigned by Sir Joseph Banks in 1800. Several years later, in 1805, the well-recognised settler farmer John Macarthur planted an olive tree on his property 'Elizabeth Farm' at Parramatta, New South Wales. The first commercial olive growing endeavour in Australia is credited to this pioneer. Over time, further plantings were made in other parts of Australia. The olive rootstock for these early Australian plantings was brought predominantly from Spain, France, Portugal and Italy.

In the 1800s in South Australia, Sir Samuel Davenport, a protagonist of commercial olive growing, was making olive oil and providing olive rootstock for others. Olive rootstock was also being distributed freely to farmers by government agencies in a number of diverse regions in Australia. The Botanic Gardens in Sydney became one of the major distribution centres for free olive cuttings. Olives were promoted for olive oil rather than as table olives and excess olives were often relegated to animal feed. The establishment of experimental olive orchards in most Australian states, for example Dookie College in Victoria and the Wagga

Figure 1.3 Olive trees planted in 1856 at the Benedictine Community at New Norcia, Western Australia.

Wagga Experimental Orchard in New South Wales, is evidence that olives were considered a potential crop species for Australia.

In Western Australia the first olive trees, brought from Cadiz (Spain), were planted by James Drummond in the Western Australian Government House gardens in 1831. These trees were used for propagating rootstock that was distributed to settler farmers in that state. One of the oldest consistently worked olive groves in Australia was planted by the Benedictine monks at their Mission in New Norcia, Western Australia, in 1856 (Fig. 1.3). Again, apart from the olives consumed by the monks, their olive activities were directed to olive oil production.

Most of the early Australian olive orchards were not seriously worked or maintained. They were abandoned because colonists and settlers, predominantly from Britain and Ireland, were unaccustomed to eating olives or olive oil. In fact, olive oil was seen as a medicine or fuel rather than a food. Furthermore, over time, as other sources of oil for cooking and heating such as whale oil and animal fat became available, olives were of little use. Nevertheless, several groves were actively worked, providing limited quantities of table olives for eating and olive oil for medicinal purposes. Neglected early settler olive orchards, particularly in South Australia, possibly provided the seeds that have resulted in significant numbers of feral olive trees invading the natural environment and farming landscapes (Fig. 1.4). In Australia, feral olives are also called 'wild olives' and opinions vary as to their worth. Environmentalists want them removed, producers are ambivalent about their usefulness, and food writers and gourmets believe that the olives have a unique taste.

More recently, olive groves were planted at Robinvale, Horsham, Dimboola and Hopetoun in Victoria as well as at Bordertown and Palmer in South Australia. The bygone Oliveholme Company at Robinvale in Victoria, operating during the 1970s and early 1980s, was the largest table olive producer in Australia at that time, producing 1000 t/year. Oliveholme had 220 ha of irrigated olive trees planted in 1946. Most of these

Figure 1.4 Feral olives along the roadside in South Australia.

were bulldozed due to a lack of technology and a slump in olive oil prices. The Robinvale olive operation has now been revived to around 100 ha of olive trees with new trees complementing some of the original trees.

Renewed interest in olives and olive oil in Australia occurred after World War II, in the late 1940s to 1960s, with the influx of southern European immigrants and more recently those from the Middle East. These new settlers, who consumed table olives as a regular foodstuff, had little impact on the local industry as they preferred to either produce their own table olives using traditional methods or buy olives produced in their mother country out of loyalty. Such immigrants, however, have had profound effects on many aspects of the Australian lifestyle, including eating habits. Traditional food in Australia, which had an emphasis on meat and dairy products, has now in part been replaced by Mediterranean and Middle Eastern foods based on salads, vegetables, pulses, fish and olive oil. Over the past 10 years as Australians have recognised the health benefits and cultural significance of Mediterranean cuisine, consumption of table olives and olive oil has increased markedly.

François Solente established an olive grove in the 1960s, Kasbah Olives, dedicated to table olive production near Loxton, South Australia (Fig. 1.5). Green olives, predominantly of the *Verdale* variety, were produced for the fresh fruit market and commercial processing. Loxton became a centre for table olive processing and today has the largest processing plant in Australia. The South Australian Olive Company, now trading under the Viva brand, took over Kasbah Olives. Unfortunately, the Solente grove was recently removed, possibly because of the lack of commercial interest in the *Verdale* variety. Evidence of the demise of the *Verdale* variety in Australia is that another significantly sized *Verdale* olive orchard in South Australia has recently been top grafted to produce *Kalamata* variety olives.

Today, many small-scale olive growers/processors are directing their activities to table olive production rather than olive oil. Activities such as stand alone table olive processing

Figure 1.5 *Verdale* olive grove, Loxton, South Australia.

facilities in the Hunter Valley and *Kalamata* variety olive orchards are examples of industry direction. Furthermore, around Australia, medium to large-scale olive growers are directing 10–20% of their crop to table olive production.

Cultural perspectives regarding table olives

With innumerable types of table olives available to consumers (see Chapter 5), cultural practices strongly influence the popularity of particular preparations. Amongst Spanish consumers green Sevillian-style olives are popular, whereas in Greece consumers prefer naturally black-ripe olives in brine or salt-dried olives. With Italian consumers, sun-dried and heat-dried olives are popular, as are those produced by traditional methods, for example Sicilian, Ligurian, Castelvetrano and Ferrandina olives. Californian-style black olives, produced by chemically treating green-ripe olives with lye, are the principle table olives produced in the USA. Generally, around the middle and eastern Mediterranean and the Middle Eastern regions, table olives are processed by fermentation in brine. Olives produced by the latter method are increasing in significance because of consumer preference for Greek-style black olives, Kalamata- and Sicilian-style olives rather than lye treated olives.

International table olive trade

World raw olive production is around 13–18 million tonnes per year depending on the season. About 10% of these olives are processed into table olives. It is likely that as the Australian olive industry matures, at least a similar proportion of the annual Australian olive crop will be processed into table olives. If Australia follows similar trends to California and South Africa, the proportion of raw olives processed as table olives could increase markedly, to possibly 30–40% of the annual crop. Countries significantly involved with table olive production are listed in Table 1.1.

Table 1.1. Countries with significant interests in international table olive trade
EC, European Community.

Major producing countries	Major importing countries	Major consuming countries	Major exporting countries	Major consumers per person
EC	USA	EC	EC	Cyprus
Spain	EC	Spain	Spain	Syria
Italy	France	Italy	Greece	Jordan
Greece	Germany	France	Morocco	Spain
Syria	Brazil	USA	Turkey	Israel
USA	Canada	Turkey	Argentina	Palestine
Morocco	Russia	Syria		Lebanon
Turkey	Romania	Egypt		Greece
Egypt	Australia	Brazil		Italy
Algeria		Algeria		
Argentina		Jordan		

World table olive production has expanded by nearly 50% over the past 15 years and is now around 1.6 million tonnes per season. According to recent International Olive Oil Council (IOOC) statistics (2005), the European Community (EC), Morocco, Syria, Egypt, Turkey and the USA accounted for over 80% of the world table olive production over that time. Consumption has also steadily increased at the same levels. People living in EC countries in and around the Mediterranean basin, in Middle Eastern countries and the USA consumed most of these olives. Reasons given by the IOOC for increased world table olive consumption include: extra availability, better presentation, enhanced quality, population increases and the increased purchasing power of consumers.

There is a delicate balance between table olive production and consumption. Any shortfall in world table olive consumption and production is generally met from olives carried over from the previous season.

World table olive exports have been climbing steadily and by 2005 reached around 450 000 t/season. However, fewer table olives relative to the quantities produced are exported or imported. Around 30% of world table olive production is exported mainly by the EC, Morocco, Turkey and Argentina. Major table olive importing countries, accounting for 65% of the world average, are the USA, EC, Brazil and Canada. Australia currently accounts for 3% of world table olive imports, mainly from Spain and Greece. Australian table olive imports account for 5–6% of the European Community table olive exports. Increases in Australian table olive production, consumption and export will, therefore, impact more on the EC table olive export sector than total world production.

There is some concern, particularly in the northern Mediterranean, as to the marginality and abandonment of olive farms. Abandonment of olive farms has also been linked to the decoupling of industry subsidies and with the socioeconomic characteristics of production. Without subsidies, variable production costs cannot be covered where olives are grown on inclined, shallow and low fertility soils. Abandoned olive orchards also pose a negative threat to the environment because of the risk of harbouring diseases. Efficient olive growing now practised within some of the traditional olive growing regions, as well as new growing regions, is advantageous to Australia, and elsewhere.

Australian table olive trade activities

Activities surrounding Australian table olives and trade involve both locally produced olives as well as imports. Most table olives imported into Australia are: Spanish green (Sevillean) and black (Californian/Spanish) styles, where lye has been used during processing; Greek/Sicilian-style green and Greek-style black olives; and Kalamata-style black olives (Figs 1.6 and 1.7). Specialty olives, including green olives in marinade (Australia), *Picholine* (France), *Bella di Cerignola* (Italy), Ligurian (Italy) and *Arbequina* (Spain), though of lesser commercial importance, have potential for the boutique sector.

In general, Australian boutique and small-scale producers of table olives favour natural processing methods. Olive varieties such as *Manzanilla*, *Sevillana*, *Jumbo Kalamata* and *Verdale* are processed as green olives in brine, whereas naturally black-ripe olives of the *Manzanilla*, *Kalamata* and *Volos* varieties are processed in brine or by the traditional water soaking methods. Larger scale Australian table olive processors produce similar products to boutique and small-scale producers except that some also treat green

Figure 1.6 Canned imported olives: *Kalamata*, Greek-style black and Spanish-style black. (Photo: Andritsos Brothers, Western Australia.)

Figure 1.7 Barrels of *Kalamata* olives imported from Greece. (Photo: European Foods, Western Australia.)

olives with lye, which result in Sevillian-type olives. Although lye treatments speed up processing there are significant drawbacks such as environmental issues, and low retail prices because of the highly competitive international market. No Australian table olive processor is currently producing Californian/Spanish-style black olives.

Australian processors mostly market their table olive products to the food services industry in bulk through specialty food outlets and, to a limited extent, supermarkets. The olive industry, like the wine industry, has adopted tourism as a major strategy in marketing table olive products with other foods in regional Australia. Uptake of Australian table olives by national supermarkets has been slow due to high prices, low levels of availability, lack of specific products such as pitted and stuffed olives, and the existing buying habits of consumers. It is expected that existing imports will persist for some time because of traditional trading patterns of importers, wholesalers, retailers and consumers. Possible competition from other southern hemisphere table olive producers could impact on the Australian table olive trade.

Currently, most Australian produced olives come from South Australia and Victoria. This will change when recently planted orchards in Western Australia, New South Wales and Queensland reach commercial production levels. Furthermore, a number of new Australian table olive enterprises are making substantial investment in table olive production facilities.

Australian table olive perspectives

Australia is now emerging as a significant table olive producing country. Australia has the physical resources, horticultural infrastructure and food processing expertise to support a modern table olive industry. Australian regions where olive varieties suitable for table olive processing exist, and regions with existing or emerging table olive processing activity, are indicated in Fig. 1.8.

To date, only relatively small quantities of Australian table olive products have been available, predominantly for domestic markets. Commonly available processed table olives in order of importance include: *Manzanilla, Verdale, Picholine* and *Kalamata*. Sustained growth of the table olive industry will depend on advances in efficient production and market development. A recent Australian Survey undertaken in 2005

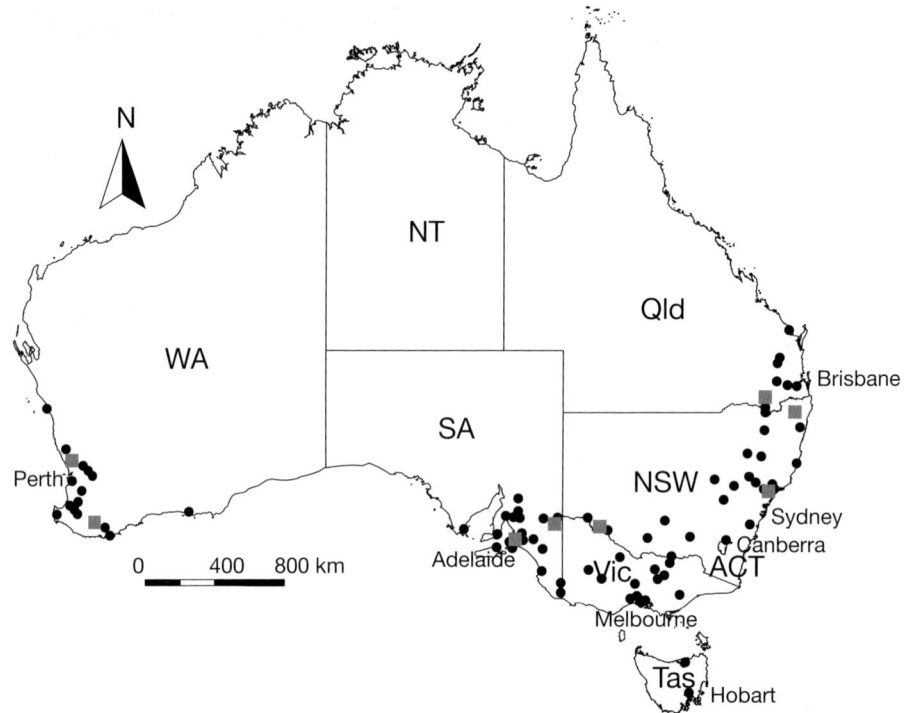

Figure 1.8 Potential table olive producing sites in Australia based on existing or proposed processing activities and sites with suitable varieties. Squares indicate emerging table olive centres.

indicated that at least 192 groups, represented in all states (but not the territories), were involved in growing and/or processing table olives. Most of these groups are also involved in olive oil production. The number of table olive/processor groups in each state is as follows: New South Wales, 68; Victoria, 50; Western Australia, 27; South Australia, 22; Queensland 19; and Tasmania 6.

Major barriers to Australian table olive production are high harvesting costs, the lack of production expertise and, perhaps, the lack of production facilities. There have been numerous attempts to establish a national table olive industry in Australia. Most have failed due to the inability to compete commercially with imported product or the lack of acceptance of the Australian product. As important is the isolation in which early promoters for table olives operated. The establishment of a network of industry bodies working towards common goals has now addressed this problem. Also, national and international food expos and fairs have provided venues to showcase Australian table olive products.

Table olive activities in Australia

Australian table olive trade

The Australian table olive trade involves all olive products – olives and pastes – including those made in Australia and those imported from overseas. This trade can be considered within three broad categories: *(1)* Australian grown olives, processed into table olive products and packed in Australia; *(2)* olives and table olive products imported into Australia, either packed in the country of origin or repacked in Australia before sale; and *(3)* olives and table olive products mainly imported that undergo secondary processing such as stuffing or marinating before sale.

Table olives packed in Australia and readily available in Australian supermarkets are depicted in Fig. 1.9.

Figure 1.9 Table olives packed in Australia and readily available in Australian supermarkets. (Photo: Swansea Supermarket, Western Australia.)

Activities undertaken by Australian table olive growers/processors

Activities undertaken by Australian table olive growers/processors can be conceptualised as follows:

- growers sell unprocessed olives to the fresh food market or to third party processors;
- growers vertically integrate their olive activities, for example orchard to plate, selling their products to restaurants, pizza bars and small-scale food outlets; and
- growers process their olives and sell them to third party processors for secondary processing and packaging.

Australian olive growers and table olive processors have a strong interest in using natural processing methods such as brine fermentation and salt/heat drying methods rather than processing with lye. This follows similar interest worldwide. With lye treatments larger amounts of water and energy are used than with traditional methods involving brine fermentation.

Specialty olives include the *Jumbo Kalamata* variety, a large olive processed at the green to turning colour maturation stage, and Ligurian-style olives using *Taggiasca* or *Frantoio* varieties. Australian growers/processors and third party processors are also undertaking secondary value-adding operations, producing specialty products such as olives in marinades, tapenades, olive pastes and antipasti.

Growing/producing table olives involves all operations from producing the fruit to delivering the olives to the processing facility. Primary processing involves any operation (soaking, fermentation, lye treatment or heating) where the olives are debittered and preserved for bulk storage.

Secondary processing involves increasing the organoleptic and financial value of the olives by adding herbs, spices, vegetables and marinades, destoning without stuffing, or destoning and stuffing with materials such as anchovies, peppers, cheeses, nuts, garlic or onion. Marketing involves packaging olive products for wholesale and sale to end users such as pizza producers, café and restaurant operators, food processors and consumers.

Australian table olive production

Australian table olive production over the period 1990–1991 to 2003–2004 has nearly doubled. Australian table olive production for the 2002–2003 season was about 4000 tonnes and can be expected to increase significantly as Australians eat more olives and the industry develops. By 2013 Australia has the potential to produce 18 000–45 000 tonnes of table olives per year, an amount that far exceeds the current level of Australian consumption. Following the resurgence of the Australian olive industry in the 1990s, there is great interest by growers and processors to produce truly Australian table olive products. Table olive processing in Australia is generally small scale by world standards, with one large plant sited in Loxton, South Australia, capable of handling up to 500 t/year. New processors are still at the pilot stage, with some planning to produce 100 t/year. Their current production is mostly less than 10 t/year, with some up to 30 t/year. There are many small-scale boutique table olive operations that are processing less than one tonne of table olives per year.

Commercially available Australian table olive products include green and black olives – whole, varietal and marinated – as well as olive pastes and tapenades. Varieties commonly processed into Australian table olives include *Kalamata, Volos, Sevillana, Manzanilla, Barouni* and *Verdale*. Olives marinated with combinations of olive oil, lemon, vinegar, garlic, chilli and other herbs and spices are popular. Olives packed with fetta cheese, capers, sun-dried tomatoes, garlic, chilli and pimento extend the olive into the gourmet market. The indicative scale of Australian table olive production as new plantings come on line is shown in Table 1.2. Information is given on the possible size of the operation and, for boutique to small-scale growers/processors, the number of olive trees required.

Half the Australian olive groves are operating at the boutique to small-scale levels with orchard areas of 6 ha (approximately 1500 trees) or less.

Table 1.2. Indicative scale of Australian table olive enterprises
*, based on an average seasonal crop of 25 kilograms/tree.

Size of operation	Capacity in tonnes/season	Olive trees required*	Equivalent orchard area
Boutique	<5	<200	1 ha or less
Small-scale	5–<100	200–<4000	1–16 ha
Medium-scale	100–<500	4000–<20 000	16–80 ha
Large-scale	>500	>20 000	>80 ha

Australian table olive imports

Australia imports table olives mainly from the European Community: Spain, Greece and to a lesser extent Italy. Imports are of the order of 12 000 t/year and valued at A$40 million, representing an approximate doubling of imports since 1992–1993. From 2002–2004 over 40 000 tonnes of table olives were imported into Australia with around 40% and 50% coming from Greece and Spain respectively. Over the same period, the cost of imported olives was around A$144 million. By extrapolation, if the amount of

Australian produced olives were included, the retail trade value of table olives in Australia is around A$190 million. An interesting feature of imported table olives is that the average price for imported Greek table olives is around A$4/kg compared with around A$2.5/kg for Spanish olives. This differential in price is reflected also at the retail level. It should be noted that most of the Greek olives are produced by traditional methods whereas those from Spain are treated with lye. Of imported table olives, around 45% are only provisionally processed and, therefore, not suitable for immediate consumption so need to be repacked before sale to consumers. The rest, made up of green and black olives, whole, pitted or stuffed, are in a form ready for consumption.

The marked increase in imports is a significant indicator of the popularity of olives in Australia and a clear signal to Australian growers and processors of demand. Imported table olives (black and green) are sold in bulk by wholesalers to the food services industry or repackaged by third parties into consumer-sized quantities to be sold at retail outlets. Most imported olives are: Spanish-style green and black, Greek-style black and green, and Kalamata-style black. These are included in all types of products, such as whole olives with or without marinades or stuffings, olive pastes and other pickled vegetables.

Australian table olive exports

The export value of Australian table olives is relatively low: currently less than $1 million per year. Those exported are either locally produced, or imported and then exported from Australia, mainly to Asian and Pacific countries. Although some boutique table olive producers are exporting table olive products to the UK and USA, a number of large-scale olive enterprises in Australia are exploring opportunities for more substantial levels of table olive exportation.

Australian table olive consumption

Consumption of table olives by Australians has increased steadily since the 1990–1991 trade year from around 0.4 kg/person/year to 0.9 kg/person/year, making them one of the largest consumers per capita outside those living in and around the Mediterranean. On a per capita basis, Australians consume more table olives than Americans or Canadians. However, up to the 1999–2000 trade year, increased table olive consumption by Australians was accounted for by imports from Spain and Greece. Since then, Australian produced table olives have contributed to the available trade pool.

Table olives are eaten with bread, cheese and wine, included in salads and incorporated into cooked foods and pizzas. Australians appear to consume mainly marinated and/or stuffed Spanish-style green olives and black olives in brine. Where Australian consumers have a choice they prefer to eat olives processed in brine, green or black. Olive varieties commonly eaten by Australians are: *Manzanilla*, *Sevillana*, *Hojiblanca*, *Kalamata* and *Konservolia* (*Volos*). Some oil varieties processed as table olives, such as *Frantoio*, *Leccino*, *Arbequina* and *Taggiasca*, are also available in small quantities.

Olive growing in Australia

Since 1995 there has been intense interest in the commercial potential of an Australian olive industry. Olive orchards have been established at numerous centres in southern and eastern Australia. Initial interest was for olive oil production but interest in an Australian table olive industry is increasing at all levels from boutique to larger scale enterprises. Areas and regions in Australia with olive activities are listed in Table 1.3.

Table 1.3. Areas/regions in Australia with an interest in table olives or that have varieties growing suitable for table olive production

Regions	Some areas/sites with olive varieties suitable for table olive production	Commonly planted varieties suitable for table olives
Australian Capital Territory	Canberra	*Kalamata, Manzanilla*
New South Wales	Bowral, Branxton, Broke, Buronga, Casino, Cessnock, Cowra, Darlington Point, Dungog, Forbes, Gol Gol, Goulburn, Hunter Valley, Inverell, Maitland, Mittagong, Moss Vale, Mudgee, Orange, Riverina, Rylstone, Tamworth, Vacy, Wagga Wagga, Wentworth	*Barnea, Correggiollo, Frantoio, Jumbo Kalamata, Kalamata, Leccino, Manzanilla, Mission, Nevadillo Blanco, Paragon, Picholine, Picual, SA Verdale, Sevillana, UC13A6, Volos*
South Australia	Fleurieu/Kangaroo Island, Strathalbyn, Meadows, Myponga, McLarenVale, Willunga, Normanville, Yankalilla, Cape Jervis, Eyre Peninsula, Port Lincoln, Yorke Peninsula, Kadina, Kulpara; Northern Region – Balaklava, Jamestown, Clare, Riverton, Southern Flinders Adelaide Plains, Virginia; Adelaide Hills, Balhannah, Stirling; Limestone coast – Coonalpyn, Parilla, Keith, Pinnaroo	*Barnea, Barouni, Jumbo Kalamata, Kalamata, Manzanilla, Picual, SA Verdale, UC13A6*
Tasmania	Devoit, Devonport, Evandale, Forth Valley, Hobart, Legana, New Norfolk, Launceston, Sandy Bay, Tinderbox	*Barnea, Frantoio, Kalamata, Manzanilla, Paragon, SA Verdale*
Victoria	Alexandria, Armstrong, Ararat, Bellarine Peninsula, Beaufort, Boort, Euroa, Elmhurst, Geelong, Goulbourn, Great Western, Heywood, Irymple, Kaniva, Macedon, Mansfield, Mildura, Mornington Peninsula, Pomonal, Redcliffs, Robinvale, Rutherglen, Stawel, Strathboogie, Sunbury, Sunraysia, Tinzanna, Wangaratta	*Barnea, Correggiollo, Frantoio, Kalamata, Manzanilla, Nevadillo Blanco, Paragon, Picual, Sevillana, Volos*
Western Australia	Baldivis, Bindoon, Bridgetown, Chittering, Dandaragan, Denmark, Donnybrook, Dongara, Esperance, Geraldton, Gingin, Frankland R., McAlinden, Margaret R., Mogumber, New Norcia, Northampton, Swan and Avon Valleys	*Correggiollo, Frantoio, Kalamata, Leccino, Manzanilla, Sevillana, UC13A6, WA Mission*
Queensland	Bunya, Dalby, Deepwater, Doonan, Frazerview, Gatton, Gin Gin, Grantham, Gympie, Harlin, Inglewood, Ipswich, Jandowae, Kingaroy, Millmerran, Mount Mee, Murgon, Ravensbourne, Stanhope, St George, Taroom, The Caves	*Correggiollo, Manzanilla, Nab Tamri, Paragon, Picual, Nevadillo Blanco, Kalamata, Sevillana, UC13A6*
Australia overall	Commonly planted olive varieties suitable for table olive production: *Manzanilla, Verdale, Picholine* and *Kalamata*. Varieties commonly used for olive oil production that can also be used for table olive production are: *Frantoio, Leccino, Picual* and *Barnea*.	

The Australian olive industry is fragmented, with most growers having between 500 and 5000 trees. Such olive groves are often found within popular wine growing regions around Australia. Major olive orchards (50 000 trees or more), accounting for around 70% of all planted olive trees, have been established in New South Wales, Victoria, South Australia, Queensland and Western Australia. Although accurate statistics on plantings or productive olive trees in Australia are unavailable, one estimate is around 8.5 million trees.

Based on an average seasonal production of 25 kg of olives per tree, the potential table olive crop is around 20 000 t/year representing around 1% of the world production of table olives.

Although table olive processing and olive oil processing are two quite different operations, at the orchard level this delineation is not as clear. For the olive grower two issues are important to table olive production: variety and fruit quality. Few orchards are specifically set up for table olive production. What is clear is that olive growers at the boutique/small-scale end of the industry producing olive oil are also interested in table olives to extend their options. In this case, dual-purpose olive varieties, for example *Manzanilla*, *Mission*, *Leccino*, *Kalamata* and *Konservolia* (*Volos*), should be planted. Owners of several olive orchards in South Australia and Western Australia have targeted the table olive industry by planting substantial numbers of *Kalamata* variety olive trees.

The following varieties that are suitable for table olive processing are commonly found at various sites in Australia: *Frantoio* (*Correggiollo*, *Paragon*), *Jumbo Kalamata* (*Giant Kalamata*), *Leccino*, *Manzanilla*, *Nevadillo Blanco*, *Mission*, *Sevillana*, *UC13A6*, *Barnea* and *Barouni*. Varieties with limited distribution suitable for table olive products are: *Nab Tamri*, *Boothby's Luca*, *Lecqure*, *Picual*, *Hardy's Mammoth*, *Azapa* and *Boutillan*.

Conceptual approach to the Australian table olive industry

In summary, the Australian olive industry is presented as a conceptual model in Fig. 1.10. The grower can undertake primary and secondary processing, package the products and then sell them to various customers, including wholesalers, representatives of the food services industry sector, the retail sector or direct to consumers. This sequence is likely with small-scale producers. Alternatively the grower can pass the olives on to third parties to complete processing and marketing. Some growers may sell fresh olives directly to processors or send them to the fresh market sector for sale to small-scale processors or home processors.

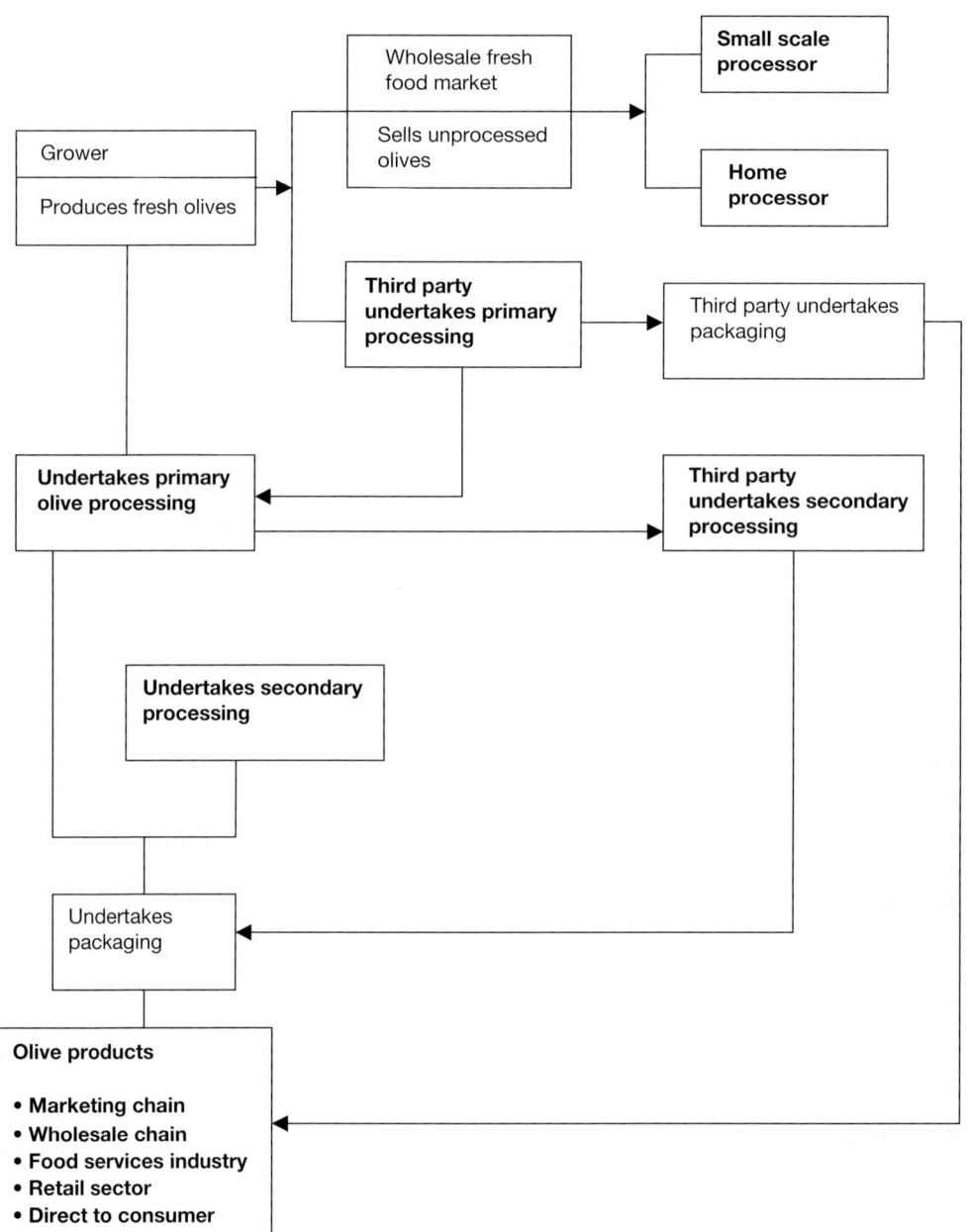

Figure 1.10 Australian table olive industry – a conceptual model.

2

The olive tree *Olea europaea*

This chapter gives a detailed account of the olive (*Olea europaea* L.) life cycle and requirements for its successful cultivation. A timetable is provided that shows the flowering and fruiting stages of the olive and the various orchard operations that have to be undertaken during the annual cycle of the olive tree. It sets out the stages involved in fruit set and problems that can occur with alternate fruit set. The properties of olive fruit and its progress to different stages of ripeness are discussed. This is important as olives are harvested at different stages of ripeness for the production of different table olive products. The physical characteristics of the olive fruit, such as size, skin pressure and Flesh:Stone ratios are examined. The general chemical composition of fresh olives are discussed, such as minerals, protein, oil (fat), carbohydrate, sugars, fatty acids, phenolic compounds and pigments. A detailed segment on olive varieties provides information on those suitable for processing as table olives. Diagrams are given of key table olive varieties.

The olive tree

The olive *Olea europaea* L. is a medium-sized evergreen shrub/tree (Fig. 2.1) that grows and fruits well under a Mediterranean climate, such as occurs around the Mediterranean Sea, southern Australia, parts of New Zealand, the Americas and South Africa, and to a lesser extent in other countries. Olive trees, depending on the variety and growing conditions, can grow to a height of 15–20 m. However, for commercial production they are best trained and pruned to a height of 3–6 m depending on the harvesting method and the available technologies. Olive trees bear fruit that are bitter mainly due to the glycoside, oleuropein.

Olive trees need sufficient winter chill to ensure fruit set and a long hot growing season to ripen the fruit, particularly if naturally black-ripe olives are required. Frosts, especially during the spring flowering time, and hot dry winds during flowering

Figure 2.1 A typical well-managed *Manzanilla* olive tree with fruit.

(anthesis) and fruit set are detrimental to olive production. The limiting factor for the reproduction of olives (fruit production) is determined by winter temperatures.

Olive trees are long lived. They will survive under highly unfavourable situations, but if given luxury they will thrive but not necessarily be the most productive. Olives will grow under tropical conditions but rarely will experience sufficient chilling hours for flower and fruit production. The olive is subject to the biennial bearing phenomenon as experienced by many fruit trees, requiring horticultural intervention to even out the crop from season to season. There are varietal differences in the degree of biennial bearing. Olive fruit are classified a drupe fruit, as they are fleshy with the seed enclosed in a stone.

Natural history of the olive tree

As a prelude to the natural history of the olive tree, a brief but relevant discussion on olive tree propagation follows. With modern olive growing, the planting stock is derived from two principal propagation technologies: self-rooted clonal cuttings, or by grafting buds or scions (shoots that are cut especially for grafting or planting; in the case of olives, these are pieces of one-year-old wood with one node: two leaves and buds). In the latter case, the desired variety is grafted onto self-rooted cuttings of an easily rooting variety, for example *Frantoio* or *Manzanilla*, olive seedlings, or wild olive rootstock such as the feral olive tree.

Micropropagation of olive trees is an emerging technology that has the potential to provide disease-free planting material. All propagating operations should be undertaken under sanitary conditions and precautions made to reduce the risk of transferring diseases such as Peacock Spot, Olive Knot, Verticillium Wilt and Phytophthora Root Rot.

Self-rooted clonal cuttings. Self-rooted clonal cuttings are produced by dipping the bottom end (1–1.5 cm) of individual, or bunches of, semi-hardwood olive cuttings in IBA (Indole Butyric Acid) hydro-alcoholic solution (4 g/l–4000 ppm), gel (3 g/l of IBA) or powder (IBA 3 g/kg) for 5–10 seconds. Treated cuttings are then inserted into trays of moist perlite placed in a tunnel with intermittent mist (constant moisture) and bottom heat, or a closed frame that provides constant humidity. Some propagators use a combined solution of IBA and NAA (Naphthyl Acetic Acid). Semi-hardwood olive cuttings are prepared from one to two-year-old vigorous shoots (6–8 mm diameter and 10–15 cm in length) collected in spring or late summer. The cuttings should have seven to eight nodes. A square cut is made just below the bottom node, and an angled cut is made above the sixth node. Leaves are left at the top two nodes and the remaining leaves removed. Three different protocols for propagating olive trees with pre-treated semi-hardwood cuttings inserted in trays of perlite are as follows: *(1)* Placing them in intermittent mist with the basal temperature of the cuttings kept at 30°C for 15 days, then gradually decreasing to 18°C until the callus and roots develop. *(2)* Placing them in intermittent mist with the basal temperature of the cuttings kept at 21°C (or even up to 26°C). *(3)* Placing them in closed frames and maintaining warm humid conditions (with no extra heat or mist). For all methods, rooting takes about two to three months.

Some table olive varieties, for example *Kalamata* and *Sevillana*, are difficult to propagate as self-rooted cuttings, and are generally grafted onto rootstock. Poor rooting of *Olea europaea* cuttings is possibly related to lignification of cell walls; that is, they turn woody. This presents a mechanical barrier to the emergence of roots or results in the release of chemical inhibitors in the tissues. Soaking cuttings in water at pH 8.5 can improve strike rates.

Once struck and the roots are of sufficient length, the rooted cuttings are removed from the perlite and planted into 3 litre pots with a well-draining potting mix containing peat, perlite, compost, lime and nutrients. After several months of growth, the young olive trees are allowed to harden in a shade house and are then ready to plant out in the orchard. Most nurseries sell olive trees to commercial olive growers when they are 12–18 months old. Younger and older olive trees are also available.

Hardwood cuttings can also be used to propagate olive trees. Micropropagation techniques have also been developed for the olive to provide rapid multiplication of new varieties as well as to produce relatively disease-free olive trees. Also, existing olive trees can be reworked by grafting them with more desirable cultivars, for example *Frantoio* to *Kalamata*. For more detail on olive propagation, particularly grafting and budding, the reader is directed to the sister publication, *Olive Propagation Manual* (Fabbri et al. 2004).

Olive seedlings. Olive seedlings are produced by germinating olive seeds. Seedlings are often found growing naturally under olive trees from fallen seed. Seedlings are not clonal, so they will display differences from the parent trees. Seedlings will produce fruit; however, as they remain in a juvenile phase for a long time, this may take up to 10 years. Once productive, the olive seedlings behave in a similar way to clonal olive trees. Olive seed germination is maximised by physically or chemically removing the flesh and oil from the stones, then storing these in a moist inert medium, such as sand or vermiculite, at 10–13°C for one to two months in the dark before planting in well-draining soil in cold frames, trays or in the open. This procedure is known as 'stratification'. Stratification of olive stones possibly reduces their abscisic acid content. Abscisic acid is a germination inhibitor within the seed coat or embryo. In some cases, nipping the stone at the radicle end (pointed end of the stone from which the root emerges) or completely removing the seed from the stone facilitates the germination process. Germination rates, which range from 5–90%, can vary with variety, the maturation state of the olive and from one tree to another. If seeds are placed in water, the non-viable ones will float. Discarding faulty stones can improve the 'apparent' germination rate.

Seedlings v. self-rooted cuttings. There have been many debates between traditional growers and scientists as to which technology produces the best commercial olive tree, for example one that is early bearing, productive, has good fruit quality or resistance to stress (climatic, edaphic or biotic). However, there is insufficient scientific evidence supporting any perceived advantage of using either method of propagation technology.

Propagation by grafting or budding onto rootstock. Difficult-to-root olive varieties such as *Sevillana* and *Kalamata* are mostly grafted onto clonal rootstock such as *Manzanilla* or *Frantoio*. In the past seedlings were used; however, such a practice produces trees of variable quality. Clonal rootstock is propagated by the method indicated above. A common method used is to cut one-year-old rootstock so as to leave about 10 cm of stem. A scion of the variety to be grafted is prepared by cutting a piece of one-year-old olive wood with one node/two leaves. The tip of the scion is cut into a sharp wedge shape. The wedge of the scion is inserted into a vertical cut at the top of the rootstock so that it comes in contact with the actively growing cambium to ensure a successful graft. The rootstock and scion should be of the same diameter so that the cambium layers of both come into contact. The scion and rootstock are then held firmly together with grafting tape or an elastic band. Grafted trees are generally kept in the nursery for a further year. Autumn and spring are the best times for wedge grafting. A bark grafting technique can also be used when the rootstock is of greater diameter than the scion.

Budding involves removing a shield-shaped piece of bark that includes an undifferentiated bud. The rootstock is prepared by making a horizontal slit then a vertical slit in the bark along the stem, of dimensions that can accommodate the scion bark. The bark on the rootstock is then carefully separated from its wood and the scion bark and the bud is inserted and the bark closed over. Note that the leaves of the rootstock are retained at this point. The bud is held firmly in place with an elastic band so that the bud graft does not desiccate (dry out). As the bark on the rootstock needs to separate easily from its wood, the best time for bud grafting is in mid-spring. After the bud has taken, the rootstock is cut back to above the graft so that the bud can grow and develop.

More elaborate grafting and budding methods are available and the reader is directed to the *Olive Propagation Manual* (Fabbri *et al.* 2004) for more detail.

Grafting or budding onto rootstock can have a number of advantages, such as: propagating difficult-to-root varieties; increasing vigour; controlling the size of the tree (for example to produce dwarfing); and influencing the tree's ability to ward off stresses, such as pests and diseases, drought and salt.

Research in California has shown that clonal rootstock of *Oblonga* induces a dwarfing effect on *Manzanilla* and greater vigour with *Sevillana*. The use of rootstocks for advantage has not been the general trend in commercial olive growing and requires more research, particularly when developing planting material for intensive olive orchards.

Natural history of olive production

Trees derived from all types of propagation technology follow a similar life cycle modulated by variety and growing environment. The cycle can be divided into four phases.

Unproductive phase. In the first few years after planting, the olive tree grows in a vegetative phase, producing leaves, branches and roots. This phase ends when flowering commences. This can take two to five years after planting a 12 to 18-month-old tree, depending on variety and growing conditions. Under favourable growing conditions the olive tree grows quickly during this phase. Seedlings are unproductive as they are in a juvenile phase that could last up to 10 years.

Early productive phase. During this phase the olive tree is productive, flowering and fruiting, but it could take 8–10 years to reach significant production. The first crop is small and the olives are often larger than expected. Once cropping stabilises, olives are generally smaller, particularly with heavy crops.

Commercial productive phase. Maximum productivity occurs during this phase with the biennial bearing patterns influencing annual yields. While well-managed trees can be commercially viable for 50 years or more, rejuvenation through radical pruning every 10–20 years will ensure healthy productive trees. Replacement with new olive trees is another option.

Declining productive phase. During this phase productivity falls, the olive tree structures deteriorate and the commercial value of the tree declines. Some historic olive trees in traditional olive growing countries such as Israel and Greece are reported to live for 2000 years or longer. Neglected olive trees, which can grow 10 m or more in height, often revert to a more vegetative state and produce only small quantities of fruit.

Principal components of the olive tree

Like other evergreen trees, the olive is made up of leaves (vegetative growth), a trunk with branches and stems, roots (both structural and functional) and the reproductive structures (flower buds, flower inflorescences, flowers, fruit and seed).

In principle the olive tree is like a continuous set of tubes, embedded in the ground (roots) and extending above the ground (trunk, branches and stems) where expanded surfaces (leaves) capture the sun's energy (photosynthesis) and participate in the uptake and release of the gases carbon dioxide, oxygen and water vapour (gas exchange). The warming of leaves by sunlight, and to a lesser extent the effect of air temperature and wind, increases water evaporation and water loss (transpiration) from leaves, thus creating a negative pressure that draws water into the tree through the roots then through the xylem vessels to the leaves. Water flow up the xylem channels is also facilitated by capillary action. The olive tree requires both nutrients and substances synthesised (synthates) within its structures, such as carbohydrates (sugars), by photosynthesis. The movement of dissolved sugars including sucrose in the phloem vessels is not energy dependent. Sugars are passed into the phloem near photosynthetic sites and delivered where required.

Grown under normal environmental conditions, most of the nutrients, dissolved in soil water, are accumulated by the roots and carried in the water stream throughout the olive tree. Transport of synthates from leaves and other photosynthetically active tissue occurs as an aqueous solution that moves through the phloem.

Olive leaves

Olive leaves have a number of functions including photosynthesis and transpiration. They are relatively small and dark green on their upperside (adaxial side) and grey-green on their underside (abaxial side). They have a protective coating (the cuticle) and their underside surface has specialised pores, the stomata, interspersed amongst overlapping peltate hairs that modulate transpiration and water loss. Olive leaves can also take up water and nutrients (the basis of foliar sprays) and lose nutrients through leaching by rain and dew. Old, senescent leaves turn a bright yellow/orange colour before they fall. This should not be confused with nutritional related chlorosis or frost damage.

Transpiration. Water is taken up by the roots and transported up to the leaves and most is lost to the atmosphere through transpiration. Water evaporates from the surfaces of mesophyll cells within the leaf into intercellular spaces (essentially air pockets). These spaces are linked to the outside of the leaf through the stomata. Water lost via the leaves passes through the stomata as water vapour. A continual supply of water via the xylem to the leaf maintains the transpiration process. Stomata have a number of functions including the egress of water vapour and intake of carbon dioxide into the leaves for use in photosynthesis. As gases like carbon dioxide and oxygen are moving in and out of leaves through stomata, water is also lost via transpiration. Stomata can respond to water stress within the tree by opening and closing. Structurally each stoma consists of two guard cells that control the size of the opening, hence the amount of water loss. Stomata open when guard cells become turgid (cells are plump and full of water and there is no water stress) and close when limp (when water stress is present), preventing water loss. Guard cells remain turgid through active pumping of ions, predominantly potassium, that helps maintain a high solute concentration within the cell.

It has been estimated that mature productive fruit trees can lose 15–20 tonnes of water per year. It has been suggested that transpiration may have physiological functions such as mineral absorption and translocation, or even reducing leaf temperature. In reality it may just accompany efficient gas exchange so when stomata are open water vapour is lost.

Because of this constant loss of water and a dependence on a constant water stream for growth and survival, water must always be available to the roots. Under extreme hot, dry conditions transpiration slows, so water losses are reduced allowing the olive tree to survive. If such conditions are prolonged, growth and productivity decrease. Olive leaves are replaced every two to three years, usually in spring, when the new growth spurt occurs.

Photosynthesis. Photosynthesis is the process by which green plants use the energy of sunlight to produce sugars. Carbon dioxide and water, the raw materials for photosynthesis, enter leaf cells and the products (sugars and oxygen) leave the leaf. Photosynthesis requires the involvement of chlorophyll pigments, complex organic molecules containing magnesium.

The process of photosynthesis occurs predominantly in leaves (but also in other green tissue including green olive fruit). It is undertaken by specialised structures within mesophyllic cells, called chloroplasts. The upper leaf surface absorbs more photosynthetically useful light than the undersurface. Relative to many other species, olive leaves have a lower photosynthetic activity, related to their thickness and the low density of photosynthetic sites. Many of the chloroplasts are distributed in the inner palisade layers, so receive low levels of light. Also, depending on the size and shape of the olive canopy, understorey leaves and those within the canopy receive less light than leaves on the outer canopy surface, markedly reducing photosynthesis to <30% compared with fully sun-exposed leaves. Furthermore, extensive shading inhibits the growth of new nodes and retards flowering. Hence the importance of developing olive trees with an open canopy and the optimum tree spacings.

Photosynthesis consists of a series of metabolic reactions occurring in three stages: *(1)* The absorption of light energy by chlorophylls that drives a number of chemical reactions resulting in the production of the high energy compound ATP (adenosine triphosphate); *(2)* the use of ATP to organically fix carbon; and *(3)* the regeneration of chlorophyll pigments.

Stage 1. The first stage, which is light dependent, needs the direct energy of sunlight (or equivalent light source) to make the energy carrier molecules utilised in the second stage. In the first stage, light initiates photochemical reactions involving chlorophylls a and b (green pigments) that absorb violet-blue (435–438 nm) and orange-red (670–680 nm) wavelengths. Chlorophyll b can also absorb light from the green part of the light spectrum, greatly increasing the amount of sunlight that can be harvested. The absorbed light energy is transformed into plant useable chemical energy (high energy ATP). When light strikes chlorophyll molecules, electrons are excited to a higher energy state and through a series of reactions the energy is transferred (lost from the chlorophyll) via an electron transport system to $NADPH_2$ and ATP. Water is also split in the reaction, releasing oxygen as a by-product. More precisely, during the first stage water acts as a hydrogen donor to Nicotinamide-Adenine Dinucleotide-Phosphate (NADP) forming $NADPH_2$ and releasing oxygen gas.

$$NADP^+ + H_2O \rightarrow NADPH + H^+ \rightarrow (NADPH_2) + \tfrac{1}{2}O_2$$
$$ADP + Pi \rightarrow ATP$$

Stage 2. In the second stage (termed the 'light independent' stage) a series of enzyme-catalysed reactions use the ATP from Stage 1 to drive the fixation of the carbon of CO_2 (carbon dioxide) into organic molecules such as sugars.

The overall reactions for photosynthesis can be summarised as follows:

$$6CO_2 + 12H_2O \xrightarrow[\text{Chlorophyll } (ATP/NADPH_2)]{\text{Light}} C_6H_{12}O_6 + 6H_2O + 6O_2 \text{ (gas)}$$

$C_6H_{12}O_6$ is the chemical formula for monosaccharide hexose sugars such as glucose and fructose. These two sugars can combine to form the soluble disaccharide sugar sucrose ($C_{12}H_{22}O_{11}$). Energy can also be stored as the insoluble polysaccharide starch formed from the polymerisation of many glucose molecules. These compounds accumulate in the leaves during photosynthesis and are drawn upon by the tree for its growing requirements and the synthesis of other carbon based organic compounds such as amino acids, fatty acids, nucleic acids, hormones and cofactors.

Similar reactions occur in developing olive fruit, where much of the organic carbon is converted to sugars and to fatty acids of different chain lengths then into triacylglycerols, especially triolein, a major component of olive oil. Some residual sugar, relevant to table olive processing, persists at all stages of olive fruit maturation.

Sugars → organic carbon → fatty acids (oleic acid) → triacylglycerols (e.g. triolein) → olive oil

Research in Italy suggests that olive fruit growth and oil synthesis are both stimulated by the availability of synthates from leaves and by exposing the fruit to light.

Stage 3. During this stage, the chlorophyll molecules that have lost an electron/molecule during Stage 1 are regenerated, otherwise the pigment would become deficient in electrons.

Olive roots

After three to four years the olive tree forms a fascicular root system that grows with age. Roots anchor the olive tree and absorb water and nutrients from the soil. Olive trees grown without irrigation have a root system that mirrors the branch system and extends outwards slightly further than the spread of the branches. With severely pruned olive trees growing under dryland conditions, the root system stretches much further. Lateral roots can grow up to 12 m long, indicating that the olive tree explores a large volume of soil for nutrients and water. The bulk of the feeding roots of the olive tree are within a metre of the soil surface. Generally, water uptake occurs over the first 1.2–1.7 m of soil depth. In heavy and poorly aerated soils olive tree roots are concentrated near the soil surface in contrast to those growing in lighter soils where roots grow much deeper. Also, large roots are commonly linked to particular branches, acting almost independently, which has nutritional and disease implications. This accounts for why one limb succumbs to Verticillium root disease when its accompanying root is infected. Olive roots associated with irrigated olive trees tend to gravitate towards the water source.

Root function. Most of the water absorbed by the olive tree is by the root hairs. Because the root hairs have a higher concentration of dissolved minerals than the soil solution, they are always turgid (plump and swollen with water) because water tends to move into them. Mineral ions are taken up by specific mechanisms even against large concentration gradients. At night when the relative humidity at the leaf surface may be near 100%, there may be little or no leaf transpiration. Many of the uptake processes require energy that is derived by respiration through the breakdown of carbohydrates and the absorption of oxygen. Roots derive their carbohydrates from photosynthesis and from starch stores in the bark and roots. In the absence of transpiration, water does not travel upward in the xylem. As a result, ions continue to accumulate in the roots also drawing water into the root hairs by osmosis. The consequence of this process is the movement of water up the xylem.

The growth of the olive tree depends on the growth of its root system. Plant bioregulators, called cytokinins, are produced in the root tips. These are drawn up through the transpiration stream to branches and stems where they promote stem growth and bud formation. Root growth is reduced during heavy cropping years, hence for table olive production the crop should be thinned out. Root trimming also reduces tree growth, a positive or negative factor depending on the objective.

Olive trunk, branches and stems

The olive trunk, branches and stems support the plant and conduct water, mineral nutrients and a number of organic substances upward to the leaves and fruit and downward to the roots through specific transport systems, the xylem and phloem. The xylem is the principle region for the upward movement of water, nutrients and some organic compounds. The phloem is mainly involved with the downward flow of synthesised plant food and other organic compounds produced in the leaves. As indicated previously, shoots arise from leaf buds on the stems in spring, reaching maximum length after full bloom. Shoots tend to grow upwards, but bend in more pendulous varieties such as *Frantoio* and *Pendolino* and in overgrown olive trees.

The woody parts of the olive tree (stems, branches, trunk and roots) consist of an outer layer of bark with an inner core of xylem (wood). The phloem, which is found below the outer layer of bark, is the tissue that transports synthates (carbohydrates) produced by leaves to the olive fruit and leaves. Both the bark and phloem are able to store carbohydrates, including starch, that can be released when there is active growth. Between the wood and bark is the cambium layer made of active tissue that rapidly divides, forming new wood on the inside and new bark on the outside. The net result is an increase in the diameter of the tree. The interior part of the trunk and stems is predominantly water-conducting tissue (xylem) transferring water and nutrients to transpiring leaves.

Olive suckers

Because olive trees are basitonic they send out vigorous shoots, called suckers, from the base of the trunk. Olive suckers are of two types. In grafted trees the suckers are from the rootstock, unless the grafts are buried below the surface when planting. In clonal trees the suckers are identical to the 'parent' and so can be used to propagate the original variety.

Suckers should be removed regularly as they compete for resources with the main stem or trunk. If clonal suckers are used for propagation, either a sucker with roots or a semi-hardwood cutting can be used. As olive suckers have some juvenile characteristics, the propagated olive tree may take a number of years to yield fruit.

Annual olive events

A series of annual biochemical, physiological and anatomical events occur in the productive olive tree around which olive orchard activities revolve. Information presented in Table 2.1 is indicative of the annual events for olives growing in the southern and northern hemispheres. Timing may vary according to the growing conditions and variety, but the sequence of events does not change.

In spring and early summer, new shoots develop (the spring flush). Ideally 30–50 cm of growth for each shoot is required to support a good crop of olives in the following season, depending on the variety and maturation stage. Early in summer, flower buds are induced on this new growth. In late summer to early autumn, flower bud initiation takes place and undifferentiated flower buds begin to develop. After a period of vernalisation (chill) in winter, flower bud differentiation takes place in late winter where the bud starts to form into the inflorescence. This is followed by anthesis in mid to late spring. Note that at the same time as the olive tree is commencing to produce fruit it is also competing with vegetative structures for water and nutrients.

Table 2.1. Annual events associated with the olive tree for table olives
*, January in the northern hemisphere.

Southern hemisphere	Jan.	Feb.	Mar.	Apr.	May	Jun.	Jul.	Aug.	Sep.	Oct.	Nov.	Dec.
Northern hemisphere	Jul.	Aug.	Sep.	Oct.	Nov.	Dec.	Jan.	Feb.	Mar.	Apr.	May	Jun.
Season	Summer		Autumn			Winter			Spring			Summer
Vegetative growth	Slow growth			Active: autumn flush		Dormancy: rest period			Active: spring flush			
Flowering	Flower bud induction		Flower bud initiation			Vernalisation: average July* temperatures of 10–12°C		Buds differentiate • Vegetative • Floral	• Anthesis • Pollination • Fruit set • Fruit development			
Fruit	Development and growth					Olive fruit have reached their maximum size						
				Ripening period Green → black → over-ripe				Fruit drop Unharvested fruit can reduce the following season's crop				
				Skin and flesh colour change								
Stone	Hardens											
Orchard operations	Undertake leaf analysis	Add nutrients		Commercial harvesting period Green → turning → black colour				Prune and add nutrients	Copper spray		Soil analysis if required	
Irrigate	Irrigate (essential period for table olives)			• In Mediterranean-like climates this period is covered by natural precipitation. • Otherwise, for regular cropping, periodic/regular irrigation may be required.					Irrigate depending on soil water holding capacity			

Critical water requirements

A lack of water has negative consequences on photosynthesis, respiration and carbohydrate levels in olive trees. Olive trees require sufficient water at critical times such as the vegetative, reproductive and fruit growth phases. Olive trees are able to withstand some degree of water deficiency because of their xeromorphic habit; that is, their ability to withstand drought, and an adaptation that allows significant amounts of water to be absorbed through their leaves from the atmosphere and by their roots via the soil. Consequences of lack of water or water stressed conditions include: reduced flower formation, incomplete flowering, low fruit set, increased biennial bearing, decreased shoot growth, decreased fruit size and fruit shrivel.

Vegetative growth in the olive

In Australia, olive trees have a vegetative growth spurt starting in spring where both apical and some lateral buds from the previous season elongate (Fig. 2.2). Unlike reproductive growth, vegetative growth does not require chilling to be initiated. Shoot elongation correlates with rising temperatures, with the rise in daytime temperatures and longer days in spring starting the vegetative cycle. Depending on the ambient temperatures, growth rates vary with a slowing of growth at low (<12°C) and at high temperatures (>30°C). Olive leaves may be photosynthetically active all year round if there are no significant temperature falls in the growing environment. Slowest growth rates occur in winter when the olive undergoes a relatively dormant period. In areas experiencing high summer temperatures growth rates are bimodal. In summer growth rates slow, and as the temperatures fall in autumn a second vegetative growth spurt occurs. In colder regions growth rates are unimodal, peaking in summer. Hence, olive trees growing under colder conditions are generally less vigorous than those growing in warmer regions (Fig. 2.3).

Figure 2.2 New vegetative growth on *Olea europaea* at the beginning of spring.

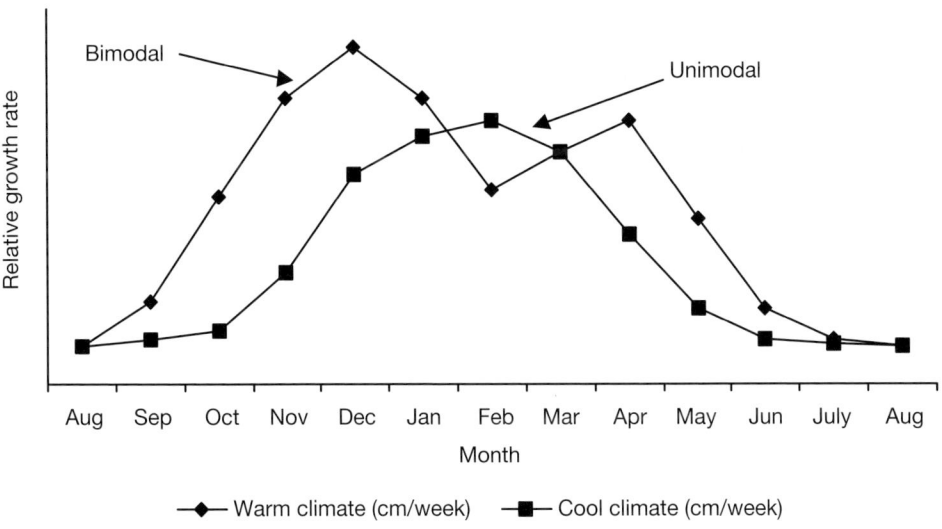

Figure 2.3 Indicative olive vegetative growth rates in the southern hemisphere over the year in hot and cool climate areas. (After IOOC *World Olive Encyclopaedia* 1996.)

At temperatures >35°C as experienced in warmer olive growing areas, photosynthesis is generally impaired. However, olive varieties acclimatised to high temperatures have the advantage of undertaking significant photosynthetic activity at temperatures up to 40°C.

A characteristic of trees, including the olive, is the presence of buds on the growing stems. These buds are active growing locations on the plant where new cells are created and are the starting point of leaves (vegetative growth) and flowers (reproductive growth). There are two types of buds: axillary (lateral) and terminal (apex) buds.

Figure 2.4 Undifferentiated axillary buds at the base of the leaves of *Olea europaea*.

Axillary or lateral buds. These buds are located just above the base of each leaf (Fig. 2.4). In the olive tree, leaves and buds occur along the branches and stems, generally in pairs, at regular intervals. The attachment of a pair of leaves and buds is called a node. The internodal length (the distance between two nodes) is influenced by the variety and growing conditions. For example *Manzanilla* olives have shorter internodes than *Olea Mission*. Axillary buds are undifferentiated but can develop and grow as side shoots that may eventually form branches, or differentiate into flower inflorescences and form olive fruit.

Terminal or apical buds. At the tip of a branch or stem of an olive tree is a terminal bud. As well as performing a similar function to axillary buds, terminal buds control the growth of the shoot. By deliberately removing terminal buds, the shape of the olive tree can be manipulated. Apparently intact terminal buds produce the hormone indole acetic acid that moves backward through the stem either slowing or stopping the growth of axillary buds. This results in orderly and regulated growth by axillary buds and as the shoots age they gradually assume the shape of the olive tree. If the apical buds are removed when training and pruning olive trees, or by animals, this control is lost and axillary buds begin to grow.

Reproductive development in the olive

Olive trees bear small, white to cream-coloured flowers with four petals borne as inflorescences on lateral buds of one-year-old wood. Flowers rarely develop from apical buds or from two-year-old shoots. Although relying on vegetative growth, the reproductive growth phases (flowering and embryo development) are partly independent of the vegetative cycle. The flowering processes are climate dependent.

Olive trees bear two types of flowers: perfect flowers containing both male and female parts, and imperfect (staminate) flowers with only stamens. Anthers are relatively large, producing copious amounts of pollen, hence facilitating wind pollination. Olive flowers open before their pollen is released from the anthers, facilitating cross-pollination before self-fertilisation is possible. Most olive varieties are self-compatible (self-fertile), but some are self-incompatible (self-sterile), and others in between.

Perfect flowers. Perfect flowers have both female (pistil) and male (stamens anthers and pollen) components. They have a large round green pistil, short style, and a green ovary and have the potential to develop into olive fruit.

Imperfect (staminate) flowers. These are male flowers with stamens, anthers and pollen. The pistil is small and rudimentary and unable to develop into olive fruit. Pistillate flowers have no stamens and are rare in the olive.

The number of flowers/inflorescences ranges from 10–35 depending on the variety. Pollen from both perfect and imperfect flowers is generally viable. In most cases only one mature fruit develops from each inflorescence, with only a few varieties with small fruit setting more than three fruits per panicle, for example *Koroneiki*. Flowering is not synchronous and maximum anthesis can take two to three weeks after the commencement of flowering.

The flowers are mainly wind pollinated and many are self-pollinating. Fruit set is generally improved by cross-pollination with other varieties. Self-incompatibility occurs with some varieties, such as *Leccino* and *Pendolino*, whereas others show incompatibility with other varieties, for example *Barouni* and *Sevillano*, and *Manzanillo* and *Mission* (Californian). High temperatures can also cause incompatibility.

The reproductive phases of development can be summarised as follows: flower bud induction, flower bud differentiation (Fig. 2.5), flower bud formation (Fig. 2.6), flowering (anthesis) and pollen release, pistil growth after pollination (Fig. 2.7), fruit set, fruit and embryo development (Fig. 2.8), and fruit growth and maturation.

Figure 2.5 Differentiated flower buds in *Olea europaea*.

Figure 2.6 Flower bud formation in *Olea europaea*.

Figure 2.7 Anthesis and pistil growth after pollination in *Olea europaea*.

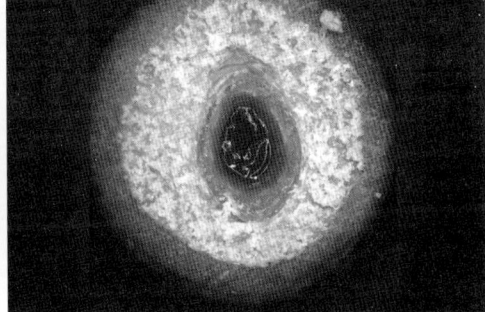

Figure 2.8 Fruit and embryo development in *Olea europaea*.

Temperature is the most significant environmental factor that limits olive growing. Critical temperatures are those that favour survival, provide sufficient chilling for flower bud induction, and sufficient heat for growth and development of the olive. Such temperatures are met in regions between 30° and 45° latitude, north or south, and to a lesser extent in areas with suitable microclimates, such as high altitudes at lower latitudes. Light is also important for flowering and extensively shaded olive trees do not produce flowers or bear fruit. If there is a reduction in light, the percentage of flower bud induction and differentiation falls.

Flower formation. Flower formation is a two-stage process consisting of flower bud induction and flower bud differentiation. Flower bud induction, a process depending on the olive's metabolic processes, is thought to occur in summer whereas flower bud differentiation requires vernalisation, a period of chilling, or day/night temperature fluctuations. During induction, biochemical and physiological changes occur in the bud with no obvious morphological changes. During differentiation early anatomical changes are observed, such as a flattening of the bud and the formation of flowers and inflorescences. The differentiation process, from the early changes to the formation of inflorescences, occurs 40–60 days before anthesis. In order to have some degree of flowering, winter temperatures must be less than 14°C for a 14-day period. A dormancy period of two months during the coldest part of the year, with average daily temperatures less than 10°C, favours flower bud differentiation at levels that will ultimately yield commercial size crops. In Australia, an average July temperature of 10–12°C (northern hemisphere January) or less provides sufficient chilling hours for most varieties to give effective cropping.

Flowering is most prolific when temperatures fluctuate between 2° and 15°C. During winter, induced flower buds accumulate sufficient chilling hours to break flower bud dormancy. Chilling requirements are variety dependent with some requiring few chilling hours, for example *Azapa* and *Kalamata*, and others, such as *Sevillana*, requiring up to 1200 hours. Varieties acclimatised to warm winters need fewer chilling hours. In years when productivity is high or harvest is late because of hormonal changes, the chilling requirements may increase.

Research undertaken at the University of California indicated that inflorescence and subsequent fruit production in the olive was, in general, directly related to the amount of winter chilling and that olive trees maintained in a warm greenhouse failed to flower. Furthermore, it was found that *Sevillano* and *Ascolana* required the maximum amount of chilling; *Mission* (Californian), *Criolla*, *Barouni* and *Manzanillo* developed some flowers with intermediate amounts of chilling, but many more with full amounts of chilling; and *Azapa* and *Rubra* produced substantial numbers of flowers with minimum chilling.

With the use of isoclimatic data, the authors believe that olive varieties originating from southern Greece, Crete, southern Italy, Sicily, the west coast of Turkey, Cyprus, Lebanon, Israel, Palestine and North Africa, which require little winter chilling to flower and fruit, would be suitable for growing in Australia, South Africa and the USA.

During the dormancy period, olive trees can tolerate short periods of frost (–6°C). Lower temperatures, such as –12°C, severely damage aerial parts of the olive, greatly reducing productivity, and possibly survival. Prolonged periods where environmental temperatures are <7°C or >15°C may impair flowering, while temperatures >20°C for several weeks may prevent release of flower buds from dormancy. Extrapolating from northern hemisphere data, differentiated buds that lead to flowering in late spring can be between three months (the autumn growth of same year) and 11 months old (the spring growth of previous year). Flower buds develop into groups of flowers called panicles or inflorescences.

Reproductive bud differentiation and associated morphological changes (inflorescences) occur just before flower buds emerge in spring. The presence of leaves and active photosynthesis are required for flowering, hence olive trees need direct sunlight to flower and fruit effectively. High light intensities increase flower bud differentiation, so attention needs to be paid to tree spacing when planting olive trees, avoiding excessive shade and applying strategic pruning. For flower bud development, spring warming is required with optimum temperatures being 15°C or more. Low spring temperatures can prolong flowering by several weeks, briefly delay flowering or even hinder cross-pollination. If the temperatures are abnormally high in spring, flower development, and therefore productivity, can also be impaired.

Inflorescence development and flowering. Depending on the variety and growing conditions, the development of inflorescences begins early in spring around 30 days after differentiation and may continue for 40–60 days or even longer in warmer climates. Flower buds and inflorescences are sensitive to spring frosts and if temperatures fall to <10°C, pollination rates are reduced. To develop a higher proportion of perfect flowers (see below), adequate light (avoid shade), moisture and nutrients are required. Shading reduces flower differentiation whereas after flowering shading can induce a high degree of sterility with ovary abortion. Leaf shading or defoliation (loss of leaves) may also inhibit flowering. The application of foliar nitrogen and boron when the buds begin to swell ensures that vascular connections are established between the stem and fruit, securing the fruit. This is important for prevention of fruit drop. Full bloom occurs in mid-spring. The flowers remain open (anthesis) for a few days and open flowers can be observed on an individual tree for one to two weeks depending on environmental temperatures. Open flowers release pollen from their anthers during the hottest part of the day.

Careful selection of varieties for hot climates using local knowledge or isoclimatic data therefore becomes critical for successful olive production. Where environmental temperatures fall below 0°C for prolonged periods, it can be too cold for commercial olive production. Furthermore, high humidity at flowering results in flower drop and increased susceptibility to fungal disease such as Sooty Mould.

In summary, early flower drop occurs with low pollination levels, nutrient deficiencies and lack of water. Late flower and fruit drop occurs when there is a lack of water and nutrients, the crop load cannot be supported with the available resources or the crop is attacked by pests.

A productive olive tree can produce over half a million flowers of which less than 5% develop into mature olive fruit.

Olive flower fertilisation and fruit set

Olive flowers are mostly wind cross-pollinated requiring a continuous source of pollen over the time that the stigma is receptive. Although pollen grains have been shown to travel up to 7 km from their original source, pollinating varieties and/or multiple varieties (about four to six) in an olive orchard are best included to increase yield. While olive flowers do not produce nectar, insect pollination by honeybees occurs but is less efficient than wind pollination. As some varieties show self-incompatibility (that is, they are infertile), the dominance of cross-pollination by the olive suggests that the mechanism of wind pollination has probably developed to maintain persistence of the species.

Note: Like all pollens, olive pollen can trigger allergic reactions in susceptible people in the form of skin irritations, hayfever or asthma. Workers in olive groves need to take appropriate precautions, such as wearing gloves and taking suitable medication to minimise these effects.

A pollen grain is a microspore that germinates (begins to grow) to form a male gametophyte made up of the pollen grain and a pollen tube. Pollination is the transfer of pollen grains from the stamens of the flower to the ovules (seed precursors). The female reproductive organ of flowering plants is the carpel. Normal olive fruit have two carpels, each containing two ovules, but only one of the four ovules develops into a seed (so there is one seed/stone). Occasionally two seeds are formed and in other cases there are none (parthenocarpy, see below).

The process of fertilisation starts when a pollen grain (male) attaches to the stigma of a perfect flower. A slender pollen tube emerges from the germinating pollen grain on the stigma, growing through the style (a slender part of the female organ) and the micropyle (a small opening in the cells enclosing the potential seed) enabling the two sperm nuclei to enter the embryo sac. One of the sperm nuclei fertilises the egg, which develops into the embryo (new plant), while the second combines with two polar nuclei eventually forming a starch filled endosperm, a food storage reserve for when the seed germinates. Some researchers believe that pollen tube growth is faster with cross-pollination than with self-pollination under the same temperature conditions. If pollen tube growth is too slow, the embryo sac degenerates so no fertilisation occurs. After successful fertilisation, the flower loses its petals and stamens and the ovary starts to grow.

The ability of a specific olive variety to self-pollinate is genetically determined, even though the growing environment and climate can strongly influence genetic expression. As indicated earlier, some olive varieties are self-incompatible (infertile) whereas others are self-compatible (fertile). The former requires cross-pollination. Furthermore, the degree of fertility can vary between regions.

Fertilisation of the perfect olive flowers results in the setting of fruit and their subsequent development. The olive fruit forms, develops and grows from the carpel under the influence of plant regulators. This explains why, when describing the parts of the olive fruit, one talks of the exocarp (skin), mesocarp (flesh) and endocarp (stone). The woody endocarp (stone) surrounds a seed that is made up of a seedcoat, and an endosperm that covers a large embryo. The embryo is made up of two seed leaves (cotyledons), with a short rudimentary root (radicle) and terminal bud (plumule). Hormones produced by the developing embryo and associated endosperm tissue prevent the development of an abscission (rejection) zone in the flower fruit stem, stopping them from falling prematurely. The fertilised ovary changes to a dark green colour (due to chlorophyll production) within 10 days and expands over time into an olive fruit that develops during summer and grows in autumn. In contrast, non-fertilised fruits are paler in colour because they contain less chlorophyll.

Under normal flowering conditions, two to three weeks after full bloom, 10–15% of flowers set fruit; however, many of these are lost as surviving fruit draw upon the resources of the olive tree. Fruit drop is significant for another two weeks then losses slow down thereafter. Only a final fruit set of 1–2% is needed for commercial olive production and this is reached six to seven weeks after full bloom. In seasons when flowering in the olive is relatively light, final fruit set can reach 6–7%.

Pollination problems can occur. Pollen can be lost under very windy conditions, so attention must be paid to the direction of prevailing winds when planning the orchard. Furthermore, rain during flowering impairs wind distribution of pollen and shortens pollen viability. Over-zealous irrigation with sprinklers at flowering time can wash away pollen and damage flowers resulting in poor fruit set. For most olive varieties perfect flowers are receptive to pollination for around 48 hours. Temperatures of 20–25°C are conducive to pollination. Excessively high environmental temperatures (>30°C) damage and reduce pollen tube growth, and hence reduce pollination. Also, under hot, dry conditions the ovaries, fertilised or not, can harden and mummify, remaining on the trees for at least two to three months before they fall. Hence both hot/dry or cool/wet climates can reduce fruit production. Olive trees subjected to high temperatures and dry winds have poor fruit set, excessive fruit drop and fruit shrivel.

Parthenocarpic fruit. Parthenocarpic olives, also called 'Shot Berries', are small under-developed fruit. They are associated with some table olive varieties including *Sevillana* (Fig. 2.9) and to a lesser extent *Verdale* (personal observation), *Barouni*, *Ascolana*, *Uovo di Piccione* and *Cucco*. Two types of parthenocarpic fruits can occur: *(1)* those that remain green in colour due to either defective pollen, defects in the pollination process or ovary abnormalities; and *(2)* those that change colour from green to black following the usual ripening process of normal-sized olives, indicating an ability to form anthocyanins. In this case the likely cause is embryo abortion (loss of the embryo). Parthenocarpic olives lose the characteristics of the variety and are round in shape.

Figure 2.9 Parthenocarpic fruit in *Olea europaea Sevillana*.

Olive fruit development and growth

Olive fruit development, growth and chemical composition depends on variety and climatic factors such as temperature and rainfall (or irrigation) and growing environment. This is illustrated by traits such as the Flesh:Stone ratio, oil quality and specific constituents such as polyphenols. The size of olives also depends on the amount of fruit on the tree. Because of the sectorial nature of olive trees, the size of fruit can differ even between scaffold branches depending on their relative load. Unlike the vegetative growth of olive trees, olive fruit continues to grow at high temperatures (32–38°C). With heavy fruit loads, particularly in unirrigated olive orchards and at high environmental temperatures, olives can shrivel, but these will recover if water is applied to the trees or the temperature falls markedly, resulting in lowered transpiration. This reversal of fruit turgor is only possible if water stress has occurred after oil accumulation. When rain occurs late in the season, shrivelled fruit may not recover. Where olive trees are heavily laden with olives, such fruit can over-winter on the tree and continue to develop in spring to full physiological maturation.

A heavy fruit load on the olive tree can have a number of negative effects. It may delay olive fruit maturation, delay or inhibit anthocyanin synthesis, or lower oil content.

As the olive is a drupe fruit, its development and maturation follows a double sigmoidal growth curve. Five stages of development have been identified from fertilisation to the black-ripe stage (Fig. 2.10). The period of fruit growth at around five to seven months varies according to variety and growing conditions and can change from season to season.

Stages of olive fruit growth

Stage 1. Fertilisation and fruit set
Initially the cells in immature fruit multiply and so increase in number (late spring). During this stage almost all cell multiplication occurs. The developing fruit are visible 10–14 days after pollination. Many immature fruit are lost during this stage, particularly if olive trees are stressed. At this stage most of the fruit volume is taken up by the endocarp tissue and a liquid endosperm, the main part of the seed.

Stage 2. Embryo and endocarp (stone) development
In Stage 2 (early summer) the fruit increases in size at a more rapid rate, mainly due to enlargement of the endocarp surrounding the developing embryo. After four weeks the

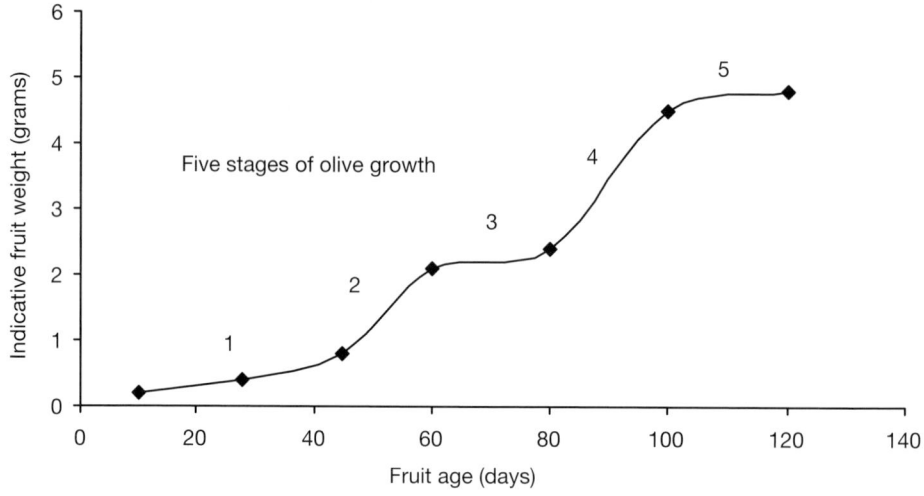

Figure 2.10 Five growth stages of the fruit of *Olea europaea*. (After IOOC *World Olive Encyclopaedia* 1996.)

parts of the olive, the exocarp, mesocarp and endocarp, are identifiable. There is little development of olive flesh or significant oil production. The vacuole of mesocarp cells increases in size and the endocarp accumulates lignin and begins to harden.

Stage 3. Endocarp hardening (sclerification)
During this stage the rate of fruit growth slows down, the embryo and endocarp reach maximum size and the endocarp hardens forming the stone. This stage is generally completed in mid-summer.

Stage 4. Mesocarp development and oil accumulation

At this stage (late summer) the olive fruit rapidly increases in size due to accelerated growth of the mesocarp and exocarp because of increased cell size rather than cell multiplication. As this is the main period for oil synthesis, oil starts to accumulate in the cells of the mesocarp. This stage continues through to autumn, where the olive changes from dark green to a yellow-green straw colour ripe stage due to a reduction in chlorophyll. During Stage 4, if there are periods of water stress, fruit growth and oil levels are compromised. Furthermore, seeds from fruit at the green-ripe stage (March/April) have a higher germination potential compared to seeds from black-ripe olives.

Stage 5. Olive fruit ripening

Fruit growth during this stage slows down although oil accumulation still occurs, but at a slower rate than in Stage 4. International studies on oil accumulation in oil and table olive varieties indicate that by the end of autumn most varieties achieve maximum oil accumulation with final levels varying between 5–30% w/w in the raw flesh depending on the variety. Furthermore, olives from an individual variety, such as *Manzanilla*, grown at different locations can show distinct variations in oil content. Except in the early stages of fruit development, the fatty acid composition in the oil fraction of the olive flesh is less affected by the maturation-state than variety and growing environment. During Stage 5 anthocyanins accumulate in the skin and flesh eventually leading to black-ripe olives. Colouration starts in the skin mostly at the tip of the olive and proceeds to the base. When the skin colour reaches the stalk (petiole), the flesh begins to colour. In some varieties, colouration starts at the base of the olive. Where fruit production per shoot is high, anthocyanin synthesis is partially inhibited with the olives taking on a uniform rose-red colour, which may or may not go to a violet-black colour. Olive seeds from fruit at the black-ripe stage have a reduced germination potential and require either stratification or a long resting period to germinate.

A period of water stress in the latter stages of Stage 2 results in a smaller stone that can improve the Flesh:Stone ratio in the mature fruit. As Stage 4 is important for fruit growth, water application during this stage increases fruit size. In warmer regions up to a two-fold increase in olive fruit size has been observed in irrigated trees when compared with olives from unirrigated trees. During the fruit growth period, olive fruit damaged by frost renders them unsuitable for table olives.

Photosynthetic activity in both fruit and the leaves is important for fruit growth and development. Furthermore, fruit photosynthesis strongly influences oil production in the flesh during the final period of fruit growth. As the photosynthetic activity of olive fruit depends on its development stage, adequate exposure to sunlight is essential. Overall, photosynthesis can be optimised by having adequate tree spacings as well as implementing training systems and pruning methods to reduce inter-canopy shading. Generally, fruit production is less on the sides of the canopy and within the canopy than at the top of the canopy. However, in regions with high levels of radiation, a degree of shading may be beneficial, protecting the tree and fruit from sunburn.

Fruit load is a determining factor of fruit size. To maximise olive fruit size, thinning procedures can be undertaken within three weeks of full bloom resulting in larger quality fruit, without compromising yield. Heavy fruit loads also retard maturation of olives as well as oil accumulation, so thinning will give a more even and enhanced ripening phase. Fruit on olive trees with heavy loads under water stress are more likely to shrivel, but if this is not severe it is reversible with water application through natural precipitation or irrigation. In any case, when olives for table olive production are grown in a Mediterranean-type climate without irrigation, harvesting should be held back until after the first rains in autumn so as to achieve a larger size fruit.

Biennial bearing (alternate bearing) and the olive tree

Biennial bearing, a feature of many fruit trees, including the olive, is characterised by small crops ('off' year) and large crops ('on' year) in alternate years. It is variety specific, but influenced by the growing environment (climate and soil) and horticultural practices such as pruning and water application. Heavy fruiting reduces the growth of new wood and fruit and eventually leads to biennial bearing. Biennial bearing is more pronounced in older trees, but is less of a problem with good soil nutrition, adequate water and early harvesting. Meta-analysis of data from a past olive trial in Mildura, Australia, illustrates the effect of biennial bearing on table olive varieties over a number of years (Table 2.2). The trees were planted at a density of 150/ha and grown under irrigation. Yield data was collected from the 5th to 14th year. The yield terms high and low, based on raw data, are relative and can be used as a guide to illustrate alternate bearing. Data collected from year nine was more consistent. The varieties *Nevadillo Blanco* and *Manzanilla* appeared to be less affected by the alternate bearing phenomenon than the other varieties.

Table 2.2. Overview of yield data of table olives grown in Mildura (Australia) (McClure pers. comm.) low, 'off' year; high, 'on' year.

Variety	Yield, years after planting								
	5 yrs	6 yrs	7 yrs	8 yrs	9 yrs	10 yrs	11 yrs	12 yrs	13 yrs
Nevadillo Blanco	low	low	high	low	high	high	high	high	low
Oliva a Prugna	high	low	high	low	low	high	low	low	high
Azapa	high	low	low	high	high	low	high	low	high
UC13A6	low	high	low	high	low	low	high	low	high
Barouni	low	high	low	high	high	low	high	low	high
Mission (Californian)	low	high	low	high	high	low	high	low	high
Nab Tamri	low	low	low	high	high	low	low	high	high
Manzanillo	low	low	low	high	high	low	high	high	high
SA Verdale	increasing yield over this period ⟶					high	low	high	low

Biennial bearing is less obvious in young productive trees, becoming more noticeable in older trees. With some cases the alternation is not as obvious because even with fewer olives in the 'off' year, the overall crop weight may be of the same order as that of the 'on' year. In the 'on' year, because of the larger crop, little new shoot growth occurs whereas in the 'off' year new shoot development is greater.

Biennial bearing is believed to be under hormonal control, particularly from developing seeds (the embryos). Auxins (hormones) produced by seeds move into the fruiting spur. It is possible that different hormonal and nutritional factors in combination depress flower formation when a large crop of olives (seeded fruit) forms, thus allowing an endogenous (internal) cycle to begin. Large crops of small olives induce biennial bearing to a greater extent than small crops of large olives. Furthermore, early harvesting favours more favourable crops in the next season, whereas late harvesting increases alternate bearing. All olives should be removed at the end of each season to prevent the problem from occurring. Unless diseased or damaged, even if the olives are not to be used for table olive production, they have value for olive oil production.

The level of winter chilling can influence biennial bearing. In the case of relatively warm regions (as experienced in most parts of Australia) a single event in which there is either insufficient chilling or a high winter chilling can start the biennial bearing process. If this occurs in a particular orchard or region, a pattern of biennial bearing develops. Such synchronisation with olive trees has also been attributed to spring frosts. Biennial bearing occurs in both irrigated and non-irrigated olive orchards; however, fruit thinning, controlled nutrition and more intensive irrigation can modulate crop size. That varying degrees of summer drought can cause low yields further complicates the biennial bearing picture.

Beneficial to table olive production, fruit thinning in the 'on' year can reduce the alternate bearing effect in the next year. Also, a specific tree or whole olive orchard can inadvertently be set into biennial bearing following severe pruning after the 'on' year.

In summary, there are several horticultural strategies to control alternate bearing: control flower formation by selective pruning, thinning of fruits by selective pruning or chemical thinning, reduce crop influence by early harvest, and undertake regular and maintenance pruning.

Olive fruit

Olive fruit consists only of carpel tissue that develops from the female reproductive organs of the flower. Raw olives are fleshy fruits ranging from 0.5–15 grams or more depending on the variety, crop load and growing conditions. Some olives are spherical but many are elongated. Shape can be distinctive with some being ovoid whereas others are heart or pear-shaped. Raw olives range from yellow-green to purple-black in colour depending on their maturation state. Fruit from some olive varieties take on a dark copper-brown colour. The consistency and texture of raw olive flesh is dependent on variety and maturation and can change with processing.

Chlorophyll levels in the flesh fall during maturation as the olive goes from dark-green to yellow-green in colour. The colour changes occurring in olives during maturation depend on the variety and growing conditions. The characteristic purple-black pigmentation of skin and flesh is due to a group of chemicals called anthocyanins. Initially these accumulate in the skin either from the base, apex or uniformly over the fruit, then appear in the flesh. Fully black-ripe olives have purple-black flesh down to the stone. After the olive fruit reaches the black-ripe stage it loses moisture and the bitter oleuropein breaks down to the point where some varieties are edible without requiring any processing. One in particular is *Thrubolea* (*Olea europaea* var. *media oblonga*), a small-fruited variety grown in Attica (Greece), some Aegian islands and Crete.

Fruit from different olive varieties can be broadly classified into freestone or clingstone. Although the former are more desirable, some consumers like to suck the stones, experiencing a nutty flavour, while removing residual flesh. Clingstone varieties include *Frantoio, Verdale, Sevillana* and *Jumbo Kalamata*. Olive stone shape and surface features are characteristic of variety and together with DNA data can be used for varietal identification or authentication.

Description of olive fruit

Olive fruit is classified as a drupe because it has a woody stone containing the embryo, or seed, surrounded by flesh (the pericarp) enclosed by a protective skin (Fig. 2.11). (See also Fig. 2.8.)

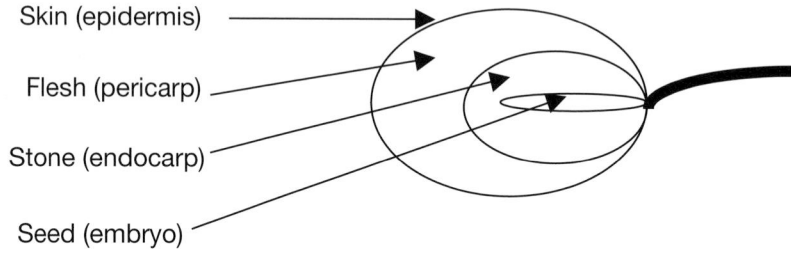

Figure 2.11 Schematic diagram of an olive fruit.

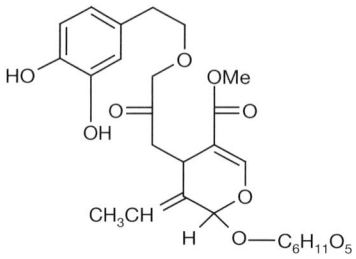

Figure 2.12 Chemical structure of oleuropein. (Image: Harris.)

Apart from olives, avocado is one of a few commonly eaten oil-rich fruits. Compared with other drupe fruits such as peaches, apricots and plums, olive flesh has a low sugar content (2–6% w/w), high oil content (10–30% w/w) and contains the unique extremely bitter glycoside oleuropein (Fig. 2.12).

The weight distribution of components varies with variety, growing environment and cultural practices.

Once olive fruit sets on the tree, it increases in size over several months. During development the olive takes on a leafy green colour. When the olive changes from a leafy green to a yellow-green colour it has generally reached its maximum size and can be picked for processing by any method as a green table olive (or if oxidised under alkaline conditions as Californian/Spanish-style black olives). As olive weight is related to water availability, targeted irrigation will ensure quality table olives of a good size with favourable Flesh:Stone ratios. Olives when green are in their most bitter state because of the high levels of phenolic compounds, oleuropein in particular.

Fruit from most olive varieties naturally ripens to a deep red to black colour. Olives with black skins and white flesh are not fully black-ripe! This problem is experienced early in the season in warmer parts of Australia, particularly with *Manzanilla* variety, where only the skin fully pigments. Oil content appears to be a better index of maturity than colour change. Olive fruit, like avocado, contains no starch (storage carbohydrate) in the flesh. Ripening of olives does not appear to be linked to insoluble carbohydrates, or be accelerated by ethylene. The significance of this is that, unlike fruits such as tomato, apricot and peaches, natural ripening of olives does not continue once picked.

Naturally black-ripe olives are less bitter than the green-ripe olives, because of lower levels of oleuropein and other polyphenols. Overripe olives, because of their lower oleuropein levels, require little processing to debitter. Olives allowed to ripen and partially dehydrate on the trees are most suitable for preparing dried olives. Of course, leaving olives on trees too long leads to poor quality fruit, crop losses due to natural fruit drop, attack by pests and diseases, as well as reduced production in the following season. Olives that are not fully pigmented when processed as naturally black-ripe olives by spontaneous fermentation give a firmer product than if fully black-ripe olives were used.

In Australia, most olives for table olive processing are harvested from early autumn to the end of winter, depending on the variety and the required level of ripeness. Olives harvested at the broad ripening stages of green-ripe, turning colour and naturally black-ripe (Plate 1) can be processed by different methods to give specific table olive products. Traditionally *Kalamata* olives are processed at the black-ripe stage. This variety can also be successfully processed at the green-ripe and turning colour stages. The resulting processed olives at these maturation states have a firm texture and an oily, nutty flavour.

Olive fruit shape and size

The size and shape of olives and the stone are often characteristic of a particular variety. Obviously with over 1000 olive varieties available around the world, there is a wide variation in the size and shape of the olive fruits. In some cases rootstock can influence size and shape of olive fruit. Varieties selected for table olives tend to be larger than oil olives, for example *Sevillana, Manzanilla, Barouni, UC13A6, Cucco, Conservolea, Mission* (Californian), *Oliva di Cerignola, Jumbo Kalamata* and *Kalamata* (see Fig. 2.16). As indicated earlier, olive size depends on the genetic characteristics of the variety as well as crop load, growing conditions and cultural practices. If raw olives are sorted and graded before processing, exceptionally large sorted olives may represent only a small proportion of the actual crop.

As moisture and oil make up a significant part of the olive, changes in the levels of these parameters will affect the overall weight of olives. The moisture content of olives is determined, to a large extent, by the availability of water to the olive tree. This is demonstrated where olives grown under irrigation are larger than those grown under dryland conditions. The moisture content of the raw olive flesh is generally between 60% w/w and 70% w/w.

Olive weight is important for table olives, because consumers show a preference for larger olives than smaller ones, except for specialty lines such as Ligurian olives (*Taggiasca*) and *Arbequina*. Around the Mediterranean local shortages of *Kalamata* olives from time to time has necessitated the processing of the much smaller oil olive variety, *Koroneiki*. As the weight of olive stones reaches a maximum well before the olives ripen, the stone weight alone does not provide useful information, except when calculating the Flesh:Stone ratio of a particular variety. Irrigation improves the commercial value of olives by increasing fruit size and the Flesh:Stone ratio. Irrigation, however, does not markedly affect olive fruit shape and only marginally affects firmness of olive fruit.

General physical features of olive fruit

In summary, the general physical features of olive fruit are as follows:

- Length 1–3 cm, diameter 1–2 cm
- Weight from 0.5–15 grams or more
- Flesh makes up 60–90% of total fruit weight
- Stone makes up 10–40% of fruit weight
- Embryo (seed) inside stone makes up 1–2% of fruit weight
- Olive shapes include pear, egg and heart; other features include asymmetry, nipples and points
- Some varieties are freestone whereas others are clingstone

Olive exocarp (epidermis and cuticle)

The exocarp is the thinnest layer of the olive fruit, and consists of the epidermis (two to three layers of epidermal cells) and cuticle on top (Fig. 2.13). The cuticle is a continuous protective layer of material, connected to the epidermal cells, which consists of carbohydrates (pectin/cellulose/hemicellulose), cutin and wax layers. Cutin is a complex mixture of fatty acid derivatives that have waterproofing properties.

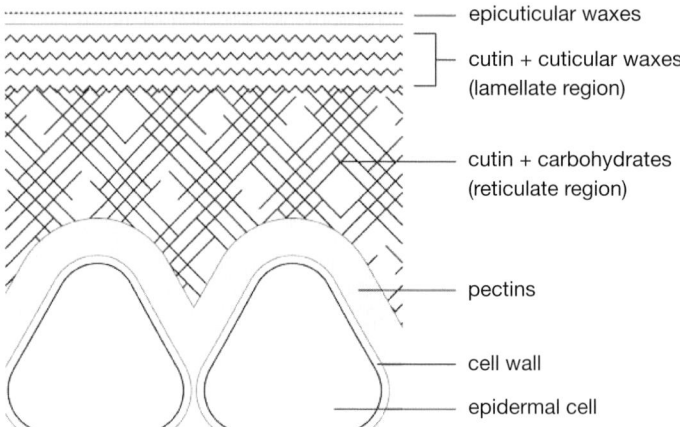

Figure 2.13 Stylised schematic of the cuticle of olive fruit (after Marsilio *et al.* 1996).

Some stomata (openings, or pores) form in the epidermal layer and later change into lenticels, possibly acting as gas exchangers. The number and size of lenticels are often characteristic of particular varieties, for example *Jumbo Kalamata* lenticels are large and diffuse whereas *Manzanilla* lenticels are pinpoint size and numerous. In many cases the lenticels disappear during ripening to the naturally black-ripe stage.

In the absence of water, natural precipitation or irrigation, the fruit adapts by increasing the thickness of its cuticle and repellent layers of wax and cutin to prevent the loss of water and nutrients. When water is readily available, cutin content decreases and thins. Skin thickness also decreases when olives are debittered with lye.

In summary, the skin protects olive fruit from pests and infection. It changes in colour from green to black during ripening, and, if damaged (particularly with green and turning colour olives), the skin darkens by oxidation due to the polymerisation of polyphenols.

When olives are treated with lye or undergo fermentation, water-soluble constituents pass from the flesh through the skin into the processing medium. When fruit from most olive varieties are processed in strong salt brines (greater than 10% w/v salt) water is lost through the skin by diffusion, resulting in shrivelled olives. Large fleshy olives such as *Sevillana*, *UC13A6*, *Ascolana* and *Jumbo Kalamata* are prone to this problem, giving less than desirable products. Of course, when preparing dried olives, shrivelling of the whole fruit is the desired outcome.

Olive fruit flesh (mesocarp)

Structurally, the raw olive flesh is made up of large irregular cells and fibrous materials such as cellulose and lignin. The mesocarp cells change in size, form and function during maturation. The water-soluble components of the cells are vacuolised, whereas the oil is complexed in the cytoplasm. Early in olive development the oil accumulates as droplets that coalesce against the vacuole when the olive matures. In all olive varieties the mesocarp cells contain chlorophyll that is gradually lost during maturation.

Olive flesh has relatively large parenchymatous (a basic cell type) cells surrounded by a rigid cell wall. A large water filled vacuole (sac), in the centre of the cell contains dissolved sugars, weak acids, tannins, water-soluble pigments and minerals. The dissolved material and the structural components of the cell contribute to the texture and firmness of the olive. Changes in these characteristics during maturation and ripening results in a softening of olive fruit. During ripening from green-ripe to black-ripe, cell walls in the flesh become thinner, and there is greater cell separation because of partial solubilisation of pectin, hemicellulose and cellulose polysaccharides within the cell wall matrix. This weakens the cell wall structures resulting in the softening of the olive. When olives are debittered with lye, most of the protopectins in the raw flesh dissolve, resulting in changes in texture.

Table 2.3. Approximate composition of raw olive flesh (% w/w), based on composite data from different varieties and at different maturation states determined by the authors

mg/kg, milligrams per kilogram weight.

Component	Levels
Moisture (%)	60–68
Oil (%)	12–28
Saturated fatty acids (%)	12–20
Polyunsaturated fatty acids (%)	5–18
Monounsaturated fatty acid (%)	60–80
Carbohydrate	
Total (%)	8–12
Soluble sugars (%)	0.5–5.5
Protein (%)	0.7–2.0
Minerals	
Phosphorus (%)	0.02–0.25
Potassium (%)	0.5–3.4
Sodium (%)	0.01–0.20
Calcium (%)	0.02–0.20
Magnesium (%)	0.01–0.06
Sulfur (%)	0.01–0.13
Boron mg/kg	4–22
Copper mg/kg	0.3–5.8
Iron mg/kg	3–95
Manganese mg/kg	0.91–5.5
Zinc mg/kg	1.5–33.0
Ash (minerals) (%)	0.4–1.1

The chemical composition of olive flesh is complex. It contains water, oil, carbohydrates, minerals, vitamins, protein, fibre and pigments. As indicated earlier, unlike other drupe fruits, olive fruits are rich in oil and low in soluble sugars. Characteristic of olive fruit is its high content of phenolic compounds and unique bitter glycoside, oleuropein. During processing, the levels of oleuropein are markedly reduced either by leaching or hydrolysis, hence the fruit is debittered. Soluble sugars in the olive flesh are essential for supporting fermentation during processing. With dried olives, some of the soluble sugars in addition to the oil content contribute to the energy and organoleptic qualities of the processed olives.

Other nutritionally important substances in the flesh are proteins, minerals and vitamins. Summary information on the chemical composition of raw olive flesh, from data generated by the authors, is given in Table 2.3.

Olive endocarp and embryo

Deep inside each olive fruit is an endocarp (stone) that is essentially made of wood and has cells showing a large vacuole and a thick parietal membrane. A small cell layer on the inner part of the endocarp serves as the internal wall that comes into contact with the olive seed (the embryo). The seed is enclosed and protected by the endocarp. The oil content of olive seeds is around 20–30% w/w and has a different chemical composition to the oil in the flesh. The olive endocarp tissue is lighter in colour and is made up of smaller cells than the mesocarp. The endocarp cells harden through lignification (turn into wood). The colour of olive stones is buff-yellow initially and changes to brown through oxidation when exposed to air. Olive stones are of different shapes, sizes and have characteristic surface markings that can be used to identify broad varietal groups, but not clones within a variety. Some varieties are prone to having split stones, an undesirable characteristic.

Moisture content of raw olives

The water content of olive flesh is around 60–70% w/w. It follows, therefore, that the lower the water content of olive flesh, the higher its nutritive and energy value. The water content of the raw olive is responsible for its turgor. This is under cellular control via a selectively permeable cytoplasmic membrane so that the olive does not dehydrate and shrivel when on the tree. Olives, however, can shrivel under extreme conditions or when the olives senesce (age).

Effects of processing. During processing of raw olives with salt, lye and pH reduction, selective permeability is lost and the water in the olive is retained as a layer around predominantly colloidal protein particles. As long as there are sufficient colloidal particles, the processed olives will retain their shape and appearance. Where the colloid levels in the olive fruit are insufficient, the fruit becomes a package of liquid enclosed by the olive skin. Processing, as well as the effects of frost, will shrivel olives if the colloidal particles are denatured or changed. During processing olives lose moisture and other water-soluble components, so their net weight is often lower (up to 10%) than that of the raw olives. With salt-dried or heat-dried olives, the water content falls to around 30% w/w.

Flesh to stone ratio of olives

The Flesh:Stone ratio is derived by dividing the weight of flesh by the weight of the stone. Ideally the Flesh:Stone ratio of table olives should be 5:1. With processed olives the Flesh:Stone ratio can be used as a crude indicator of the nutritional quality of the olives. An indication of the required processing time can be obtained knowing the weight of the olives and the Flesh:Stone ratio. Smaller olives with a low Flesh:Stone ratio (*Verdale*) process faster than larger olives with a high Flesh:Stone ratio (*Jumbo Kalamata*), especially by traditional methods. Large olives with high Flesh:Stone ratios should be slit, stoned, cracked or bruised to facilitate processing.

Oil content of raw olive flesh

The fruit of all olive varieties contains oil (lipid), most of which (98%) is concentrated in the flesh. In some cases, table olive varieties have lower oil levels in the flesh than oil varieties. Some olive varieties are arbitrarily divided into oil, table or dual purpose; however, within reason, all olive varieties can be used for either table olive or olive oil production. The percent oil content of the olive depends on the variety, growing conditions and its maturation state. Most of the oil fraction in raw olives consists of triacylglycerols (98%), combinations of fatty acids and glycerol, as well as some diglycerides (1.1%) and free fatty acids (0.3%). Two other types of lipids incorporated into the olive fruit cell membranes are phospholipids and galactolipids. These increase as the olive fruit develops. Other oil soluble compounds in raw olive flesh include sterols, triterpenic acids and tocopherols (Vitamin E).

In triacylglycerols, three molecules of a fatty acid are esterified (combined) to one of three positions on a molecule of glycerol. Oleic and linoleic acids generally occupy 'position 2' of the glycerol molecule. As naturally occurring oils contain a number of specific fatty acids, in excess of 70 different combinations of triacylglycerols can occur in the oil fraction of olive flesh. When consumed, triacylglycerols are broken down to individual fatty acids and glycerol and these are utilised by the body for energy and for the synthesis of body chemicals, with excess fatty acids stored as fat. The proportions of the more commonly occurring triacylglycerols in the oil fraction of raw olive flesh are presented in Table 2.4.

Table 2.4. Indicative proportions of common triacylglycerols in the oil fraction of raw olive flesh

Proportion (%) of total triacylglycerols	Number of fatty acids combined with glycerol			
	Oleic	Palmitic	Linoleic	Stearic
40–60	3	–	–	–
12–20	2	1	–	–
12.5–20.0	2	–	1	–
3–7	2	–	–	1
5.5–7.0	1	1	1	–

Oil (olive oil) is synthesised and accumulates in the olive during development at a rate depending on the growing environment. For many olive varieties most oil accumulation occurs before fruit maturation, whereas in others it may continue even into the black-ripe stage. The composition of oil in olive flesh differs between varieties, but is relatively consistent for individual varieties grown under similar conditions. Higher levels of oleic acid are present in the oil fraction when olive trees are grown in cool regions compared with hot regions. At late maturation stages some decomposition of the oil fraction in the olive occurs, with amounts of free fatty acid increasing, particularly when harvested olives are stored.

Indicative ranges of the main fatty acids in the oil fraction of raw olive flesh at maturity are:

Oleic acid (MUFA)	70–80%
Linolenic acid (PUFA)	<1.5%
Linoleic acid (PUFA)	5–10%
Palmitic acid (SFA)	10–15%
Stearic acid (SFA)	2–3%

Oil starts to accumulate in the olive flesh when the stone hardens, reaching maximum levels when the olives reach maturity. Changes occur in the oil levels of olive flesh during growth and development and follow a pattern as indicated in Table 2.5. The oil composition also changes with maturation. Initially, the relative quantities of saturated fatty acids (SFA) and the polyunsaturated fatty acids (PUFA), linoleic and linolenic acid, are higher and the monounsaturated fatty acids (MUFA), oleic acid, are lower. As the olive matures this distribution is reversed.

Table 2.5. Indicative pattern of oil levels during olive fruit development

Season	Days	Oil (%)	Moisture (%)
Summer	Start	1	73
	20	2	76
	40	6	74
	86	16	68
	109	19	68
Winter	143	20	70

The oil in table olive fruit is the same as olive oil including its nutritional and health benefits. Unlike extra virgin olive oil there are no set limits for the different levels of fatty acids in raw or processed olive flesh. A major difference between the oil composition in raw olive flesh and olive oil is that the olive seed contributes to the fatty acid profile of the extracted oil. Research has shown that for the table olive varieties *Kalamata*, *Ascolana Tenera* and *Nocellara del Belice*, the fatty acid composition is not affected by irrigation and the quality of the oil varies between olive varieties according to the temperature of the growing environment.

Effects of processing. Generally, the oil content of the olive is retained during processing because only water-soluble compounds can pass through the flesh and skin during soaking and fermentation operations. Some investigators have observed loss of oil content in olives during lye treatment. A possible cause is a reaction between lye (sodium hydroxide) and oil yielding water-soluble salts of fatty acids (soaps) that are lost during the soaking and washing steps during processing. Sometimes lye treated olives have a soapy taste for the same reason. A small amount of oil is also lost into the brine from olives that are slit or cracked prior to processing. The brine takes on a cloudy appearance, a state that should be differentiated from microbiological causes of cloudiness (a signal of possible microbiological problems). The waxes impregnated in the olive skin tend to remain intact during processing.

Carbohydrate levels in raw olive flesh

Total carbohydrate levels. Raw olive flesh contains soluble sugars (simple sugars) and sugar polymers (complex sugars) including cellulose, hemicellulose or pectins, and lignin. Total calculated carbohydrate values for raw olive flesh determined by the authors was found to be between 8% w/w and 12% w/w. Lignin is concentrated in the stone. Hemicellulose and cellulose are structural components of the cell walls involved in linking adjacent cells, hence contributing to flesh texture. As these compounds contribute to the structural characteristics of olive flesh, changes or a reduction in these polysaccharides during ripening or processing can influence the organoleptic characteristics of processed olives. Cellulose (3–6% w/w) makes up a substantial amount of the fibre in olive flesh. Quantitative and qualitative differences between varieties are discernible. Also, processed olives tend to have less fibre than raw olives.

Pectin is composed of the specialised sugar, galacturonic acid, forming a cement-like substance that contributes to the structure of cells. When pectins are hydrolysed or broken down the flesh loses texture and softens. This occurs when: lye treatment is used to debitter olives; the temperature increases, for example in poorly stored olives; spoilage microorganisms grow during fermentation (softening is the result of the action of released enzymes); and when the fruit is overripe (internal enzymes in the flesh break down pectin).

Where enzymes are involved in the deterioration of olive flesh, they include varying amounts of polygalacturonases and pectinesterases. The effect of these enzymes is more likely to occur in fruit during maturation, storage or prolonged soaking in water/weak brines rather than when processing by brine fermentation or with lye treatments. An explanation for this is that the enzymes have low activity in brines and lye solutions.

Soluble sugars. The main soluble sugars in olive flesh are glucose, fructose, sucrose and mannitol (sugar alcohol). In the olive tree, mannitol is a translocating sugar (a form of sugar that moves from one part of the plant to another) whereas glucose and fructose (reducing sugars) are utilised in metabolic processes. The levels of soluble sugars in raw olive flesh range from around 0.5% w/w to over 5% w/w depending on variety and growing conditions, with glucose predominating. Soluble sugars decrease as the olive fruit develops and oil synthesis begins. They are twice as high in green-ripe olives compared with black-ripe olives and their levels are inversely related to the oil content. Soluble sugars can also form esters, and in the case of the olive, are a component of oleuropein (a beta-glucoside) and other related glycosides.

Glucose, and to a lesser extent fructose, are important substrates when raw olives are processed by fermentation. On a dry-weight basis, sugar levels are lower in olives grown under irrigation than if grown under rainfed conditions, which has implications for the effectiveness of fermentation during processing.

Points to note are that the soluble sugar concentrations in the raw olive flesh changes with moisture content and if olives are subjected to prolonged soaking and multiple washing steps, much of this intrinsic sugar is lost and additional soluble sugar (glucose or dextrose) may need to be added if fermentation is required during processing. It is generally accepted that olives with higher levels of fermentable sugars process easily. A problem that can occur with olives picked late in the season is that they may have insufficient quantities of sugars to support fermentation, particularly if they are to be treated with lye (Spanish-style green olives) then required to undergo fermentation. This is not a problem for olives that are processed in brine by natural fermentation.

Effects of processing. Reducing sugars in raw olive flesh provide the main energy source for fermentative microorganisms during processing. As microorganisms consume these sugars during fermentation, the olive flesh is left practically sugar free. The numerous glycosides in olive flesh (including oleuropein) release soluble sugars when hydrolysed that can be an additional source of sugar for fermentation. In naturally processed olives (that is, non-lye treated olives), hydrolysis of these compounds is slow, and can even occur after the olives are packed, resulting in secondary fermentation and stability problems.

If the level of fermentable substrate (sugars) in the olive flesh of green-ripe olives is around 4%, when processed (as in Spanish-style green olives) the expected titratable acid and pH levels in the brine are 0.8–1.0% w/v (lactic acid) and pH 3.8–4.1 respectively. With untreated naturally black-ripe olives, if the fermentable substrate level in the flesh is 2.0–2.5% w/v, when processed by spontaneous fermentation the expected titratable acid levels and pH in the brine will be 0.5–0.6% w/v and pH 4.5 respectively. Differences in acid production and pH can be influenced by factors such as olive variety, maturation state of the olives, the initial sugar concentration of the flesh/brine and the types of microorganisms involved.

Sun-dried and heat-dried olives retain some of their sugar content, which becomes relatively more concentrated due to the loss of moisture from the flesh during processing. Salt-dried olives, however, 'leak' moisture during processing and some loss of sugars and other water-soluble substances occurs.

Phenolic substances in raw olive flesh

Phenolic compounds, also called polyphenols, are secondary metabolites present in all plant tissues and are involved in protective mechanisms against external stresses. They have antioxidant activity, so when consumed in the diet they have beneficial effects on health. Polyphenols in olives include oleuropein, hydroxytyrosol, caffeic acid and tyrosol.

Phenolic compounds make up 2–3% w/w of olive flesh, with oleuropein (a bitter tasting compound) the most abundant polyphenol. Oleuropein accumulates during fruit growth and is slowly converted to elenolic acid glucoside and demethyloleuropein as the fruit ripens. When some olive varieties, for example *Leccino* and *Kalamata*, are at the completely black-ripe stage the oleuropein levels fall to near zero, which facilitates processing. In a recent Italian study, researchers found that total polyphenol levels (mainly oleuropein) in *Frantoio* olives fell from 1.6% to 0.2% over summer (green) to winter (black).

Three stages have been identified in relation to oleuropein levels in the olive: *(1)* the growth stage, when oleuropein accumulates; *(2)* the green maturation phase, when oleuropein levels fall; and *(3)* the black maturation phase, in which oleuropein levels continue to fall (and anthocyanins accumulate).

Oleuropein levels in raw olives also fall during storage after harvesting even if the olives are stored under controlled conditions.

Research has shown that irrigation results in a decrease in the polyphenol levels of raw olives in the table varieties *Kalamata*, *Ascolana Tenera* and *Nocellara del Belice*. This is most noticeable with green-ripe olives. It is not unreasonable to expect that this effect would occur in all table olive varieties.

Phenolic compounds other than oleuropein found in olive flesh include: hydroxytyrosol-4-beta-D-glucoside, hydroxytyrosol, tyrosol, verbascocide, luteolin 7-O-glucoside and rutin. In *Manzanilla*, oleuropein is the major phenolic compound through all stages of ripening. Relevant to table olive processing, glycosides (such as oleuropein and hydroxytyrosol–4-beta-D-glucoside) provide sugars for fermentation when hydrolysed. Furthermore, both these compounds, together with verbascocide, contribute to the amount of hydroxytyrosol in fruits and brines. Hydroxytyrosol, a small active molecule, well absorbed by the gastrointestinal tract compared with oleuropein, provides useful levels of antioxidant activity when consumed via processed olives.

Figure 2.14 Cracked raw green-ripe olives showing darkening of skin due to the oxidation of polyphenols.

Injured green-ripe or turning colour olives or olives exposed to air undergo a browning of skin or flesh because of the presence of polyphenols (Fig. 2.14).

Effects of processing. As oleuropein is water-soluble, it is extracted from the raw flesh by diffusion when olives are soaked in water or brine during processing. Slitting, cracking, bruising or destoning the olives facilitates extraction of oleuropein. However, important to processing, oleuropein is degraded by lye but not in the acid environments created by fermentation. Recent research has identified some strains of fermentative lactic acid bacteria that can breakdown (hydrolyse) oleuropein through enzymatic reactions, hence debittering the olives as well as participating in the fermentation process. This biotechnology could replace the use of lye in table olive processing.

One of the goals in processing raw olives into table olives is to reduce the levels of oleuropein in the flesh, hence debittering the olives. If polyphenols occur in high enough concentrations, they can inhibit fermentative organisms due to their antibiotic properties, thus slowing down fermentation and prolonging processing times. Oleuropein and its hydrolysis end products (elenolic acid and aglycone) inhibit the activity of a number of microorganisms involved in the lactic fermentation of green-ripe olives. Such organisms include *Lactobacillus pentosus* (formerly *Lactobacillus plantarum*), *Lactobacillus brevis*, *Leuconostoc mesenteroides*, *Pediococcus cerevesiae*, *Geotrichum candidum*, *Rhizoctonia solani* and *Rhizopus* spp. Another potential problem is that oleuropein can also be used as a source of carbon (energy) for contaminating organisms during fermentation, leading to spoilage of the olives.

When olives are treated with lye, they debitter easily because all the oleuropein is destroyed by an alkaline hydrolysis reaction. When olives are subject to soaking and/or natural fermentation without prior lye treatment, oleuropein is diluted and/or partially destroyed so the final products retain some level of bitterness. Some have suggested that the attractiveness to consumers of processed 'feral olives' is their significant residual bitterness.

Protein levels in raw olive flesh

Olive flesh contains low levels of soluble and insoluble protein at concentrations of around 1.5% w/w protein. Soluble proteins can transfer into the brine, providing amino acids for fermentative organisms and possibly the growth of spoilage organisms. Major amino acids in raw olives include arginine, alanine, aspartic acid, glutamic acid and glycine. Other amino acids such as histidine, lysine, methionine phenylalanine, tyrosine, leucine, isoleucine, threonine, valine, proline and serine are also present.

Effects of processing. Olives processed with lye have lower levels of flesh protein when compared with raw olives. With directly brined olives, green-ripe olives have higher protein levels than naturally processed black-ripe olives. Directly brined green-ripe olives have the highest concentrations of amino acids compared with directly brined naturally black-ripe olives or lye treated olives. With naturally black-ripe olives the decrease in amino acid levels is possibly due to degradation of the protein in the raw olive flesh and/or to soaking in water or brine. Regardless of the processing method, commonly found amino acids in processed table olives are phenylalanine, isoleucine, leucine, methionine, tryptophan, valine, aspartic acid, glutamic acid, alanine, glycine and proline.

The amino acids lysine, threonine, serine and tyrosine are also present in Spanish-style green olives, whereas directly brined green-ripe olives have serine and tyrosine but lack lysine and threonine. Processed naturally black-ripe olives contain threonine but lack lysine, serine and tyrosine whereas Californian-style black olives (lye treated) lack lysine, serine, tyrosine and threonine.

Pigments in raw olive flesh

Olive flesh contains chlorophyll a and b (green), carotenoids and triterpenic hydrocarbons (yellow) and anthocyanins (purple-black). The green and yellow pigments are oil soluble, whereas anthocyanins are water-soluble. Initially chlorophyll is the major pigment in the olive fruit that plays an important role in photosynthesis. As the fruit matures and ripens chlorophyll levels decrease while other pigments, beta-carotenes and the purple-black anthocyanins increase.

Anthocyanins give naturally black-ripe olives their characteristic purple-black colour. The major anthocyanin in olives is cyanidin, but a number of others have been identified. Anthocyanins develop initially in the olive skin and progressively appear in the flesh; in fully ripe olive fruit the flesh is blue-purple down to the stone. In some varieties such as *Leucocarpa margareta*, anthocyanin synthesis is blocked resulting in 'white' olives when fully ripe. Accumulation of anthocyanins in skin and flesh can filter out harmful light rays, hence protecting the nuclear material in the embryo from potential damage. Anthocyanins make up about 0.5 mg/100 g of raw olive flesh. The variety, ripeness stage, and level and degree of solar radiation influence the synthesis of anthocyanins.

Effects of processing. When processing naturally black-ripe olives, care must be taken to ensure that the flesh is also pigmented otherwise pale insipid looking olives result. During processing of naturally black-ripe olives, anthocyanins leach into the fermentation brine resulting in rose to magenta-coloured brines. The resulting olives can have a washed out pink-buff colour. Exposing such pale coloured olives to air for 24–48 hours oxidises polyphenols in the flesh, which darkens them. The colour of anthocyanins are also affected by pH. They are pink at low brine pH and darken as the pH increases.

Lye treatment of green-ripe olives denatures chlorophyll in the flesh, releasing water-soluble compounds into the lye solution. Under some circumstances, lactic acid removes magnesium from the chlorophyll in the flesh, leaving residual pheophytins and pheophorbids, which give the olives a grey-green colour. Carotenoids, on the other hand, are more resistant to processing procedures.

Mineral content of raw olive flesh

The mineral content of raw olive flesh (defatted and digested in perchloric acid) is quantified by inductively coupled plasma-atomic emission spectroscopy. The nitrogen content of defatted raw olive flesh is quantified by the Dumas combustion procedure or by the Kjeldahl procedure that involves digestion in sulfuric acid. These nitrogen levels can be used to determine protein levels in olive flesh by multiplying the nitrogen levels by a conversion factor of 6.25. The levels of eleven minerals (both macro and micro) in raw olive flesh are presented in Table 2.3. The macro-elements are phosphorus, potassium, sodium, calcium, magnesium and sulfur and are expressed in percentages or g/100 g. The microelements (trace elements) are boron, copper, iron, manganese and zinc and are expressed as mg/kg or parts per million (ppm). Variations in mineral levels of raw olive flesh are a function of the growing conditions, such as soil quality, level of water application, and the amounts and composition of manures and fertilisers used.

Nutritionally, table olives are a good source of minerals. Raw olives contain reasonable amounts of potassium, phosphorus, nitrogen, copper, calcium, boron, magnesium, manganese, iron, potassium and sulfur. The levels of these minerals depend on the olive variety, maturation state and processing method.

Effects of processing. Some minerals, especially potassium, are lost when prolonged soaking methods are used during processing. Processing also increases the sodium content of olives when carried out in brine. Where iron salts (ferrous gluconate, ferrous lactate) are used to stabilise the colour of Californian/Spanish-style black olives, iron levels are markedly increased. Furthermore, minerals are lost when they diffuse out of the olive flesh into the processing brines where they can be utilised by fermentative organisms. Sun-dried and heat-dried olives retain their minerals, which become relatively more concentrated due to the loss of moisture from the flesh during processing. As salt-dried olives 'leak' moisture during processing, some loss of minerals occurs.

Further components in raw olive flesh

Vitamins. Both water- and oil-soluble vitamins are present in raw olive flesh and can be added to by microbiological activity during fermentation. In general, water-soluble vitamins are mostly lost during processing whereas the oil soluble ones are retained. Water-soluble vitamins include: Ascorbic acid (Vitamin C), Thiamine (Vitamin B1), Riboflavine (Vitamin B2) and Niacin (Vitamin B6). Oil-soluble vitamins include carotenes (Vitamin A precursor) and tocopherols (Vitamin E group). The latter group are antioxidants. The addition of stuffing material into olives will add to the vitamin levels of olive flesh.

Organic acids. Small quantities of organic acids such as citric, oxalic and malic acids are present in raw olive flesh. Actual quantities depend on the variety, maturation state and growing conditions. When raw olive flesh is ground into a paste it is slightly acidic with a pH of around 5. With untreated olives, diffusion of free organic acids from the flesh into the brine contribute to the initial acidity of the brine, especially when the olives are slit or cracked. The organic acids play no part in processing when olives are processed in lye.

Natural microbiological flora on olive fruit

A variety of microorganisms, including bacteria, yeasts and fungi are found on the surface of raw olive fruit. The types and mix of microorganisms is influenced by climatic conditions, olive variety and maturation stage of the olive.

It is these naturally occurring microorganisms that initiate spontaneous fermentation when olives are processed in brine using natural methods. However, other organisms on the raw olives, mostly contaminants from the environment, have the potential to cause spoilage or food poisoning. Microbiological blooms (white/grey film) are often visible on raw olives, especially *Kalamata* olives (Plate 1) and should not be mistaken as chemical spray contamination. Nevertheless, all olives should be washed before processing.

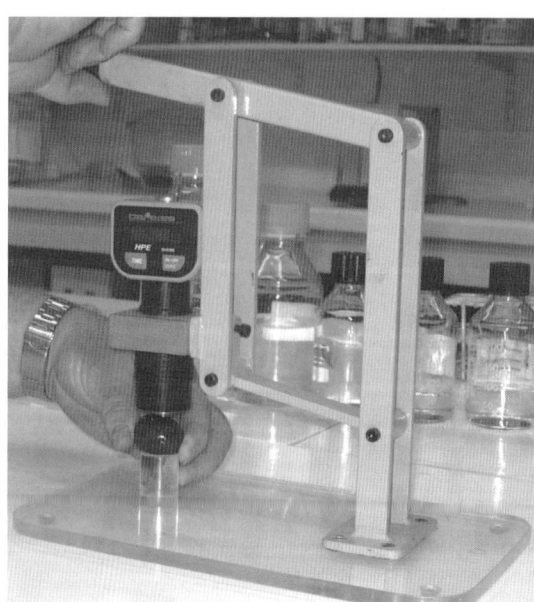

Figure 2.15 Table olive assessment using Shor 00 hardness tester.

Changes in olive flesh resistance during maturation

Olive flesh, composition and texture changes during maturation and ripening. With a simple hardness test (Shor 00) differences can be assessed between olives of different varieties at different maturation states, and after processing. Hardness is the resistance that a sample offers to induration.

Hardness is a characteristic of a material and not a fundamental physical property. In the case of olive fruit hardness has been equated to firmness.

The Shor 00 (Fig. 2.15) has an indentor (penetrometer), which, when placed in contact with the surface of the material and activated, measures firmness. The instrument is activated for a short period during which time the indentor depresses the material and a value appears on the digital display. A suitable measuring time for olives is four seconds. Using identical testing conditions, resultant values can be compared.

Using the Shor 00 tester, unripe green olives of all varieties give values of 100+ Shor units. Green-ripe olives, however, generally give values between 80 and 90 units, whereas naturally black-ripe olives with pigmentation to the stone give values between 50 and 70 units. Turning colour olives give values in between. This test can be used as an indicator of flesh integrity, and for an individual batch of olives can be used as a preprocessing baseline value. Processed olives when tested give lower pressure values than the original raw olives. More elaborate equipment for testing olive texture and firmness is also available. For more information the reader is referred to Fernández *et al.* (1997).

Internationally important table olive varieties

There is a myriad of olive products available worldwide, with each olive growing country or region producing local and regional styles.

Some examples of varieties and their final products are:

Manzanilla – Spanish-style green olives
Californian *Mission* – Californian black-ripe olives and Californian green-ripe olives
Kalamata – Kalamata-style olives
Volos (*Conservolea, Konservolia*) – Greek-style naturally black, lye treated green-ripe olives
Hojiblanca – Spanish-style black olives, natural black olives, Kalamata-style olives
Sevillana (Spanish Queen) – Spanish-style green olives
Nocellara del Belice – Castelvetrana-style olives, Sicilian-style green olives
Chalchidikis – Also called Greek donkey olives, lye treated green-ripe olives
Taggiasca – Ligurian olives
Thrubolea – Greek black dried olives (Thrumba-style)
Maiatica di Ferrandina-style – Italian dried olives (Tostata)
Picholine – Olives de Nimes

Some internationally recognised trade products have also emerged and these include: untreated black olives in brine (untreated black-ripe olives that undergo spontaneous brine fermentation), Spanish-style green olives (lye treated green-ripe olives that undergo lactic fermentation in brine) and Californian/Spanish-style black olives (green-ripe/turning colour olives treated with lye and darkened by oxidation).

Most producing countries use only a few varieties as raw material for table olive production and these are processed by a handful of methods. It is daunting to the prospective table olive processor trying to unravel this situation. Ideally the raw olives should firstly be debittered during '*primary processing*' then secondly embellished by '*secondary processing*'. Embellishment involves adding other components to the olives or brines, such as olive oil, vinegar, herbs, spices and aromatics. Other secondary processing can include destoning the olives with or without stuffing, for example pimento, anchovy or almonds, and preparing olive pastes and tapenades from processed table olive flesh.

Note: Starting processing with raw olives, brine and embellishments all at the same time is likely to give products of variable quality and anomalous characteristics and should not be practised in commercial processing. However, this is commonly practised by some home processors.

Important table olive varieties

As the olive is not indigenous to Australia, all Australian olive trees have been derived from imported olive trees or their progeny. The domesticated olive, *Olea europaea*, originates from a group of varieties and semi-wild forms that have evolved by hybridisation and selection from several tropical and subtropical wild olive species. The European olive has been cultivated around the eastern Mediterranean and Middle East since ancient times. Over 2000 named olives are recorded but these are generally clones selected from spontaneous, uncontrolled crosses or are, at most, a few generations removed from 'wild' forms.

Generally, local or regional names rather than botanical names are used when referring to a particular olive variety. It is not unusual for a specific variety to have several different names within the same country or even in other countries. Furthermore, the same variety can develop different organoleptic characteristics when processed by different methods. It is well recognised that olive fruit characteristics can change when some varieties are grown outside their traditional growing region. Research is needed to evaluate varieties brought in from international sources for Australian growing conditions. Work undertaken as part of the NOVA (National Olive Variety Assessment) in South Australia revealed that only some of the varieties sold by Australian nurseries matched international standards. Of 100 NOVA accessions (planted at Roseworthy, South Australia), which were supposedly 87 different varieties, only 55 different genotypes were detected. Twelve differently named varieties were of the same genotype as the Italian variety *Frantoio*.

Scientific breeding programs are underway in several countries, including Australia. Such research could deliver varieties with improved fruit yield and/or fruit quality, disease resistance and other horticultural benefits such as tree dwarfing and easy harvesting. International researchers are developing some new varieties suitable for table olive production and several examples are provided below.

Three new semi-dwarf varieties have been bred and evaluated in Israel. All three varieties have a medium to weak vigour, enabling harvesting from ground level, and require minimal pruning. They can be used for green and black olives. The three varieties are:

Kadeshon   7–8 g and slightly elongated and freestone
Sepoka   4–6 g and elongated and freestone
Masepo   7–8 g large round olive and freestone

A previous introduction was the low-fat olive variety, *Kadesh*.

In Italy, three new dual-purpose medium vigour olive varieties are under investigation: *Arno* (6.2 g), *Tevere* (5 g) and *Basento* (5.9 g). Eleven table-olive clones selected from the Sardinian cultivar *Nera* are also under investigation.

Selected characteristics for table olive varieties

Information on important varieties used for table olive production is given in Table 2.6. Most of the varieties are available in Australia, although authenticity should be established at the time of purchase. Fertility information provided in this section is based on local and northern hemisphere information. The sizes and shapes of numerous olive varieties in two planes (A and B) are presented in Fig. 2.16 at the end of this chapter.

Definitions of terms relevant to the olive profiles are as follows (for more information on olive varieties the reader is directed to the International Olive Oil Council's *World Catalogue of Olive Varieties* (2000).

Flowering time. This can vary from region to region so the terms *early*, *intermediate* and *late* are comparative.

Fertility. Most olive varieties will set some fruit in a monoculture (single variety) environment. Olive fertility can be divided into three categories: self-incompatible requiring cross fertilisation, partially self-compatible, and self-compatible. Many olive varieties are partially incompatible, with the degree of incompatibility dependent on climate and, in particular, high temperatures, which are detrimental. Fertility can vary from region to region so the terms described below are comparative.

Self-incompatible (self-infertile). Fruit set is not possible by self-pollination so open pollination is required (pollinating varieties need to be planted).

Partially self-compatible (partially self-fertile). Fruit set may occur occasionally by selfing.

Self-compatible (self-fertile). There are small differences in fruit set between self-pollination and open pollination.

Ripening. The time of olive ripening depends on the variety, growing environment and crop load. Ripening can be divided into three categories: early (late April, or in the northern hemisphere late October), intermediate (early winter) and late (late winter).

Oil content. Oil content is categorised for whole raw olives including the stone. Note that the oil content of the flesh is higher than for whole olives and overall levels depend on the moisture content. Raw olives from non-irrigated olive trees (of the same variety and maturation state grown under the same conditions) can have higher oil levels than those from irrigated olive trees. The oil content can be described as: low <18%, medium 18–22%, and high >22%.

Start of bearing. This is the interval in years between planting and cropping, with early, intermediate and late bearing corresponding to three, four and five years respectively.

Productivity. Productivity, based on a single olive tree, depends on the growing environment and cultural practices and the descriptive terms are relative. Ripening can be divided into the three categories low, medium and high.

A 14-variety olive trial was conducted in Mildura (Victoria, Australia) to assess productivity. Olive trees were planted under irrigation in 1964 and five years later.

Productivity was recorded from five years after planting over nine years. Data from this trial are included with the information on particular varieties (see Table 2.2).

Bearing. Bearing depends partly on the variety, growing conditions and cultural practices. The relative descriptive terms can be divided into two categories: *(1)* alternate (the variety shows clear bearing irregularities under normal growing conditions), and *(2)* constant (the variety shows only small fluctuations in annual crops when trees have reached full bearing).

Rooting ability. Rooting ability is based on olive varieties where scientific data is available and where IBA has been used as the auxin. Rooting ability is divided into the following three categories: high, 66–100%; medium, 33–60%; and low, 0–33%.

Tolerance to abiotic and biotic factors. Although olive is considered a very tolerant species, in some circumstances some varieties are more or less susceptible to cold, drought, salinity, scale infestation, Peacock Spot, Olive Knot and Verticillium Wilt attack. With the latter problems, growing conditions and the state of the tree are contributing factors. The severity of these factors is classified low, medium and high tolerance. Only selected information is presented in the variety profiles and the absence of information on specific tolerances should not be taken as not being a problem.

NOVA (National Olive Variety Assessment – Australia). Relevant information from the NOVA trial has been included in the variety profiles. Where the term 'matches the international standard' is used, the accession in the NOVA trial has matched at least one similarly named variety in an overseas country. Furthermore, the harvesting time (if indicated) is based on a Maturity Index of 3 (MI = 3).

Table 2.6. Important varieties used for table olive production

Arbequina (Plate 2)

Alternative names	*Arbequi, Arbequin, Blancal*
Original growing region	Spain: Catalania
Rooting ability	High
Tree characteristics	Hardy, low vigour with drooping, spreading, medium canopy Susceptible to lime induced chlorosis
Flowering time	Early to intermediate
Fertility	Self-compatible
Fruit shape	Spherical to heart-shaped
Typical size range	1.5–2.0 g (500–660 fruit/kilogram)
Ripening	Unevenly maturing early to mid-season
Flesh:Stone ratio	4.6:1
Stone characteristics	Small elongated clingstone stone
Oil content	High
Start of bearing	Early
Productivity	High
Bearing	Alternate to constant
Tolerances	High to cold, humidity, salinity, Black Scale Medium to drought, Olive Knot, Verticillium Wilt Low to Black Scale, Olive Fruit Fly
Uses	Processed at turning colour stage in brine and for olive oil
Comments	Low levels of oleic acid and polyphenols Processed green to turning colour in brine popular in Spain NOVA: matched the international standard. Harvesting time (MI = 3)

Ascolana Tenera

Alternative names	*Oliva Dolce* (contains little bitterness)
Original growing region	Italy: Le Marche and central Italy
Tree characteristics	Prefers cooler rather than hotter growing environments but widely adapted
Rooting ability	High
Flowering time	Late with high pistil abortion
Fertility	Self-incompatible, requires pollinators, e.g. *Manzanilla, Mission* (Californian)
Fruit shape	Round to ovoid and slightly asymmetric
Ripening	Early
Typical size range	Large 8.3–9.1 g (110–120 fruit/kilogram)
Flesh:Stone ratio	6–8:1
Stone characteristics	Freestone
Oil content	Low to medium
Start of bearing	Early
Productivity	Medium, but high under ideal growing conditions
Bearing	Constant, heavy when young but develops alternate bearing habit in mature trees
Tolerance	High to cold, Peacock Spot, Sooty Mould Medium to drought, Olive Knot Low to Olive Fruit Fly, Black Scale, Verticillium Wilt
Uses	Green table olives
Comments	Skin is delicate and flesh has soft texture easily bruised during handling and processing NOVA: matched international standard

Azapa

Alternative names	*Sevillana de Azapa*
Original growing region	Chile: Azapa
Tree characteristics	Hardy, resistant to salinity and drought, fruits well in climates with warm winters
Rooting ability	Medium
Flowering time	Intermediate
Fertility	Partially self-compatible
Fruit shape	Asymmetric, ovoid, slightly pointed at the apex
Ripening	Late
Typical size range	Variable 4–12 g (80–250 fruit/kilogram), average 8 grams
Flesh:Stone ratio	May reach 11:1, i.e. very fleshy
Stone characteristics	Clingstone
Oil content	Low
Start of bearing	Early
Productivity	Medium Mildura trial: 53.2 kg/tree/year averaged over nine years, five years after planting
Bearing	Alternate (increased effect with *El Nino*)
Tolerance	High to drought Medium to cold Low to Black Scale
Uses	Green and black table olives and sometimes for oil production
Comments	Easy to pit when processed with lye Bears well in Chile under conditions of low winter chilling NOVA: Harvesting time: late (MI = 3)

Barnea (Plate 3)

Alternative names	K18 (Breeding Code)
Original growing region	Bred in Israel: Galilee, coastal and southern plains
Tree characteristics	Moderately hardy, vigorous, with erect growth
Rooting ability	High
Flowering time	Intermediate with moderate pollen
Fertility	Partially self-compatible; pollinators *Picholine*, *Picual*, *Frantoio* and *Manzanilla*
Fruit shape	Asymmetric elongated with a nipple like point at apex
Ripening	Early to intermediate
Typical size range	2–4 g or more (250–500 fruit/kilogram)
Flesh:Stone ratio	7:1
Stone characteristics	Freestone
Oil content	Medium
Start of bearing	With irrigation, bears from year 3
Productivity	High constant under irrigation
Bearing	Constant with irrigation, alternate without irrigation
Tolerance	High to Peacock Spot Low to frost, Verticillium Wilt
Uses	Green and black table olives, Greek-style in brine, also for oil production
Comments	Irrigation is important Shape can be confused with *Kalamata* olives When processed in brine, *Barnea* olives have a granular flesh NOVA: matched the international standard. Harvesting time: intermediate (MI = 3)

Barouni

Alternative names	Several different names in Tunisia. The variety *Uovo di Piccione*, although considered a separate variety (*Bidh el Hamman*), is often included in the *Barouni* group.
Original growing region	North African origin: Tunisia
Tree characteristics	Relatively small tree with a spreading habit that is easy to harvest.
Flowering time	Early
Rooting ability	High
Fertility	Self-compatible
Fruit shape	Ovoid to elongated ovoid
Ripening	Harvest mid to late season
Typical size range	7.4 g or larger (130–140 fruit/kilogram)
Flesh:Stone ratio	6.8:1
Stone characteristics	Clingstone
Oil content	Low
Productivity	Intermediate Mildura trial: 66.1 kg/tree/year averaged over nine years, five years after planting
Bearing	Alternate
Tolerance	High to drought (heat resistant) Low to cold, Peacock Spot, Olive Knot, Verticillium Wilt
Uses	Table olives
Comments	Suitable for home processing Green-ripe olive fruit travels well NOVA: matched nursery standard. Harvesting time: intermediate (MI = 3)

Chalchidikis

Alternative names	*Chondrolia Chalchidikis, Gaidourelia* (Donkey-olive), *Halkidiki, Khalkidiki*
Original growing region	Chalchidiki region: near Thessaloniki
Tree characteristics	Medium hardiness
Rooting ability	Medium
Fertility	Partially self-compatible
Fruit shape	Asymmetric, ovoid, elongated with a nipple at apex Similar appearance to *Ascolana Tenera*
Ripening	Harvested early, does not turn completely black
Typical size range	6–10 g or more (100–165 fruit/kilogram)
Flesh:Stone ratio	10:1
Stone characteristics	Freestone
Oil content	Medium
Start of bearing	Intermediate
Productivity	Medium
Bearing	Alternate
Tolerance	Low to salinity
Uses	Spanish-style green olives in brine, Sicilian-style green olives
Comments	Unsuitable for naturally black olives in brine, off odours can develop during processing because of its low sugar content, e.g. *Zapateria* defect A red ring that can develop during processing is considered a serious defect Has a buttery flavour when processed

Conservolea

Alternative names	*Chondrolia, Konservolia, Amphissis, Volos*
Original growing region	Central Greece: most common variety grown in Greece (up to 85% of all trees), Amphissa, Volos and Ebia
Rooting ability	Medium
Tree characteristics	Quick grower under irrigation, adapts to many environments Can grow at higher altitudes up to 600 m
Flowering time	Intermediate
Fertility	Self-incompatible
Fruit shape	Asymmetric round to ovoid with pointed apex
Ripening	Intermediate to late, fruit matures slowly
Typical size range	5.0–5.5 g or more (180–200 fruit/kilogram)
Flesh:Stone ratio	8–10:1
Stone characteristics	Freestone, large
Oil content	High
Start of bearing	Within three to four years with irrigation
Productivity	Intermediate to high (depending on horticultural practice)
Bearing	Alternate
Tolerance	High to cold, Olive Knot Medium to drought Low to Verticillium Wilt
Uses	Table olives used for naturally black olives and oil production
Comments	Raw olives have thin, elastic skin, firm flesh, travel well, and can withstand shrinking in highly concentrated (12–19%) salt brines Sugar content is 2–3%

Cucco

Alternative names	*Cucca* and many local names
Original growing region	Italy: Abruzzi, Molise, Chieti and Pescara regions

Tree characteristics	Hardy and vigorous
Rooting ability	Medium
Flowering time	Early with little pollen
Fertility	Self-incompatible
Fruit shape	Slightly asymmetric, ovoid to heart-shaped
Ripening	Early
Typical size range	Large
Flesh:Stone ratio	4:1
Stone characteristics	Freestone
Oil content	Medium
Start of bearing	Late
Productivity	High
Bearing	Alternate
Tolerance	High to cold conditions Medium to Peacock Spot Low to wind, Olive Knot and Olive Fruit Fly
Uses	Green and black Greek-style olives, Spanish-style green olives
Comments	Subject to fruit drop

Frantoio (Plate 4)

Alternative names	*WA Mission, New Norcia Mission, Correggiolo, Paragon, Mediterranean*
Original growing region	Italy: central Italy including Tuscany
Rooting ability	High
Tree characteristics	Adaptable to different growing conditions, cold sensitive, medium vigour, pendulous branches
Flowering time	Intermediate with low pistil abortion
Fertility	Self-compatible with productivity increasing if cross pollinated
Fruit shape	Slightly asymmetric, ovoid to pear-shaped towards the apex
Ripening	Late and phased; may not reach black-ripe stage in some areas
Typical size range	2–4 g (250–500 fruit/kilogram)
Flesh:Stone ratio	5:1
Stone characteristics	Clingstone
Oil content	Medium
Start of bearing	Intermediate
Productivity	High
Bearing	Constant and high in well-managed trees
Tolerance	High to salinity Medium to drought, Verticillium Wilt Low to cold, fog, Peacock Spot, Olive Fruit Fly, Olive Knot, Black Scale
Uses	Predominantly oil, but can be used for Greek-style olives in brine
Comments	Final products are similar to Ligurian olives (*Taggiasca* variety) NOVA: matched the international standard. Harvesting time: intermediate (MI = 3)

Hojiblanca

Alternative names	*Ojiblanco, Barquillero* (Malaga)
Original growing region	Spain: several sites, Andalucia, Cordoba and Malaga regions
Rooting ability	Medium
Tree characteristics	Moderately vigorous with an erect medium to dense canopy Tolerant to calcareous soils

Flowering	Intermediate to late
Fertility	Self-compatible
Fruit shape	Ovoid
Fruit ripening	Late
Typical size range	Variable 1.5–4.5 g (220–670 fruit/kilogram)
Flesh:Stone ratio	5–6.5:1 depending on fruit size
Stone characteristics	Clingstone
Oil content	Low
Start of bearing	Intermediate
Seasonal bearing	Late ripener
Productivity	High
Bearing	Alternate
Tolerance	High to drought Medium to cold, salinity Low to Peacock Spot, Olive Knot, Black Scale, Verticillium Wilt
Uses	Table olives and oil
Comments	Fine skin and firm flesh, prone to gaseous spoilage with Greek-style method When at the turning colour stage they are prone to shrivelling during storage Used for Spanish/Californian black-ripe olives NOVA: matched the international standard. Harvest time: intermediate (MI = 3)

Kalamata (Plate 5)

Alternative names	*Nychati Kalamon, Calamon, Calamata*
Original growing region	Greece: Kalamata region in the Peloponnese, Messina, Lakonia and Lamia Second most important table olive variety in Greece
Rooting ability	Low to medium (often grafted)
Tree characteristics	Medium-sized tree, medium hardiness Grows vigorously when grafted onto strong rootstock Possibly sensitive to acid soils
Flowering time	Intermediate
Fertility	Self-compatible
Fruit shape	Elongated asymmetric with pointed apex but no nipple
Ripening	Late and generally harvested black-ripe
Typical size range	4.0–5.5 g (180–250 fruit/kilogram)
Flesh:Stone ratio	8:1
Stone characteristics	Freestone
Oil content	Medium
Start of bearing	Intermediate
Productivity	High
Bearing	Alternate to constant (depending on horticultural practice)
Tolerance	High to Olive Fruit Fly, Verticillium Wilt Low to wind, Black Scale Moderately resistant to cold, sensitive to climates with excessively high temperatures
Uses	*Kalamata* olives, culls and rejects used for oil production
Comments	Thin skin, good flesh characteristics, copes well during handling and processing Sugar content is 3–5% NOVA: matched the international standard. Harvesting time: intermediate (MI = 3)

Leccino (Plate 6)

Alternative names	Leccio (note that there are at least seven clones of *Leccino*)
Original growing region	Italy: central regions, Tuscany and Umbria
Rooting ability	High
Tree characteristics	Vigorous, upright with dense canopy (some clones may be pendulous) and adaptable to different growing conditions
Flowering time	Late
Fertility	Self-incompatible, has low quality pollen, requires pollinator: *Frantoio/Pendolino*
Fruit shape	Slightly asymmetric ovoid
Ripening	Early with all fruit ripening at the same time
Typical size range	Medium 2.0–2.5 g (400–500 fruit/kilogram)
Flesh:Stone ratio	4.8:1
Stone characteristics	Freestone
Oil content	Low/medium
Start of bearing	Early
Productivity	High
Bearing	Highly productive and constant but may be a low performer in hotter growing areas
Tolerances	High to wind, fog, Olive Knot, Peacock Spot Medium to drought Low to salinity, Olive Fruit Fly, Black Scale, Sooty Mould, Verticillium Wilt
Uses	Usually used for oil, but can be processed by Greek-style method in brine
Comments	Some low cold tolerant clones have been identified NOVA: matched the international standard. Harvesting time: intermediate (MI = 3)

Manzanilla (Plate 7)

Alternative names	*Manzanilla de Sevilla, Manzanillo* (Italy, USA)
Original growing region	Spain: Seville region
Rooting ability	High
Tree characteristics	Low vigour, low spreading and medium dense canopy Sensitive to prolonged cold periods, e.g. average July temperatures of less than 8°C or frosts below –5°C
Flowering	Early to intermediate
Fertility	Partially compatible, may require pollinator outside of Spain, e.g. *Sevillana, Ascolana* Fertilisation affected by high temperatures
Fruit shape	Spherical (apple like)
Fruit ripening	Early but later than *Sevillana*, harvest early (green) to prevent frost damage to fruit
Typical size range	3.5–5.0 grams (200–280 fruit/kilogram)
Flesh:Stone ratio	6–8:1 or more
Stone characteristics	Small, smooth, freestone
Oil content	Medium
Start of bearing	Early
Productivity	High, crops well in areas with warm winters Mildura trial: 93.1 kg/tree/year averaged over nine years, five years after planting
Bearing	Alternate

Tolerance	Medium to salinity, drought
	Low to cold, Peacock Spot, Olive Fruit Fly, Olive Knot, Black Scale, Sooty Mould and Verticillium Wilt
Uses	Table olives (all styles) and oil
Comments	Fine skin and good texture, not prone to gaseous spoilage or yeast spots
	Sensitive to blistering and peeling with lye
	Store for two to three days under ambient conditions before lye treatment to prevent sloughing of the skin
	Prone to *Zapateria* defect because of low acid production during fermentation
	NOVA: matched the international standard. Harvest: intermediate (MI = 3)

Mission

Alternative names	*Californian Mission, Mammoth*
Original growing region	USA: California. Introduced via Mexico in 1769
Rooting ability	High
Tree characteristics	Hardy, erect growth habit
	Can grow up to 15 m tall so needs training/pruning
Flowering time	Early to intermediate
Fertility	Self-compatible
Fruit shape	Slightly asymmetric, ovoid to elongated ovoid
Ripening	Mid–late in the season
Typical size range	Medium, approximately 4 g (250 fruit/kilogram)
Flesh:Stone ratio	6.5:1
Stone characteristics	Freestone
Oil content	High
Start of bearing	Intermediate
Productivity	Medium
	Mildura trial: 80.5 kg/tree/year averaged over nine years, five years after planting
Bearing	Variable alternate to constant
Tolerances	High to cold (at least −13°C) but early autumn frost can damage fruit
	Medium to drought, Olive Knot
	Low to salinity, Peacock Spot, Black Scale, Verticillium Wilt
Uses	Green and black table olives and olive oil
Comments	Flesh is very bitter
	NOVA: widely mislabelled in Australia. Harvest time: intermediate (MI = 3); California, late.

Nocellara del Belice

Alternative names	Many local names, e.g. *Oliva di Castelvetrano*
Original growing region	Italy: western Sicily
Rooting ability	High
Tree characteristics	Moderate growth
Fertility	Self-incompatible, needs pollinator (*Giarrafa* in Italy)
Fruit shape	Asymmetric spherical
Ripening	Late
Typical size range	6–8 g (125–160 fruit/kilogram)
Flesh:Stone ratio	6–7:1 or more
Stone characteristics	Freestone
Oil content	Low to medium

Start of bearing	Early
Productivity	Intermediate to high
Bearing	Constant
Tolerances	High to humidity
	Medium to cold, Peacock Spot, Olive Fruit Fly
	Low to drought, fog, salinity, wind, Olive Knot, Verticillium Wilt
Uses	Green table olives
Comments	Firm flesh

Oliva di Cerignola

Alternative names	Various local names: *Bella di Cerignola, Grossa di Spagna, Oliva a Prugna*
Original growing region	Italy: Puglia
Rooting ability	Low to medium
Tree characteristics	Sensitive to cold
Flowering time	Late with high pistil abortion
Fertility	Partially self-compatible but needs pollinator
Fruit shape	Asymmetric, elongated void with a pointed apex
Ripening	Mature early and are not cold resistant
Typical size range	Variable sized large fruit (100 fruit/kilogram)
Flesh:Stone ratio	3:1
Stone characteristics	Clingstone
Oil content	Low to medium
Start of bearing	Early
Productivity	Medium
	Mildura trial: 51.6 kg/tree/year averaged over nine years, five years after planting
Bearing	Alternate
Tolerance	Medium to Olive Fruit Fly
	Low to cold, drought, fog, Peacock Spot, Olive Knot, Black Scale, Sooty Mould
Uses	Table olives: Spanish-style or Greek-style green
Comments	Tough, hard and fibrous flesh
	Has similar shape to *Jumbo Kalamata*

Picholine (Plate 8)

Alternative names	*Picholine Languedoc, Olive de Nimes*
Original growing region	France: No. 1 olive, grown in Provence, southern coastal and hillside areas
Rooting ability	Medium
Tree characteristics	Medium vigour with spreading canopy
	Hardy and can adapt to different growing environments, moderately tolerant to cold and drought and resistant to Peacock Spot
Flowering time	Intermediate with high germinating pollen
Fruit shape	Asymmetric, elongated with pointed apex but no nipple
Fertility	Partially self-compatible
Ripening	Mid (green) to late season (red-black)
	Turning colour olives are a red wine colour
Typical size range	Medium and even size range 2–4 g (250–500 fruit/kilogram)
Flesh:Stone ratio	Approximately 5–6:1
Stone characteristics	Freestone
Oil content	Medium but difficult to extract
Start of bearing	Early

Productivity	High with irrigation
Bearing	Constant with irrigation
Tolerances	High to Peacock Spot, Black Scale Medium to cold, drought, Olive Fruit Fly Low to Olive Knot
Uses	Green table olives, olive oil
Comments	Used for *Olives de Nimes* Flesh has a firm texture NOVA: matched international standard

Sevillana (Plate 9)

Alternative names	*Gordal Sevillana, Spanish Queen, Gordal* and *Sevillano* (Italy, USA)
Original growing region	Spain: Seville and Andalucia
Rooting ability	Low (often grafted)
Tree characteristics	Vigorous when grafted but not when self-rooted, prefers warm mild climates, spreading growth habit
Flowering time	Early to intermediate
Fertility	Self-incompatible, needs pollinator, e.g. *Manzanilla, Mission* (Californian), *Ascolana*
Fruit shape	Ovoid/heart-shaped and grows singly
Fruit ripening	Early: green to purple-black
Typical size range	8–10 grams (100–120 fruit/kilogram) or more
Flesh:Stone ratio	High, 7:1 or more
Stone characteristics	Clingstone
Oil content	Low
Start of bearing	Intermediate
Productivity	Low
Bearing	Alternate
Tolerance	High to cold, Peacock Spot Medium to salinity, Olive Fruit Fly Low to drought, Olive Knot, Black Scale, Verticillium Wilt, Sooty Mould Susceptible to Soft Nose (too much nitrogen fertiliser or fungus: Anthracnose), high chilling requirement
Uses	Table olives; can be destoned easily; used for Sicilian green olives and lye treated olives
Comments	Thick skin with compact flesh, sugar 4–6% Prone to shot berries, thin skin, soft texture, sensitive to lye and Fish Eye, easily bruised, also possibility of split stones Flesh difficult to remove from stone NOVA: matched international standard; also had identical fingerprint to *Cucco* and *Nab Tamri*. Harvesting time: intermediate

Taggiasca

Alternative names	Many local names: *Olivo di Taggia, Tagliasca, Gentile*
Original growing region	Italy: Liguria in Imperia
Rooting ability	Low to medium
Tree characteristics	Large, adapts to coastal and hilly areas, susceptible to cold and drought
Flowering time	Intermediate
Fertility	Partially self-compatible with high fruit set
Fruit shape	Symmetric ovoid
Ripening	Late
Typical size range	1.5–2.0 g (500–670 fruit/kilogram)

Flesh:Stone ratio	4–5:1
Oil content	High
Start of bearing	Early
Productivity	High
Bearing	Constant
Tolerance	Low to cold, drought, fog, wind, Peacock Spot, Olive Fruit Fly, Olive Knot
Uses	Oil and table olives; used for Ligurian-style olives
	Developed by Benedictine monks in the town of Taggia

Additional table olive varieties grown in Australia

Several additional olive varieties are grown in Australia for table olive processing. These include *Hardy's Mammoth*, *Jumbo Kalamata*, *Nab Tamri*, *UC13A6* and *Verdale*. Limited scientific information is available on these varieties in publications or in the public domain. Further studies need to be undertaken to verify anecdotal information from growers and processors.

Hardy's Mammoth. *Hardy's Mammoth* (Plate 10) is believed to have its origins in Queensland, Australia. There are 15 specimens growing in the Yanco Olivetum (NSW). Not all of the trees show DNA matches against commercial standards, with some showing clustering with *Correggiollo* and others with *Verdale Aglandau*. This variety is under investigation at Blackwood (South Australia) and Gatton (Queensland). RAPDs (Random Amplification of Polymorphic DNA) analysis in the NOVA trials indicates a possible link to the French *Verdale Aglandau*. Some growers have reported that *Hardy's Mammoth* is slow growing and a late bearer of fruit. Fruit are large with weights around 7 g and have a Flesh:Stone ratio around 6–7. Oil content is medium. The fruit is ovoid in shape and clingstone. The stones are elongated and pointed. Australian processors consider this variety suitable for both green and black table olives.

Jumbo Kalamata. This large-fruited clingstone olive is of uncertain origin (Plate 11). Trees are of medium size, intermediate flowering, exhibit alternate bearing habit and are harvested late in the season. They are also known as *Giant Kalamata* and *King Kalamata* and apart from a superficial likeness to *Kalamata*, they are unlikely to be related. The fruit of this variety is of similar size and shape to the Italian variety *Oliva di Cerignola*. Individual fruits weigh 10 grams or more, but there is significant weight variability between fruit on the same tree. Distinctive features of the skin are the large white patches or lenticels. This variety is used predominantly for table olives as it has low oil content. Because of the relatively high Flesh:Stone ratio (around 5:1), processing this olive variety needs careful attention to ensure complete debittering and prevention of spoilage such as gas pocket formation. To overcome this, turning colour olives should be used. Possible disadvantages of this variety are its very large size, the long sharply pointed stone and coarse flesh. Slitting the olives longitudinally facilitates the processing of this variety.

Nab Tamri. This variety (Plate 10) has its origins in North Africa. The fruit, which are ovoid to heart-shaped, are large, weighing around 10 g. Fruit are clingstone with a Flesh:Stone ratio of around 7–8. Trees are large and moderate to heavy regular producers of fruit. This variety is used predominantly for table olives. In the Mildura trial productivity was 73 kg/tree/year averaged over nine years, five years after planting. A biennial bearing pattern was observed.

The olive tree *Olea europaea* | 65

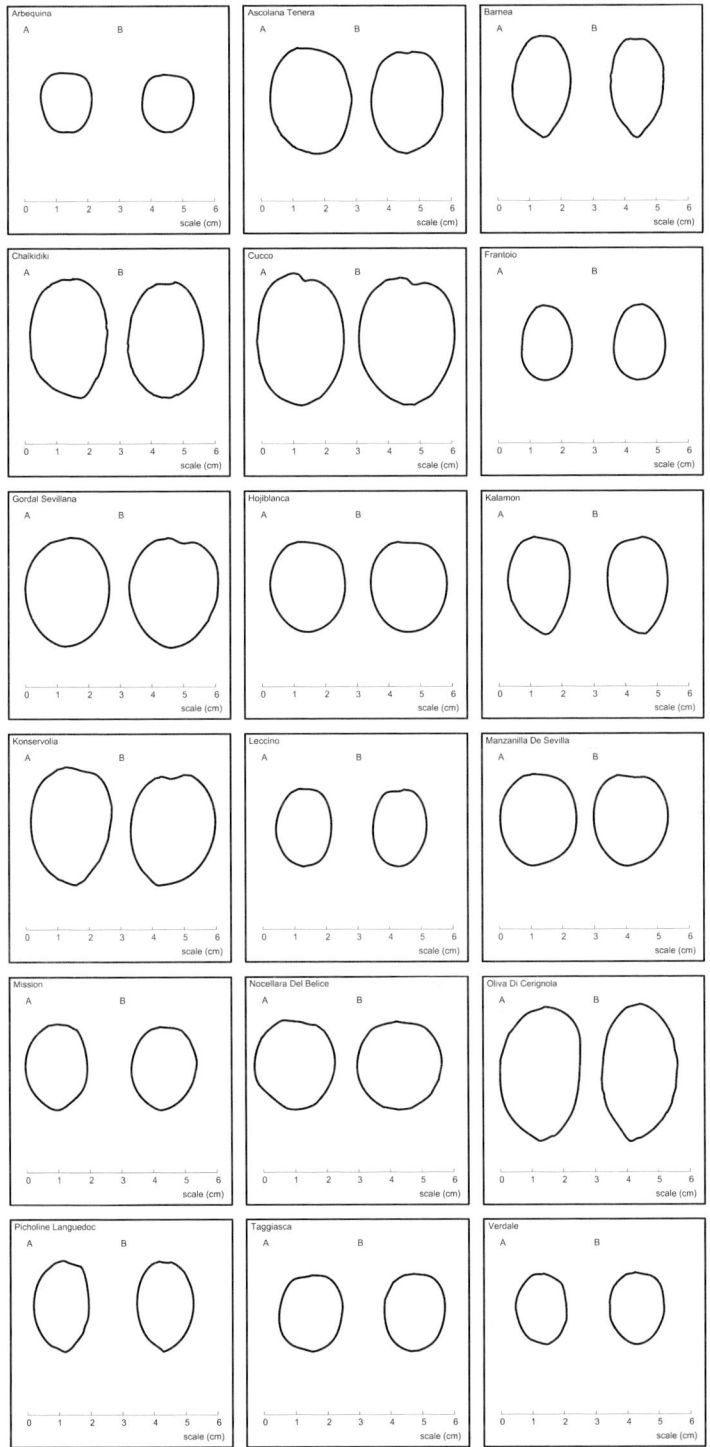

Figure 2.16 Diagrammatic representations of table olive varieties. (The base where the pedicel is attached is at the top of the diagram and olive tips are at the bottom of the diagram.)

UC13A6 and related varieties. This large-fruited olive (Plates 10 and 12) was developed at the University of California from a selection of open-pollinated seedlings of the Egyptian *Tafahi* variety and is also known as *Californian Queen*. Little is known of this olive even in California and it appears to be grown mainly in Australia. It is an early flowering variety, and a medium bearer with intermediate alternate bearing characteristics. The fruit is spherical to oval in shape, clingstone, with a typical size range of 9–11 grams. The oil content of the flesh is low, but the sugar content is high. The fruit can suffer from environmental stress leading to flesh breakdown and browning that is only visible when affected olives are cut open. Fruits do not mature evenly. This variety is suitable for green olive production and heat-dried black olives. The latter have a distinct sweet liquorice and slightly bitter taste. In the Mildura trial productivity was 89 kg/tree/year averaged over nine years, five years after planting. A biennial bearing pattern occurred as the trees aged.

Several other University of California varieties worthy of mention here were included in the Mildura trial.

UC6A7. This is an open seedling of the Egyptian variety *Tafahi*. The tree is very large, has a weeping habit with heavy yields. In the Mildura trial productivity was 66.3 kg/tree/year averaged over nine years, five years after planting. The somewhat elongated fruit weighs about 6 g. A biennial bearing pattern was observed as the trees aged.

UC23A9. This was a cross made in 1952 between *Manzanilla* and *Sevillana*. Trees from this cross produced large fruit. In the Mildura trial productivity was 53.5 kg/tree/year averaged over nine years, five years after planting. A biennial bearing pattern was observed as the trees aged.

UC22A11. This was a cross made in 1952 between *Columella* and *Sevillana*. Trees from this cross produced large fruit. In the Mildura trial productivity was 60.6 kg/tree/year averaged over nine years, five years after planting. A biennial bearing pattern was observed as the trees aged.

UC23A13. This was a cross made in 1951 between *Ascolana* and *Sevillana*. Trees from this cross produced large fruit.

Verdale and related varieties. There are several olive varieties given the name *Verdale*: *Verdale*, *South Australian Verdale* and *Wagga Verdale*. The variety *Verdalion* is another variant. The *South Australian Verdale* is commonly grown around Australia, both in commercial orchards and in home gardens. Flowering is early to intermediate and harvest is mid-season (Plate 13). The genetic basis of the *South Australian Verdale* is still unclear, but it could possibly be related to *Verdale de L'Herault*. Its origin is probably French and it may be related to the French *Verdial*. It is a hardy, relatively small, compact tree with drooping branches and easy to harvest by hand. It has adapted well to a wide range of growing environments in Australia as it is resistant to cold and drought. The fruit, which has a large clingstone, is ovoid in shape with a size range of 5–6 grams. Oil content in the whole olive is low to medium because of the large stone. It is a medium producer with an alternate bearing habit that can be modulated by picking the crop when green-ripe and by pruning. In the Mildura trial productivity for *South Australian Verdale* was 59.5 kg/tree/year averaged over nine years, five years after planting. A biennial bearing pattern was observed. For *Wagga Verdale* productivity was 69.7 kg/tree/year averaged over nine years, five years after planting. In this trial biennial bearing was not obvious.

3

Producing quality raw olives

Orchard siting and the requirements for growing high quality table olives are discussed in this chapter. Site preparation and climatic considerations, particularly with respect to moisture requirements, are outlined. Varietal selection, relevant to the types of final table olive products to be produced, is discussed. We also deal with the factors to be considered in selecting cultivars: olive size and shape, Flesh:Stone ratio, flesh detachment, firmness, colour and stone shape. Soil requirements, light, canopy and training development, olive nutrient requirements and tree row placement are examined. Information is also provided on the visual diagnosis of specific problems, requirements for chemical testing, and overcoming plant nutritional problems. Common pests and disease problems are reviewed; however, if sprays are used the authors consider that a record of type and time of application are essential requirements. The maturation states of olives are discussed and suggestions for the best time of harvesting are made. Methods of handling olive fruit during harvesting and post-harvest are detailed. The undesirable qualities of raw olives with respect to table olive production are presented.

Introduction

Good quality olives are required to produce quality table olive products. Hence, quality starts at the olive orchard. The grower needs to understand the olive tree and the qualities of the fruit produced (see Chapter 2). Poor selection of varieties, particularly if grown under sub-optimal conditions, can influence the properties of raw olives, which will impact on finished table olive products. Grower, consumer and quality considerations for table olives are summarised in Table 3.1.

Table 3.1. General considerations for table olives

Growers concerns	Consumer concerns	Factors affecting quality
Simultaneous ripening	Safe products	Growing environment
Resistance against pests and diseases	Colour	Variety
Spacing	Flavour	Growing technologies
Availability of water	Fruit size	Disease and pest control
Well-draining soils	Choice of olive product	Available nutrition
Early productivity	Price	Local weather
Canopy control		Crop load
Fruit yield		Ripening stage
Biennial bearing		Harvesting method
Fruit size		Transporting in/out of orchard
Fruit removal force		Pre-processing operations
		Processing technology
		Packaging technology

Basic requirements of raw olives

As olives have a fleshy structure, if bruised while on the tree, during harvesting or when transporting and storing, physical, chemical and biological damage of skin and flesh results. Because of the high moisture content of the flesh, enzymes released in damaged olives are able to rapidly degrade the flesh, hence softening the olives. Tissue enzymes (lipases) also react with the oil in the flesh resulting in the release of free fatty acids to levels greater than 3–4% w/w; similar reactions occur in over-ripe olives. Thus, good crop management is essential to ensure the quality of raw olives used for processing.

Table 3.2. Basic requirements for raw olives for table olive processing

Basic requirements	Comment
Plan the table olive enterprise	Select growing site: climate, land features, availability of services, availability of labour force.
Select appropriate varieties	Ensure authentic varieties are planted. Ensure variety is appropriate to the style of table olive product(s) to be made.
Grow olive trees under optimal environmental conditions	Select sites with an appropriate climate. Ensure soils are well draining. Check soil pH and ensure soils are slightly acid to neutral. Plant olive trees to give maximum radiation.
Grow olives under good horticultural practices (GHP)	Undertake leaf and soil nutrient analysis. Apply nutrients on the basis of nutrient analyses. Apply irrigation water as appropriate. Prune to manage canopy and for production. Implement an integrated pest and weed management program. Restrict the use of chemicals: herbicides, pesticides and fungicides.
Use effective harvesting methods	Harvest olives at the correct ripening state for style/method. Ensure that olives are not damaged during harvesting. Protect harvested olives from the sun and heat in the field. Ensure olives are stored under clean hygienic conditions.
Use most effective post-harvest handling methods	Sort olives according to state of ripeness and size and remove small and defective olives. Store and transport olives under cool, clean and hygienic conditions.
Ensure efficient delivery of olives to processor	Store olives under cool, clean and hygienic conditions at the processing plant.

Only the best quality fruit should be processed as table olives. Rejected fruit, for example small and/or misshapen fruit, can be used for olive oil production. Quality includes ensuring authentic varieties are grown and using sound fruit of an appropriate size and at the correct maturation stage. Common table olive varieties include *Manzanilla, Kalamata, Chalchidikis, Sevillana, Hojiblanca*, and to a lesser extent *Verdale, Picholine* and *Barouni*. Some oil olive varieties, such as *Arbequina, Taggiasca, Frantoio* and *Leccino*, are also used for table olive processing. Consumers prefer medium to large-sized olives, with a delicate skin and a Flesh:Stone ratio of around 5:1 or greater. Freestone varieties are best as the flesh separates easily from the stone. Also, olives with small, smooth stones are preferred. Small olives (*Arbequina, Frantoio, Taggiasca*) and extra large olives (*Jumbo Kalamata, Oliva di Cerignola, Cucco*) are better suited to specialty markets. The basic requirements for producing raw olives for table olive processing are listed in Table 3.2.

Planning for table olive production

To be strategic, the potential grower must determine the size of the orchard to be established and which varieties to plant. Two limitations, apart from the growing conditions, in establishing an olive orchard include: the varieties available from the nurseries and their authenticity, and the available financial, physical and human resources available to the grower at the time. The latter is important at harvesting time because olives for table olive processing are mostly picked by hand. When olive trees become commercially productive, crop yields can reach 15 t/ha. For a typical olive orchard of 20 ha with 6000 olive trees this translates to a level of production that can reach 300 t/year, which is the equivalent of 150 000 two-kilogram packs of olives.

Planning the olive orchard also requires an understanding of the end products to be produced – table olives or olive oil. In many situations these two end objectives coexist. Growers who wish to concentrate their activities on the table olive sector can plant suitable varieties, for example *Manzanilla, Kalamata, Sevillana* and *Conservolea*, whereas those wanting to produce oil and table olive products can add *Frantoio, Leccino, Picual, Barnea* or *Arbequina* (see also Chapter 2).

New entrants to the industry need to consider the following when planning their table olive operation:

- market segment for their products: fresh fruit sales, processed olives, packaged olives;
- quantity of olive product to be processed;
- olive products to be produced: table olives, tapenade, antipasti;
- available growing technologies: irrigated, non-irrigated, organic;
- quantity of trees/olives required to meet production goals;
- extra sources of raw or processed olives in case of shortfalls in production goals;
- disposal of culled and defective olives for olive oil production; and
- environmental implications: disposal of liquid and solid wastes, minimisation of obnoxious smells and noise.

Starting to grow table olives

Planning the table olive orchard

The principal objective of the table olive grower is to maximise the viability and profits of the enterprise without compromising quality or by using dubious practices. For planning purposes the following points summarise the growing requirements for olives that require consideration (see also Chapter 2):

- choose varieties that can adapt to the local climate;
- a suitable climate with preferably a wet winter and long, hot, dry summer; sufficient chilling hours for vernalisation; adequate radiation and sufficient heat degree-days for complete ripening (naturally black olives);
- soil suitable for olive growing should be a depth of 80 cm or more, friable (<40% clay), pH near neutral (6–7) and contain adequate organic matter;
- slope less than 1:5 incline;
- tree spacing (this should be adequate between trees to provide sufficient radiation and to undertake all operations); and
- water (good quality and quantity for key times during the olive cycle).

Factors to consider when siting the olive orchard include:

- local government and agency requirements and regulations;
- climate (temperature, rainfall, frost periods, wind, extreme hazards, microclimates);
- site characteristics (slope, aspect, soil quality, drainage);
- biological risks (pests and diseases);
- existing vegetation (root competition, shading);
- water availability (either as rainfall and/or irrigation);
- cost of land;
- access to site;
- existing facilities and assets associated with past farming activities;
- availability and reliability of services (community, shire, transport, electricity and gas);
- availability of workers;
- air quality; and
- bio-security.

Local government and agency requirements for growing olives

It is essential when exploring the option of establishing a table olive orchard that dialogue be opened with all necessary agencies such as shire and local councils, and agricultural, environmental, health and food agencies. Particularly important are the establishment of water rights, an understanding of environmental requirements, and the health and safety implications associated with olive growing and table olive processing.

Climatic considerations

The ideal climate for olives is one that has a mild wet winter and a long, warm, dry summer typical of a Mediterranean-type climate. Such a climate is found in southern Australia. The olive, *Olea europaea*, produces fruit when average daily temperatures in

June/July are around 10–12°C or less. Unless there is enough water and sufficient chilling hours available at a particular site, olive growing is futile.

As indicated in Chapter 2, olive trees do not tolerate very high or very low temperatures. To avoid frost damage to the trees and fruit, olives should not be planted in regions where temperatures fall to below –5°C for prolonged periods. As the olive tree also suffers under very hot dry conditions, especially during flowering and fruit set, dry arid regions, even if land is cheap, should be avoided. Olive trees grown in regions experiencing high humidity levels (fog or summer rain) with poor air circulation in the orchard are prone to fungal diseases such as Peacock Spot.

In summary, marginal olive growing areas are those with heavy soils, steep slopes, extremes of climate as well as low and/or irregular rainfall. Regardless of the quality of the site, poor cropping over three successive seasons indicates a major problem with the grove and if this cannot be corrected, signifies poor commercial prospects. As with other fruit trees, paying attention to nutrition, pest and disease management, and irrigation increases the viability of the olive orchard.

Potential olive growing areas in Australia have been broadly defined by Nix (Fig. 3.1). Areas within which current olive growing is taking place are superimposed over the Nix predictions. However, it is important to examine local climatic conditions and weather patterns to determine the suitability of a site for olive growing. Climatic information can be obtained from local agricultural agencies or national meteorological organisations. Outside a Mediterranean-like climate, where some olive growing occurs or has been suggested for future olive growing – for example China, South Korea, Japan, India, Pakistan, Texas (USA) and north-eastern Australia – olive performance is likely to be variable with mixed commercial success.

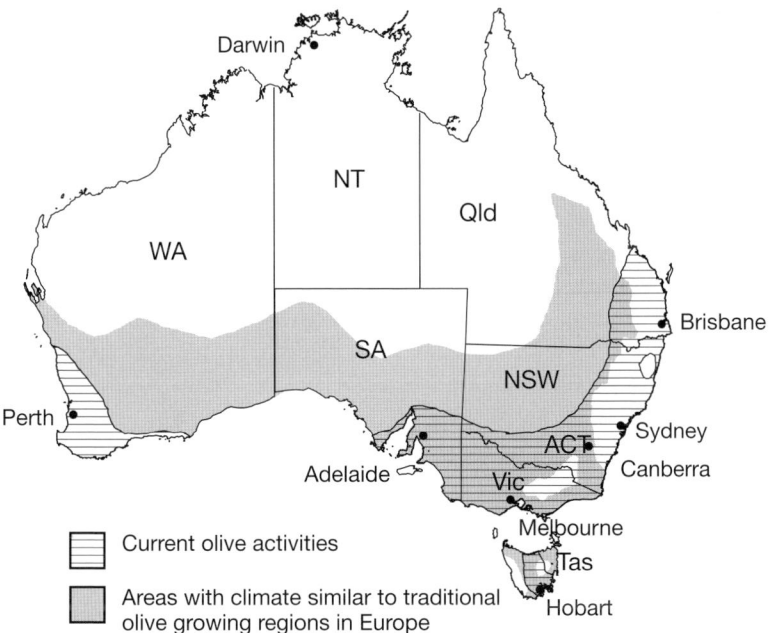

Figure 3.1 Relationship between areas of current olive activities reported (by Kailis and Sweeney) and predicted growing areas (by Prof. Henry Nix).

Chilling hours

Sufficient chilling hours for vernalisation (flower bud development) is essential in the olive otherwise it will persist in a vegetative state and will not produce fruit. Examples of approximate chilling hours available at various regions in southern and south-eastern Australia, where olives are growing, are as follows: Western Australia, 600–950 hours; South Australia, 600–1200 hours; Victoria, 1200–1400 hours; Tasmania, 1400–1500 hours; New South Wales, 900–1400 hours; Australian Capital Territory 1300–1400 hours; Queensland, 350–750 hours; and Northern Territory (Alice Springs), 600 hours.

A similar range of chilling hours occurs in olive growing areas especially around the Mediterranean Sea and the Middle East. Olive trees, especially local varieties growing in parts of southern Greece and Italy, northern Africa, and countries along the eastern Mediterranean basin, mostly require only a few hundred chilling hours to produce commercial crops.

Approximate chilling hours are calculated on the basis of days with minimum temperatures below 10°C for the months April–August less six hours for each day where the maximum temperatures exceeded 22°C during the period. Generally, fewer chilling hours are available in areas with warmer climates than in cooler ones.

Heat degree-days and radiation

Sufficient heat and radiation is required to ripen the olives. A measure of available heat is the number of 'heat degree-days'. In some olive growing regions, such as in Tasmania, or at elevated sites on the Australian mainland, there may be insufficient heat degree-days and radiation, particularly in late autumn/winter, to fully ripen the olives. In this case, some olive varieties may lock into the green-ripe or turning colour stages. This is not disadvantageous, as olive oil produced from such fruit is generally high in the nutritionally beneficial oleic acid and antioxidants. To overcome growing condition problems for olives in Tasmania, most olive orchards are found scattered near the north, east and south coasts of the island.

Frost prone areas and extremely low temperatures

Olive trees and fruit can suffer severe damage at temperatures of –5°C to –10°C, making sites that experience such temperatures unsuitable for table olive production. Prolonged frosts have similar effects. Newly planted olive trees suffer heavily under severe frosts as do young shoots and flowers on more mature trees. Frosts occur at inland sites on clear, cold nights without wind or at higher altitudes. Frost is less likely to occur in coastal areas, particularly those that are slightly elevated.

In localities that are frost prone, to improve the growing environment, one or more of the following strategies need to be employed (all will add to production costs):

- preparation of the orchard floor (smooth, weed free and moist) so the soil acts as a heat reservoir;
- spraying the olive trees with water (this raises the temperature);
- selecting sites that have good air circulation (remove impeding hedges or trees); and
- using mechanical devices such as tower mounted fans (to mix cooler and warmer air).

Other strategies that have also been tried to reduce frost damage include:

- avoiding nitrogen fertiliser application (that promotes new growth) in late autumn;
- using kelp sprays every 10–14 days;
- regular use of copper sprays (Bordeaux); and
- using anti-transpirant sprays (those that inhibit transpiration).

Hot dry areas

Without irrigation under hot dry environmental conditions olive trees undergo physiological shut-down to conserve moisture. Production is reduced and if hot conditions are prolonged the olive trees become stressed as does the fruit, making them less suitable for processing into table olives.

Hot dry winds at the pollination stage can reduce fruit set and productivity. Furthermore, under these conditions young trees and fruit desiccate and mature trees can lose limbs. Long dry summers, however, do reduce the risk of fungal diseases as well as ensuring the olives will ripen, particularly when naturally black-ripe olives are required. In regions with dry winters irrigation is required during winter–spring, as this time is critical for flower bud differentiation.

Water availability

The amount of available water is a major limiting factor in olive growth and production. Most Australian olive groves are irrigated or can be sustained by winter rainfall (600–800 mm/yr.) and supplementary irrigation. Estimates suggest 500–800 litres of water is required to produce one kilogram of olives. In areas with low rainfall, irrigation is required to offset any water shortfall. Olives can withstand drought, though fruit production is reduced. Under rainfed conditions the effective growing period for olives in southern Australia is mainly between March–November. Rain at pollination time can be a problem as fruit set and productivity are reduced. Therefore, a site with sufficient rainfall should be sought, as the cost of installing irrigation or trucking in water is expensive. With respect to rainfall a number of factors need consideration and these are listed in Table 3.3.

Table 3.3. Natural precipitation – considerations in growing olives

Aspects to be considered	Definition	Implications for olive growing
Distribution	When the rain falls	Rainfall in winter is more effective than in summer.
Variability	Rainfall consistency from year to year	Important for planning water storage and supply.
Frequency	How often it rains	Important for water storage planning.
Intensity	Total annual rainfall divided by the number of wet days with a rainfall greater than 0.2 mm	Rainfall intensity is greater on the east coast of Australia than southern Australia.
Evaporation	Evaporation of surface water	Soil water loss is greater at high environmental temperatures, low humidity and increased wind.
Effective rainfall	Calculated as rainfall minus evaporation over one month	In southern Australia rainfall is generally effective from March to November.

All olive orchards will benefit from irrigation water in addition to the immediate rainfall. Irrigation improves and regularises yield by increasing both olive fruit size and vegetative growth. The amounts required depend on the level of effective rainfall and the needs of the olive trees based on the quality of the growing environment. In Australia, irrigation water for olives is obtained from the following sources: on-site storage in tanks and dams; groundwater; commercially available water sources (dams, rivers, streams); and reticulated mains or town water systems.

Water is important both for growing olives and processing them into table olive products. As many boutique and small-scale olive growers vertically integrate their olive activities by also pickling olives and selling them in the tourist trade, they need quality water to sustain their enterprise. For growing olives, water should be non-toxic with low salt levels. Poor quality water with particulate matter and high levels of solutes also block pipes and sprinklers. For table olive processing the water used, even rainwater, must meet potable water standards to prevent spoilage and food poisoning. Seawater is an unsuitable processing medium because its composition is variable and it can have significant loads of chemical and microbiological contaminants.

Natural precipitation varies markedly within the area identified by Nix (Fig. 3.1), ranging from less than 100 mm/yr. to over 1000 mm/yr. As table olives need irrigating over summer and autumn, and in spring in regions with summer rain, additional sources of water are necessary for commercial production in low rainfall areas.

Olives have a favourable crop coefficient ($Kc = 0.4$–0.7) indicative of their particularly low water requirement. However, when estimating irrigation requirements for table olives and sufficient low cost water is available, a Kc value of around 0.7 should be used. This is based on research undertaken in California that has shown that this level of Kc yields larger and more profitable olives.

Site requirements

Olives will grow in most soil types as long as there is adequate drainage and the soil is not too shallow. Information on soil types and their distribution can be obtained from local, regional or national agricultural agencies. It is essential that olive trees be planted in healthy soils. Olive trees depend on the soil for the uptake of water, nutrients and oxygen. Poor soils can not only retard the growth of olive trees, but may also harbour detrimental bacteria, fungi and parasites. Therefore, for a particular site the soil should be evaluated prior to planting, or even before considering the acquisition of land, using soil pits and laboratory tests. As part of a sustainable management plan for the olive orchard, soils should be tested regularly and the necessary additions made, for example lime, organic matter and fertilisers.

Many Australian soils, particularly if they have been used for past agricultural or horticultural purposes, may be deficient in some nutrients or may contain toxic residues. Before purchasing properties or entering into other legal arrangements, particularly if organic table olive production is envisaged, the soil should be sampled and tested by a registered laboratory.

Well-structured friable (crumbly) soils are easy to cultivate and will permit planting of olive trees without the need for expensive, time-consuming remedial operations. Slightly acidic to near neutral (pH 5.5–7.0) friable soils should be selected for table olive growing as the need for soil amendments, hence establishment costs, are reduced. Soil structure can be improved by incorporating crushed limestone, dolomite, gypsum or organic matter. This is particularly important as relatively acid soils tend to become less fertile over time through leaching. Also, elements such as aluminium and manganese are more soluble in acid soil and may produce toxic effects to the tree when taken up.

When purchasing a property for olive growing, it should be examined especially during winter when waterlogged soils and effects of flooding are more likely to be revealed. Planting sites with slight slopes facilitate both air movement and water drainage, whereas flat areas with poorly structured soils are susceptible to water logging. If olives are planted on sites with poor draining soils prone to water logging, olive production is compromised. The olive tree suffers in poorly drained soils, causing stunted growth and death. Well-aerated soils are preferred, allowing roots to take up essential gaseous oxygen. When there is insufficient oxygen, soil microbes produce toxins such as hydrogen sulphide and methane that damage the roots. To overcome poorly drained sites, olive trees can be planted on graded ridges of 50 cm height and 2 m width. Olives are planted in the centre of the ridges. Dry organic matter placed along the ridges protects against erosion as well as helping to keep the soil cool and reduce water losses.

A favourable soil pH ensures efficient nutrient uptake to meet the requirements of the olive. Also, soils with high levels of organic matter are desirable. Very acidic or very alkaline soils should be avoided as amending these soils will be expensive and difficult. Lime and dolomite are used to increase the pH of acid soils, whereas sulfur, ammonium sulfate or superphosphate can be used to acidify alkaline soils. Although sandy soils have the advantage of good drainage, they mostly contain little organic matter and have poor water retention qualities, and will, therefore, require more effective irrigation and the addition of nutrients.

Sloping sites generally have good drainage but are prone to erosion and deterioration, particularly where there is high surface run-off. To minimise erosion, olives should be planted along contour lines. Although many of the olives grown in Greece and Italy are on slopes, cultivation and orchard operations are difficult. Developing orchards on sloping sites or hillsides may require expensive earthworks and terracing. Expensive equipment that can track up slopes may be required otherwise it may need to be done by foot or donkey (Fig. 3.2)!

Figure 3.2 Sloping olive orchard in Spoleto, Italy.

Other considerations

Boutique and small-scale olive growers/processors should set up their activities close to markets or be on a tourist route to reduce costs to customers (see also Chapter 4) and maximise profits. If cheaper suitable land is sought, it should provide some economic advantage, such as perfect growing conditions, rainfall and all weather access. A previously established orchard or farm may already have significant assets, such as roads, power, bores, water rights, pumps and tractors, which will reduce capital outlay for new equipment. If the orchard is to be sited near an industrial area or conflicting enterprises, then the risk of olive trees taking up pollutants or pollutants being deposited on the olives must be considered. If this happens there is the potential problem of chemical residues in the fruit. Furthermore, pungent and obnoxious chemicals in the air may affect fruit as well as deter potential customers. Biosecurity can be a problem if the orchard has no permanent resident or manager.

Varietal considerations for table olives

Choice of variety is important. Those planted for table olive production are selected according to different criteria to those for oil production. In practice a grower can specialise in one variety, such as *Kalamata* or *Verdale*, or several including *Manzanilla*, *Kalamata*, *Sevillana* and possibly *UC13A6*. Varieties with features such as early productivity, adaptation to growing conditions and resistance to environmental stress are advantageous in maximising financial returns.

Information on varietal suitability of olives under different Australian growing conditions is limited. Basic information derived from international practice and Australian research is given in Chapter 2. Researchers and industry stakeholders in Australia and elsewhere are actively seeking quantitative information on old and new varieties. Much of the data collecting has been directed to olives for olive oil production and there is a drastic need to systematically collect information on varieties relevant to table olives.

Of major interest is the authenticity of olive varieties being supplied to growers. Recent reports emanating from research underway at Roseworthy, South Australia, indicate that trees of the following varieties suitable for table olive production, when sourced from reputable nurseries, matched international standards: *Arbequina*, *Barnea*, *Frantoio*, *Hojiblanca*, *Kalamata*, *Manzanilla*, *Picual* and *Sevillana*.

Figure 3.3 Comparison of processed *Kalamata* (top) and *Barnea* (bottom) olives.

It is the responsibility of nurseries to provide planting stock of authentic varieties, the grower to deliver olives of authentic varieties for processing and processors to process and package olives of authentic varieties. With increasing table olive consumption, consumers are becoming familiar with the varietal names. Although databases are being developed using RAPD, AFLP (Amplified Fragment-Length Polymorphism) and micro-satellite technologies with olive leaf DNA to authenticate varieties, there appears to be a shortage of information on DNA typing using fresh or processed olives. Currently, if there is a dispute involving the authenticity of processed table olives, morphological evaluation is the only practical identification technique. A case in question is differentiating processed *Kalamata* and *Barnea* variety olives, which to the inexperienced observer show similar morphology (Fig. 3.3).

A systematic approach is required. There is a need to develop an Australian centre that is able to deliver to nurseries authentic certified disease-free propagating material from which mother trees true to variety can be developed. One such centre in Pescia, Italy, together with universities in Tuscany, ensures that local growers are able to source authentic varieties propagated under phytosanitary conditions. It will then be up to individual nurseries to implement quality management procedures that ensure growers receive the correct planting material.

The enigmatic varieties in Australia (*Verdale*, *Hardy's Mammoth*, *Jumbo Kalamata* and *UC13A6*) still elude authentication and verification of their original sources. Olive varieties currently available from Australian olive nurseries and sold for table olive production are listed in Table 3.4.

Table 3.4. Table olive varieties available for growing in Australia

Olive weight: medium, 2–4 grams; high, 4–6 grams; very high, >6 grams. Green, olives picked at yellow-green stage; black, olives picked at naturally black-ripe stage; Californian-style, olives picked at the turning colour (TC) stage.

Variety	Country of origin	Use	Fruit weight	Processing type
Ascolana Tenera	Italy	Table	Very high	Green
Azapa	Chile	Table	Variable high to very high	Green, black
Barnea	Israel	Dual	Medium	Green, black
Barouni	Tunisia	Table	Very high	Green, black
Californian Mission	Mexico/Spain	Dual	Medium	Green, black
Cucco	Italy	Dual	High	Green, black
Hardy's Mammoth	Uncertain	Dual	High	Green
Hojiblanca	Spain	Dual	High	Green, Californian-style black
Kalamata	Greece	Dual	High	Black
Kalamata Jumbo	Unknown	Table	Very high	Green/TC
Leccino	Italy	Oil	Very high	Black
Manzanilla	Spain	Dual/Table	High	Green, black, Californian-style black
Nab Tamri	North Africa	Table	Very high	Black
Oliva di Cerignola	Italy	Table	Very high	Green
Picholine	France	Dual	Medium	Green
Gordal Sevillana	Spain	Table	Very high	Green
UC13A6	North Africa/USA	Table	Very high	Green, black (dried)
Verdale	France	Dual	Medium	Green, black
Volos (Conservolea)	Greece	Dual	High	Green, black

Some of the most sought-after varieties for table olive processing are *Manzanilla*, *Sevillana*, *Barouni* and *Hojiblanca* for green olives, and *Conservolea*, *Kalamata* and *Hojiblanca* for naturally black-ripe olives. Other internationally recognised varieties that should be encompassed by the Australian table olive industry are *Chalchidikis* (Greece), *Nocellara del Belice* (Italy) and *Picholine* (France). Even oil olives such as *Frantoio* can be processed as Ligurian-style olives and *Leccino* can be processed for naturally black olives in brine or Kalamata-style olives. The very small fruit of *Arbequina*, popular amongst Spanish consumers, are seen as an alternative snack to salted peanuts. The use of olives from *Barnea* olive trees, a variety recently utilised for processing by the Kalamata method, is gaining interest in Australia, particularly if there is a shortage of *Kalamata* olives.

Varieties referred to in Table 3.5 reflect the olive germplasm available in Australia and include most of the internationally important varieties for table olive processing. *Verdale* has been a favourite of Australian olive growers for years because of its adaptation, productivity and ease of processing. *Kalamata* olives are well recognised for their firm flesh, organoleptic characteristics and their easily recognised shape. Large and very large olives, processed by the Spanish Green-style method, are generally called 'Queens' and include *Sevillana* (Spanish Queen), *Barouni*, *Azapa* and *UC13A6* (Californian Queen). There is scope for the introduction of new varieties for Australian olive growers, particularly from Greece, southern Italy, Turkey, North Africa and the Middle East.

Olives with a high Flesh:Stone ratio generally have greater acceptance with consumers, whereas over-sized olives, for example *Jumbo Kalamata* and those obtained by size sorting of other varieties, are suited for the boutique market. Olives with a high Flesh:Stone ratio absorb more salt from brines than those with low ratios, so that monitoring salt concentration of brines becomes more critical to ensure safety and prevent spoilage.

Olive pollinating varieties

Some olive varieties are better pollinators than others. Specific olive pollinators, for example *Manzanilla* for *Sevillana*, *Giaraffa* for *Nocellara del Belice*, *Manzanilla* for *Barouni*, or non-specific pollinators such as *Pendolino* or *Frantoio*, can be planted in the olive orchard to improve productivity. In some countries *Barouni* is used as a non-specific pollinator. In many traditional olive-growing regions where orchards have only one variety, specific pollinators are not generally used because of the large number of olive trees in the areas or the particular variety may be self-fertile. Where olive orchards are isolated from each other (such as in many parts of Australia) effective pollination is important. A potential problem is that if the olive orchard experiences high environmental temperatures, the level of sterility increases, hence productivity decreases. In practice, orchards with several olive varieties experience fewer pollination problems. Pollinating olive varieties should be planted at a minimum distance of 30 m from the recipient trees and make up 10% of the planted olive trees. Cross-pollination also helps reduce the proportion of parthenocarpic flowers in affected varieties.

Olive varieties and table olive processing

Australian growers/processors need to be strategic in their approach to table olive production and select olive varieties that have favourable growing and processing characteristics. Olives from heritage trees may have a local advantage or be sought after by consumers, for example *New Norcia* olives. Although most olive varieties can be processed as table olives, it is important that commercially viable varieties are chosen that can deliver consistent characteristics from season to season. Olive size, shape, Flesh:Stone ratio, ease of pitting, colour and texture are all very important selection criteria. Table 3.5 lists the factors to be considered when selecting olive varieties for table olive production.

As examples, *Manzanilla*, *Kalamata* and *Conservolea* have favourable characteristics for table olives.

Table 3.5. Factors to be considered in selecting varieties for table olive production

Feature	Description
Olive size and shape	Medium to large, 2–6 g, the olive size should be broadly uniform. Shape should be uniform ranging from spherical to elliptical dependent on variety.
Flesh:Stone ratio	Ideally this should be around 5:1; however, minimum ratios are 3:1 for black olives and 4:1 in green olives. Values are lower for naturally black-ripe olives left to dehydrate (shrivel) on the tree.
Flesh detachment from stone	Easier detachment of flesh from stone is advantageous for ease of eating and de-stoning.
Texture of flesh	The olive flesh should be non-granular and non-fibrous. Olives should be free of internal flesh damage such as browning due to infestation or environmental stress.
Olive firmness	Olives should be harvested so that they are firm enough to resist damage during harvesting and post-harvest handling.
Skin and flesh colour	Olives should have the characteristics required for the particular method/style of processing (green-ripe, turning colour and naturally black-ripe olives). The olive skin should be thin, fine and delicate.
Stone size, shape and surface	Olive fruit should have a stone that is small, round/elliptical and smooth without sharp protuberances and flesh that is easily detached from the stone (freestone).
Overall appearance	Olives should have a clean appearance with no injury or defects.

There is a great diversity of table olive varieties growing in traditional olive areas. Studies on those cultivated in Portugal, Spain, France, Italy, Greece, Turkey and Tunisia indicate links with ancient types. New varieties can also be developed through breeding and evaluating the fruit from feral olives or seedlings. A problem around the world has been varietal authentication. Current research is addressing this with the development of international libraries linking morphological characteristics, such as shape and size of olive leaves, fruit and stones, with DNA profiles. The large fruit of *Jumbo* or *King Kalamata*, which shows similarities to *Oliva di Cerignola* (Fig. 3.4), is worthy of further scientific investigation because of its popularity with Australian consumers.

Figure 3.4 Comparison of *Jumbo Kalamata* (left) and *Oliva di Cerignola* (right) olives.

New varietal considerations for table olives in Australia

Important olive varieties originating in traditional olive growing countries are listed in Table 3.6. However, at the international trade level, the important table olive varieties are

Kalamata, Conservolea, Manzanilla, Sevillana and *Hojiblanca*, and to a lesser extent *Oliva di Cerignola* and *Ascolana Tenera, Barouni, Chalchidikis* and *Mission* (Californian). Table 3.6 contains some varieties that are generally not available outside the country of origin, for example *Azeradj, Alfonso, Tanche, Carolea* and *Gemlik*, that should be considered for the Australian table olive industry.

Table 3.6. Table olive varieties growing in traditional olive growing countries

Olive weight: medium, 2–4 grams; high, 4–6 grams; very high, >6 grams. Green, olives picked at yellow-green stage; black, olives picked at naturally black-ripe stage; Californian-style, olives picked at the turning colour stage.

Variety	Countries	Use	Fruit weight	Processing type
Azeradj	Algeria	Dual	High	Green, black
Sigoise	Algeria	Table	Medium	Green, black
Arauco	Argentina	Dual	Very high	Green, black
Alfonso	Chile	Dual	Medium	Black
Oblica	Serbia and Montenegro (former Yugoslavia)	Dual	High	Green, black
Zutica	Serbia and Montenegro (former Yugoslavia)	Dual	Medium	Green, black
Picholine Languedoc	France	Dual	Medium	Green
Tanche	France	Dual	Medium	Black
Kalamata	Greece	Dual	High	Black
Karydolia	Greece	Dual	High	Green, black
Conservolea (Volos)	Greece	Dual	High	Green, black
Kothreiki	Greece	Dual	High	Green, black
Megaritiki	Greece	Table	Low	Black shrivelled
Ascolana tenera	Italy	Table	Very high	Green
Carolea	Italy	Dual	High	Green, black
Itrana	Italy	Dual	High	Black
Nocellara del Belice	Italy	Table	High	Green
Soury	Lebanon	Dual	Medium	Green, black
Picholine Marocaine	Morocco	Dual	Medium	Green, black
Carrasquenha	Portugal	Dual	High	Green
Galega Vulgar	Portugal	Dual	Medium	Black
Redondal	Portugal	Dual	Medium–high	Green
Manzanilla Cacerena	Spain	Dual	High	Green, black
Hojiblanca	Spain	Dual	High	Green, Californian-style
Manzanilla de Sevilla	Spain, USA, Israel	Table	High	Green
Al-Doebli	Syria	Dual	High	Green, black
Meski	Tunisia	Table	Medium–high	Green
Domat	Turkey	Table	Very high	Green
Gemlik	Turkey	Dual	Medium	Black
Memecik	Turkey	Table	High	Green, black
Californian Mission	USA	Dual	Medium	Green, black

As mentioned earlier, three popular table olive varieties in the Mediterranean region, *Arbequina, Chalchidikis* and *Nocellara del Belice*, are also worthy of consideration.

Arbequina. This small fruited, pea-like variety (Plate 2), generally used for olive oil production, is popular in Spain when processed in brine. This variety is of interest for olive oil production, because of its growth habit and its potential for mechanical harvesting.

Chalchidikis. This olive variety originates in the Chalchidiki Peninsula in northern Greece. It is also known as *Chodrolia*, and has similar morphological features to *Ascolana Tenera*. Although some difficulties have been experienced in processing this variety because of low levels of fermentable substrates and its pale colour, it has some advantages such as a high oil content (20%) and the large fruit size (6–10 g or more).

Nocellara del Belice. This is an Italian olive variety usually processed as a green olive. It can be grown under irrigation or under dryland conditions where the rainfall is around 600 mm/yr. The fruit is of similar shape to *Manzanilla*, weighs around 6–8 g and has a Flesh:Stone ratio of around 6–8:1.

Establishment of the table olive orchard

Olive orchard design

The planting density of the olive orchard depends on: soil quality (drainage and fertility); water availability (natural precipitation, irrigation); variety (size, competition) and growing techniques (pruning and fertilisation).

In practice, several different planting spacings have been developed:

- traditional: 100–300 trees per hectare (7 m × 6 m to 10 m × 10 m);
- dense planting: 400–500 trees per hectare (5 m × 4 m to 6 m × 4 m); and
- experimental: up to 1000 trees per hectare (6 m × 4 m to 4 m × 3 m).

Traditional planting densities for olives are more compatible with hand harvesting for table olives than denser plantings. Denser plantings are favourable when olive trees are less than 10 years old or where they have been trained for mechanical harvesting. The suggestion of dense plantings initially followed by removal of trees to reduce density as the olive orchard matures could be more theoretical rather than practical. Olive trees are planted in either square/rectangular or diamond patterns (Fig. 3.5). With diamond patterns, more trees per area can be planted in a specified area, but each tree has less soil volume from which to gather water and nutrients, and canopies may cause shading. An olive orchard planted in a rectangular pattern is shown in Fig. 3.6.

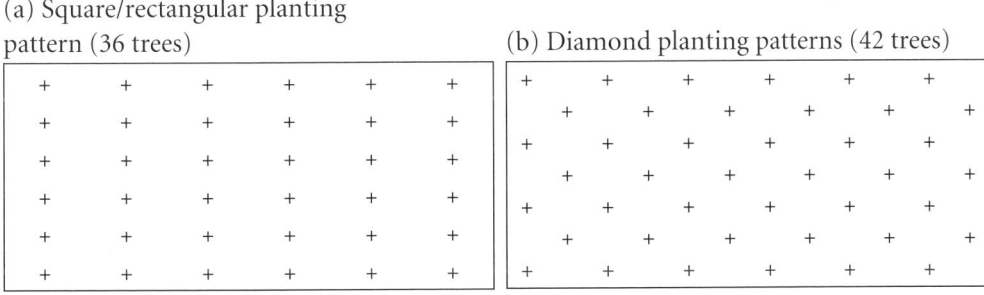

Figure 3.5 Planting patterns for new olive orchards.

Figure 3.6 An olive orchard planted in a rectangular pattern in Greece.

Olives can be grown effectively under rainfed conditions with limited water availability as long as the planting density is adjusted to the situation. Where annual rainfall is around 200 mm/yr. the spacing should be 24 m × 24 m (approximately 100 trees/ha) or less, while at higher rainfall levels, for example 600–800 mm/yr., the spacings are 6 m × 6 m (approximately 250 trees/ha). Planting in a rectangular pattern improves yield, possibly due to more efficient use of sunlight and other resources.

Olive trees require direct sunlight for growth, initiation of fruit buds, fruit yield and quality. Radiation levels in most parts of Australia are more than sufficient for olive fruit production and problems only occur when trees receive substantial amounts of shade. To maximise radiation at susceptible sites in the southern hemisphere, olive trees are best planted on north to north-east facing slopes with the tree rows in a north–south orientation. For northern hemisphere sites, olive trees are best planted on south to south-west facing slopes with the tree rows in a north–south orientation. Olive trees should not be planted adjacent to belts of eucalypt or pine trees to avoid shading and competition.

The commercial olive grower needs to manage the prevailing sunlight for the production of quality olives. Although most olive growing areas have adequate levels of sunlight, the tree canopy needs to be managed to ensure the amount of light interception is optimised, light distributes evenly within the canopy at levels that do not damage stems and branches, and that transient shade within the canopy does not affect bud development and fruit growth.

Soil requirements

The olive, although considered to be a hardy species, requires a high level of management to yield well. Soils should be assessed for pH, nutrient and organic matter levels and corrections made ideally before planting.

Knowing the *Water Holding Capacity* (WHC) of the soil is important as it can be used together with the rainfall data and evaporation rates to plan the irrigation requirements and schedules for the olive orchard. Not all the water in the soil is available to the olive trees. The water actually available to the olive tree is the *Readily Available Water* (RAW). When the roots of the olive tree take up the readily available water from soil, drawing on the remaining water requires much more work by the tree. So if the water in the soil profile is not replenished, eventually the olive tree becomes water stressed. Clay and sandy soils have the least RAW whereas loam and silt loam have the highest, with other soil combinations having intermediate values. As clay soils are fine-grained, they have a high percentage of air spaces that can hold water, hence creating a high surface tension requiring the olive tree to work hard to remove water. Sandy soils have a low RAW because of their low water holding capacity. The following RAWs can be used as a guide for different broad classes of soils.

Sand	40 mm/m
Loamy sand	55 mm/m
Sandy loam	65 mm/m
Sandy clay loam	70 mm/m
Loam	75 mm/m
Silt loam	80 mm/m
Clay loam	70 mm/m
Heavy clay loam	65 mm/m
Light to medium clay	55 mm/m
Heavy clay	40 mm/m

Further management aspects for common soil types are given below. Olive trees will grow on nutrient-poor sandy soils if well drained, but struggle in clay soils. In between sandy and clay soils are soils made up of mixtures of sand and clay in varying proportions, together with grit and humus. Soils that are crumbly have a desirable texture for growing olives as they are made up of 'crumbs' interspersed with air spaces. Such soils are easy to wet, allow water to pass through freely, drain easily, allow roots free movement and allow entry of air and ventilation for toxic gases.

Deep sandy soils. Organic matter needs to be added as well as small quantities of slow release fertilisers three to four times a year as excessive nitrogen or phosphorus applications can induce trace element deficiencies. Olive orchards with sandy soils, in the absence of natural precipitation, should be irrigated/watered lightly and frequently with micro-sprinklers rather than drippers.

Loam over clay soils. These types of soils have poor drainage and are prone to water-logging. Soil structure and fertility can be improved by mounding the soil along the planting line.

Heavy clay soils. Soils of this type require deep ripping, for example one metre along the planting line. Organic matter (straw, compost) or gypsum is placed in the rip lines. Olive trees are then planted in the rip line.

Poorly structured loams/clay loams. Soils of this type are managed by increasing the organic matter with animal and/or green manures and adding crushed lime (acid soils) or gypsum (neutral to alkaline soils). This soil type is best irrigated slowly with a dripper system.

Saline soils. Olives are moderately salt tolerant. Saline soils in Australia are commonly found along coastal strips and inland areas because of excessive clearing of native woodlands and poor soil management. Furthermore, rising water tables lift salt up the soil profile to the surface. This type of soil can be managed by adding gypsum to improve subsoil drainage and mounding the planting line. Olive trees planted in saline soil should be irrigated with a trickle system and the soil should be flushed annually to remove accumulated salt around the roots. Although some olive varieties have been reported to be salt tolerant, in time all varieties succumb.

pH adjustment of soils. Soil pH can be adjusted with agricultural lime or dolomite for acid soils and elemental sulfur for lightly alkaline soils. Superphosphate and ammonium sulfate can also be used as acidifying agents. The pH of the first 10 cm for an acid soil can be raised or lowered by one pH unit by making additions as indicated in

Table 3.7. Addition of sulfur is most effective if the initial soil pH is below 7.5. At soil pH values greater than 7.5 it is almost impossible to acidify the soil.

Table 3.7. Protocols for changing soil pH levels

Soil type	Raise pH one unit: add agricultural lime or dolomite grams/square metre	Lower pH one unit: add elemental sulfur grams/square metre
Sand	100	25
Loam	200	50–70
Clay	300–400	100

Nutritional requirements of the new orchard. Olive trees respond to fertilisers even though the nutrient requirements of olives are lower than fruit trees such as citrus, grapevines and other stone fruit. If there are nutrient deficiencies, major physiological effects can occur. Basal fertilisation should only be considered with poor soils. The macronutrient requirements for olives are met if the soil content of nitrogen is >0.1% w/w; phosphorus >0.013% w/w (or P_2O_5 >0.03% w/w); and potassium >0.025% w/w (or K_2O >0.03% w/w). At these levels, other macronutrients should be sufficient as long as the soil pH is favourable. In some centres phosphorus is measured as P_2O_5 and potassium as K_2O, so dividing values in these units by the factors 2.3 and 1.2 respectively gives the % levels of the individual elements.

Water requirements – the practicalities. Olive trees will suffer under water stress conditions that can occur in Mediterranean olive growing regions especially in summer/autumn. The minimum water requirement for the olive is believed to be around 200 mm/yr.; however, for mature olive trees to be commercially viable, a winter/spring rainfall of at least 600–800 mm is required. Unlike some other species growing under similar conditions, olive trees do not have a dominant tap root, especially when grown from self-rooted cuttings, but an extensive lateral adventitious root system that is able to collect water outside the canopy footprint when grown without irrigation. However, with irrigation the root system of olive trees is modified; irrigated olive trees have less than half the root system of trees that are non-irrigated. If olive trees grown under dryland conditions for a number of years are to be irrigated, care must be taken to ensure that the entire root system receives water, otherwise a stress signal from the non-irrigated portion of the root may retard growth. Because of the modified root structure, changing from irrigated to non-irrigated olive growing can be detrimental to the trees. With limited availability of water, supplementary watering is the best strategy.

Under non-irrigated conditions, olive yield is dependent on rainfall, the rainfall distribution, water and the water holding capacity of the soil. Orchard soils with good water holding capacity can support olive growing at lower levels of rainfall. At higher rainfall levels production increases as long as drainage is not a problem. As a rule of thumb the optimal water requirement of olive trees for commercial table olive production is the equivalent of 1000 mm/yr. of natural precipitation. That translates to 10 ML/ha/yr. So if the rainfall at a particular orchard site is 600 mm/yr., the equivalent of 400 mm of rainfall as additional irrigation (4 ML/ha/year) given at appropriate times is required to maximise production and olive size.

Saline water irrigation of olives. Olive trees are moderately salt tolerant so using saline irrigation water is possible. Although some olive varieties show more salt tolerance than others, eventually all olive trees are affected, exhibiting reduced growth, smaller crop and eventually death. High salt levels in the soil reduce water absorption by the olive tree roots and with increasing salt concentrations, shoot length, leaf area and leaf differentiation are reduced. At high levels of salinity the amount of leaf drop increases and photosynthesis is reduced. Salt accumulates in basal leaves and to a lesser extent in apical leaves. When olive trees are irrigated with saline water of 180 mS/m (1080 mg/l), olive fruit production is unaffected whereas at higher salt levels, 260 mS/m (1560 mg/l) and 560 mS/m (3360 mg/l), production falls by 10% and 20% respectively.

Saline water has been used successfully in Israel to irrigate commercial olive orchards. From a practical point of view, maximum safe levels of salt for irrigating olive trees without compromising production are 270–635 mS/m (approximately 1500–3500 mg/l or 1500–3500 ppm). Water with a high salt content is better suited to well-drained soils, such as sandy soil, rather than heavy textured soils and in areas where high seasonal rainfall can leach out accumulated salt. Nevertheless, with large fluctuations in moisture on sandy soils the risk of salt burn increases. Irrigation practices also need careful attention. Saline water delivered by trickle irrigation is less problematic than using fine sprays or intermittent sprinklers because of the continuous moist soil around the roots when using trickle irrigation, while spraying and sprinkling result in high evaporation rates. Sprinklers also deposit salt onto the leaves. More saline water can be used at each application if irrigation is infrequent or only used over short periods. Olive orchards growing close to the coast have an added problem in that salt is deposited by offshore winds onto trees and in the soil.

A visible symptom of salt damage to plants is burnt leaf tips and/or margins because of salt accumulation. First signs are a yellowing of the affected part of the leaf, progressing to dark brown (Plate 14). In severe cases defoliation and dieback occur. As similar leaf symptoms are a sign of other nutrient deficiencies, leaf analysis can be used to establish a cause.

Management strategies when using saline irrigation water include: avoiding the use of spray sprinklers, installing windbreaks, avoiding irrigation operations during periods of strong winds, decreasing the frequency and increasing the volume of water at each application, and improving drainage of heavy soils by incorporating gypsum (5 kg/m^2).

Preparing the orchard floor. Depending on the nature of the proposed orchard some basic works may need to be undertaken such as clearing, levelling, ripping, ploughing and contouring. Care must be taken to remove residual roots to reduce the likelihood of root disease being transferred to newly planted olive trees. Where necessary, the orchard floor is prepared by deep ripping, especially with duplex and heavy soils (Fig. 3.7). Based on soil tests, nitrogen and phosphorus fertilisers are

Figure 3.7 Deep ripping the orchard floor in preparation for planting olive trees. Notice the parallel rip lines.

applied into the rip lines, ploughed into the soil or spread around individual trees. One year after planting, leaf analysis should be undertaken to guide further nutrient applications. As indicated earlier, soil drainage can be improved by adding organic matter and/or gypsum. Most soil types can be improved, but of course soil modification can be slow and expensive. Ideally, major amendments should be made before planting the olive trees.

Deep ripping. Ripping to around 0.5 m or more is essential on most agricultural soils. It should be carried out in the summer months when a maximum shattering effect is possible. The ripping device should have a narrow tine with a broad/point boot to ensure maximum lift and shattering. Soil ripping has a number of functions. It dislodges rocks and roots that may interfere with growth, facilitates planting, provides ways by which roots can seek water, increases soil aeration and increases plant growth.

Weed control. Weeds in the olive orchard need to be controlled as they compete for resources with young olive trees and, if severe, can reduce productivity. Several methods of weed control are available: hand removal, mowing, cultivation, cover crops, mulches, heat sprays (fire, steam) and herbicides. Whichever methods are used, care must be taken not to injure the olive trees. Weeds are less of a problem in established olive groves as they are generally restricted to the inter-row spaces. This is because the olive is an *allelopathic* species; olive leaves and leaf litter release chemicals that inhibit the germination of weedy species that accumulate under the tree canopy. Thus, weed control under the olive tree canopy is less of a problem than in some other fruit orchards. (As an aside, it is an interesting observation that olive seedlings are less affected and are often seen emerging from the olive leaf litter.)

Current practice in Australian olive groves is mowing/slashing in the inter-rows, with hand weeding and/or chemical sprays along the rows. Non-chemical methods are used with organically grown olive trees.

Regarding cultivation, various mechanical devices can be used, such as hoes, disks and tillers. The weeds need to be turned over several times, preferably as soon as they emerge. Care must be taken not to cut or damage the olive tree roots or expose irrigation lines. If cover crops are planted in the inter-rows, these should not compete with the olive trees, otherwise additional water and nutrients will need to be applied. Organic and synthetic mulches reduce weed growth by blocking out light and are effective particularly for annual rather than perennial weeds. Herbicides should be used with extreme care, if at all, with young olive trees. If herbicides are used around olive trees, the trunks need to be protected. Glyphosate (a non-selective systemic herbicide that interferes with photosynthesis) is useful to control broad-leafed weeds once they have emerged. Other herbicides used in olive groves in traditional olive growing countries include simazine pre-emergence, and paraquat, diquat and glufosinate ammonium post-emergence. Paraquat and diquat, used either on their own or together, are contact herbicides that poison the green parts of plants, but do not translocate to roots.

Orchardists working with chemical pesticides should be trained to handle chemicals and chemical emergencies. When handling concentrates of herbicides, especially paraquat and diquat, gloves and face masks should always be worn.

Planting olive trees. After deciding on the planting densities, the growth habits of individual varieties, and, if possible, future growing and harvesting technologies, the orchard can be planted. Densities of 250–300 trees per hectare (8 m × 5 m or 6 m × 7 m) are currently the most suitable for table olives as these facilitate manual harvesting. At this planting density, each tree can utilise around 40 m² of orchard floor and the soil below.

Olive trees can be planted into 0.5 m × 0.5 m to 1 m × 1 m holes when dug manually on small sites, or in rip lines on large sites, and supported with the bamboo stakes supplied by the nursery. The sides of the planting hole should not be left glazed after digging; sides should be roughened to allow adequate movement of water and air. Planting stock, usually one to one-and-a-half year old, should be in good health and disease-free (Fig. 3.8). Heavy staking of olive trees is not always necessary but should be considered if the trees are tall, heavily foliaged, planted in sandy soil or in an excessively windy situation. Most small trees and shrubs do not require staking and will develop strong trunks if allowed to move freely in the wind.

For home gardeners small applications of blood and bone or a slow release fertiliser at the time of planting, particularly with sandy soil, is beneficial. Deep mulch, watering and the occasional application of manure or nitrogen fertiliser promotes growth in young trees.

Figure 3.8 Olive planting stock – 18-month-old trees.

Preferred planting times for olive trees are late autumn and early spring. However, as long as water (natural precipitation or irrigation) is available olives can be planted at most times of the year but not at times when frost is expected. The planting procedure should ensure that roots are not damaged either mechanically (by machinery) or chemically (by concentrated fertilisers) that can slow growth. The roots of the planting stock should be examined before planting. Damaged roots (smelly and dark brown) should be removed and suspect plants should be discarded. Circling roots of pot-bound olive trees should be redirected or cut off to prevent girdling of the plant later (this restricts transport of materials through the plant). The root balls should be dunked in water before planting then placed in the hole, making sure that the roots are not damaged. Soil is then added to the hole so that soil level is the same as in the original pot or bag. The tree is then watered. Covering the soil with straw after planting the olive trees will keep the soil cool and minimises loss of soil moisture.

Initial irrigation requirements. Newly planted olive trees require around 10 L/tree/week of good quality water delivered by bucket, tanker, sprinklers (sandy soils) or drippers (heavy soils) when rainfall is insufficient. Where possible, if irrigation lines are used, they should be installed after ripping and before the olive trees are planted. Initially one irrigation line is sufficient, but as the trees mature a second line can be installed (Fig. 3.9). Specially designed highly durable olive sprinklers are available that have insect-proof spinners providing several diameters of spray area (0.4, 1.0 and 4.5 m). The sprinklers are designed to compensate for undulating terrain. They have a number of disadvantages, such as water loss due to evaporation, and are prone to damage during orchard operations. Above- and below-ground dripper systems are also available. Above-ground drippers can be external or built in-line, with the latter less prone to damage. Underground drippers are also available and although very efficient can be impeded by root intrusion and are difficult to service.

Irrigation systems can be controlled manually, electronically or by computers. A number of systems are available that can give an indication of the level of moisture in the soil. Tensiometers and gypsum blocks are commonly installed at different soil depths, providing information on the movement of water, particularly during irrigation. They are

Figure 3.9 Established olive orchard in Western Australia with twin irrigation lines.

inexpensive and easy to use. A complete monitoring system uses a number of monitoring sites in the area to be irrigated. If required these sensors can be linked electronically to the irrigation system so the system is activated when the soil water levels fall.

Nutrient application for young olive trees. It is generally unnecessary to apply fertilisers to newly planted olive trees as their nutrient demand in the first year is low. In the season following planting, at the beginning of the vegetative cycle of the young olive trees, nitrogen fertilisers can be applied, for example 20–30 g ammonium nitrate (or equivalent), at three to four intervals and watered in to ensure uptake by the roots. Nutrient additions can be based on the results of leaf analysis for greater precision. For home growers an autumn application of citrus manure or a few handfuls of Dynamic Lifter™ watered in will suffice.

Thereafter, applications of fertilisers and manures need to be made on the basis of nutrient analysis of soil and leaf samples. Leaf samples for testing should be collected around mid-summer (if fertigation is being practised leaf analysis can be undertaken as required). The results obtained can then be used to ensure the correct balance of macronutrients and micronutrients in the olive trees. Soil analysis can be undertaken at any time.

Training young olive trees. Olive trees specifically planted for table olive production are trained as a bush or bush vase (Fig. 3.10) to facilitate hand harvesting of the fruit. With forms suitable for manual harvesting, the main objective is to bring the crop to the level of the picker. Radical pruning of the olive trees may be required to achieve this. However, severe pruning should be avoided in the first few years after planting, as the onset of fruiting will be delayed. After planting, the young olive trees should be cut back to a height of 60–80 cm in late autumn/winter to encourage the development of side branches around a central trunk. The lower the side branches the more the olive tree is encouraged to form a bushy shape. Allowing three to five side branches to develop from a height of about 30–60 cm above the ground allows the olive tree to take on this shape. For table olives the bush shape has a number of advantages over other tree shapes, including earlier fruiting, higher yields and reduced labour costs because the olives can be picked from the ground. With young olive trees, fingers as well as manual or pneumatic secateurs can be used to remove unwanted growth. Regardless of age, olive tree roots should not be kept moist and never be allowed to dry before, during and after pruning, particularly at times when rainfall is low.

Figure 3.10 *Kalamata* olive tree pruned to a vase shape in an Australian olive orchard. Notice the single irrigation line.

Olive tree maintenance for quality olive production

Raw olives for table olive processing should be produced according to good horticultural practices (GHP). The trees should be grown under conditions ensuring quality fruit production with chemical residue levels that meet the appropriate health standards, and harvested at the appropriate maturation states for the type of table olive product to be produced.

Soil management

Olives grow successfully on almost any well-drained soil between pH 6 and pH 8.5 and are tolerant of mild saline conditions. Normally olive trees take nutrients from the soil through their roots although nutrients can also be taken from solutions sprayed onto leaves (foliar sprays). Olive trees require a number of nutrients for healthy growth and production. Many of the nutrients come from the soil where they are either bound to soil particles, are within the organic matter or dissolved in soil moisture. Most soils contain adequate amounts of inorganic nutrients such as iron, calcium and magnesium for plant growth. Where deficiencies occur, corrections can be made by adding lime or gypsum to increase nutrient bioavailability, and/or trace elements can be added to depleted soils. Lime also improves the bioavailability of many nutrients especially in acidic soils. Potassium, found in most clay soils, becomes available to the plant only when released into the soil moisture. Sulfur, commonly found in the sulphide form, is not available to plants unless converted to the water-soluble sulfate form by *Thiobacillus* bacteria. Where sulfur availability is a problem, additions of gypsum can correct the deficiency. Nutrients in the sulfate form can also provide useful quantities of sulfur.

As nutrients need to be in solution to be absorbed, uptake is negligible during dry periods. When manures or fertilisers are applied, they must be watered in by natural precipitation or irrigation water, otherwise there will be little effect on the olive tree. Organic matter facilitates the uptake of nutrients as it decomposes and improves the water holding capacity of the soil. Microorganisms in the soil, such as bacteria and fungi, also facilitate nutrient uptake. Good soil drainage is just as important as soil fertility. Very fertile soils with high nitrogen levels promote excessive vegetative growth at the expense of fruit that are often fewer and smaller.

As nutrient availability is a function of soil pH, extremely acid and alkaline soils can reduce absorption of the nutrients (Table 3.8), hence the need to optimise the soil pH. Generally soils with a pH of around 6 give optimum uptake levels of most nutrients. A number of minerals are available at pH 5.5–7.5 (nitrogen, calcium, magnesium, phosphorus, potassium, sulfur, molybdenum and boron) whereas iron, manganese, zinc, copper and cobalt have greater availability between pH 4–6 and poor availability above pH 6.5. Iron chlorosis is commonly seen in olive trees growing on calcareous soils where the iron availability is reduced because of the high soil pH. Above pH 8.5 the availability of phosphorus and boron are greatly reduced. Nutrients that require regular replenishment in the olive tree include nitrogen and potassium, as these are lost through pruning and fruit production. Other minerals may not require replenishment as often as they are only used in small quantities and many are reabsorbed from ageing leaves before these fall off naturally.

Table 3.8. Nutrient bioavailability to plants at different pH levels
+, poorly available; ++, available; +++, readily available.

pH	4	5	6	7	8	9
Nitrogen	+	++	+++	+++	+++	+++
Calcium	+	++	+++	+++	+++	+
Phosphorus	+	++	+++	+++	+	+
Potassium	+	++	+++	+++	+++	+++
Magnesium	+	++	+++	+++	+++	+
Iron	+++	+++	++	+	+	+
Manganese	+++	+++	++	+	+	+
Zinc	+++	+++	++	+	+	+
Copper	+++	+++	++	+	+	+
Cobalt	+++	+++	++	+	+	+
Molybdenum	+	+	++	+++	+++	+++
Boron	++	+++	+++	+++	++	++

Excessive salt (sodium and chloride) and boron are toxic to the olive, both markedly affecting growth. Soils rich in aluminium oxide or manganese dioxide yield toxic levels of aluminium and manganese in very acid soils.

Olive roots. Three to four years after planting, olive trees develop substantial root systems that grow over time. Root systems from seedlings are dimorphic (initial tap root plus adventitious roots) whereas self-rooted cuttings develop only adventitious roots. The olive root system penetrates into the first metre of soil and exceeds the footprint of the canopy, possibly providing more effective water absorption when natural precipitation is light and intermittent. Olive trees respond to low levels of water (natural precipitation) by expanding their root systems to the point where there is a marked disproportion of roots compared with above-ground parts. This ensures that olive trees grow and bear fruit when cultivated in dry areas. Under irrigation, the distribution and structure of the olive root system tends to undergo modification, generally following the water depending on the different methods of water application, for example drippers or sprayers.

Adventitious roots form from xylopodes, the tissue between the trunk and the root crown. With heavy, poorly aerated soils most roots are found near the soil surface, whereas with lighter soils roots are found much deeper in the profile. Olive trees can search through a large volume of soil for water and nutrients with some of their lateral roots extending up to 12 m in length. With soils low in some nutrients, such as phosphorus, mycorrhizal filamentous fungi increase the efficiency of nutrient uptake.

General olive nutrition. Chemical fertilisers (NPK) with or without trace elements are commonly used to provide nutrients to olive trees. Organic olive growers generally use compost, well-composted animal manures or blood and bone, together with approved 'natural' nutrients such as rock phosphates, lime, gypsum and dolomite.

Regardless of whether chemical or organic fertilisers are applied to olive trees they will only promote increased growth and fruiting if the existing soil provides nutrients at pertinent rates. If there are nutritional deficiencies, abnormalities can occur in all parts of the tree including leaves and fruit. The nutritional status of olive trees is assessed and managed through leaf analysis and appropriate amendments made. Soil analysis data is more difficult to use to correct olive tree nutritional imbalances as soil structure and pH have a marked effect on the bioavailability of nutrients. However, soil analysis can be used to determine pH abnormalities that can impact on nutrient uptake as well as identify salinity and mineral toxicity problems.

The same processes regulate the uptake of nutrients whether they are in the form of fertilisers, manures or pure chemicals. When olive fruit are dried, 95% w/w of their dry weight is carbon, oxygen and hydrogen. The remaining 5% w/w consists of mineral elements. The more important ones are listed in Table 3.9. Olive trees grow reasonably well in nutritionally poor soils. If growing in nutritionally rich soils they are more vigorous, producing excessive foliage with less fruit. When olive trees are experiencing problems it is more likely that the cause is drainage, water logging, weed competition, or pests and diseases than a nutritional deficiency.

Table 3.9. Important mineral elements required by olive trees

Macronutrients	Micronutrients
Nitrogen (N)	Iron (Fe)
Phosphorus (P)	Manganese (Mn)
Potassium (K)	Zinc (Zn)
Magnesium (Mg)	Copper (Cu)
Calcium (Ca)	Molybdenum (Mo)
Sulfur (S)	Boron (B)
	Chlorine (Cl)

Nutritional losses through harvesting. Under Australian growing conditions it has been estimated that every tonne of harvested olives represents a nutritional loss of 9 kg of nitrogen, 2 kg of phosphorus and 10 kg of potassium. Published international estimates from Spain, France and Tunisia are of the same order as these, although phosphorus losses were less than 1 kg/t of olives. The authors of this publication found a wide range in the values of nutrients in raw olives, and their data presented in Chapter 2 (which also includes some micronutrient information) concurs with the above nutrient losses.

Fertilisers and manures

Nutrients can be supplied to olive trees as manures and solid fertilisers or as soluble fertilisers applied through the irrigation system or as a foliar spray, for example urea, boron or potassium. Olive trees store nutrients that they draw upon at the start of the growing season. For instance, research has found that leaf levels of nitrogen, phosphate and potassium are high before heavy cropping. Some investigators have found that the optimal levels of the principal macronutrients in olive leaf dry matter are: nitrogen 2.1%, phosphorus 0.34% and potassium 1.05%. These values are consistent with reported nutritional tables.

When selecting nitrogen fertilisers, attention must be paid to the existing soil pH as they can acidify the soil with long-term use and infiltrate into the groundwater system creating environmental problems. The greatest acidifying nitrogen fertiliser is ammonium sulfate, whereas urea has a lesser effect and calcium ammonium nitrate has little effect on soil pH.

Animal fertilisers and organic matter have much lower levels of nitrogen and potassium than formulated fertilisers (NPK). Fertilisers and manures are generally applied at least twice a year for rainfed olive trees (Autumn and Spring) depending on the soil quality. If the olive trees are irrigated, fertilisers and manures can be applied periodically during the growing season (see also the discussion on fertigation below). As a rule of thumb, depending on the variety and growing conditions, new shoots respond to manures and fertilisers by growing about 20–50 cm in length. Some common fertilisers (organic and synthetic) available for olive growing are listed in Table 3.10.

Note: If olive trees have adequate nutrition through conventional application, nutrient foliar sprays are of little value on biennial bearing, shoot growth, flower numbers, perfect flower numbers, fruit set, fruit yield, fruit size, shot berries or preventing fungal diseases such as Peacock Spot.

Organic manures

There is increasing interest around the world in producing organically grown table olives. Numerous Mediterranean olive growers and some in Australia have taken up this technology, with many of them applying organic manures regularly to their orchards. Unless high analysis organic fertilisers are used, the amounts of organic manures applied are up to 2 t/ha. An advantage with organic manures is that they are less likely to cause pollution by leaching nutrients into groundwater, lakes or waterways than the more soluble chemical fertilisers. With organic fertilisers, nitrogen is bound organically in plant material, particularly the protein components. Therefore, it takes time and under microbial activity the protein-bound nitrogen is released in a useful form that can be taken up by olive trees.

Protein-bound nitrogen → Amino acids → Ammonium and nitrate forms
Ammonium and nitrate forms → Absorbed by the olive tree roots

A number of organic manures are available to the olive grower and some of these are listed below. Iron induced chlorosis is less likely when olive trees are grown on soils rich in organic matter, because of the formation of iron chelates that are better absorbed.

Compost. Compost is basically a mass of rotted organic matter. Unless fortified, composts have a low nutrient content, containing around 2% nitrogen, 0.5–1% phosphorus and 2% potassium. For compost nitrogen to be taken up by the roots, composts must decompose, which can take up to 15 years. The level of nitrogen uptake should be assessed on the basis of leaf analysis, especially as annual applications of compost may cause a build-up of nitrogen levels. Apart from adding nutrients to the soil, the value of compost is improving soil structure, resulting in better aggregation, pore spacing and water holding capacity. Compost also provides humic acid, a mixture of compounds that stimulate plant growth.

Blood and bone. Blood and bone contains nitrogen (5%) and phosphorus (5%) but little potassium. Some special blood and bone products are supplemented with potassium and trace elements but these combinations are expensive.

Animal manures. Chicken manure contains nitrogen (2–3%), phosphorus (2–3%) and potassium (1%). In contrast to chemical fertilisers (NPK), these do not acidify soils. Apart from adding nutrients to the soil, the value of composted chicken manure is that it improves soil quality by increasing its aeration, water holding capacity and ability to support soil biota. Other animal manures, such as sheep, cattle and horse can also be used. Manures can have a high salt content that could be disadvantageous in salt prone soils or if used in large quantities.

Table 3.10. Commonly available fertilisers/manures for olive growing (after Baxter 1997)

Fertiliser/manure	Approximate % nitrogen	Approximate amount (kg) to deliver 1 kg of nitrogen	Additional nutrients
Ammonium nitrate	34	3	–
Ammonium sulfate	20	5	Sulfur (24%)
Potassium nitrate	13	7.7	Potassium (38%) Preferred form for fertigation
Calcium ammonium nitrate	25	4	Also adds calcium
Superphosphate	0	0	Phosphorus (8.8%) Sulfur (11.5%)
Mono-ammonium phosphate (MAP)	12.5	8	Phosphate (>20%)
Di-ammonium phosphate (DAP)	21	5	Phosphate (>20%)
Urea	46	2	
Potassium chloride	0	0	Potassium (50%)
Potassium sulfate	0		Potassium (42%) Sulfur (18%)
NPK Blue™	12	8.3	Phosphorus (5.2%) and potassium (14.1%) Contains some trace elements: sulfur, iron and zinc
NPK Blue Plus™	12	8.3	Phosphorus (5.2%) and potassium (14.1%) Contains some trace elements: calcium, magnesium, boron, sulfur, iron and zinc
Blood and bone	5	20	Phosphorus, potassium and calcium Contains some trace elements
Chicken manure	2–3	33–50	Phosphorus and potassium Contains some trace elements
Dynamic Lifter™	4	25	Phosphorus (3.2%) and potassium (1%) Contains some trace elements: calcium, magnesium, manganese, boron, copper and zinc
Compost	1–2	50–100	Phosphorus and potassium Contains some trace elements

Dynamic Lifter™. Standard Dynamic Lifter™ is pelletised organic slow-release manure based on chicken manure and blood and bone. It has an alkaline pH but differs from raw chicken manure in that levels of nitrogen, phosphorus and potassium are higher. Enriched Dynamic Lifter™ is also available as a blend of fast acting nitrogen and mineral fertilisers with slow-release organic nutrients (Long Life Complete™ 5–3–5 and

Orchard Lifter™ 8–3–8). In these products composted poultry manure and blood and bone is enriched with fishmeal, zeolite, nitrogen and potassium. The manufacturer suggests using the following as a guide for olive trees: Standard Pellets (1.5–2.5 kg/tree) or Long Life Complete™ (1.0–1.5 kg/tree) as a pre-planting application followed by Orchard Lifter™ in their first year (2–3 kg/tree) and for bearing trees (1–2 kg/tree). With commercial olive orchards, the need and quantity of these manures to be added should be checked by leaf nutrient analysis.

Green manure (leguminous crops). Leguminous cover crops, such as clovers and lupins, can be planted in autumn within the inter-rows of the olive orchard (instead of persisting weedy plants and grasses) and turned over in spring. Legumes are able to provide nitrogen to the soil in addition to providing green mulch when mown. With intercropping techniques it is important to keep the trunks free of growing plants to reduce competition and the risk of pests. This can be undertaken by hand weeding, steam sprays or flame burners in 'organic' orchards. Herbicide sprays in addition to these can be used in 'non-organic' orchards.

Fertigation of olive trees. Fertigation is the process whereby water-soluble nutrients are supplied to trees through the irrigation system. Application by this method is more precise than by providing granular fertilisers or manures to the soil. With fertigation, nutrients can be applied according to the horticultural needs of the tree. As nitrogen and potassium are heavily utilised in productive olive trees, demands can be monitored by several leaf analyses over the season and replacement made at rates relative to use. Of course all nutrients supplied by fertigation must be in a soluble form or a form that will pass through the irrigation system and be available to the olive tree. Although other macronutrients such as calcium, magnesium and sulfur are available in forms suitable for fertigation, because of cost and incompatibilities, conventional forms such as lime, gypsum and dolomite are spread in the usual way.

Note: Although irrigation can be delayed if there is sufficient rainfall in late spring, fertigation should commence before spring budding.

Foliar sprays. Nutrients applied as a spray can be taken up quickly by leaves. Foliar sprays can be used as an emergency or supplemental measure at key times during growth and development. They are particularly beneficial where winter injury has damaged the conduction system impeding the movement of nutrients from the roots to the rest of the tree. Foliar sprays are useful for the management of nitrogen deficiency as well as trace element deficiencies (boron, zinc and manganese). Foliar spraying has a number of disadvantages: the effects are temporary, uptake may be low, losses can occur through run-off or washing off by rain, translocation rates are limited and leaf damage may occur. Also, where the olive is replete with a particular nutrient, foliar sprays have little additive effect. Spraying nutrients during periods of high temperatures and very dry conditions may cause leaf damage so should be avoided, as should spraying during blossom time as this may interfere with pollination, hence affecting yield.

Nitrogen fertilisers

Ammonium sulfate. Ammonium sulfate, available in crystalline and granular forms, contains approximately 21% nitrogen and 28% sulfur; thus, 5 kg of ammonium sulfate delivers 1 kg of nitrogen. It is acidifying, so can be used to modulate the pH of alkaline soils but is a disadvantage when the soil already has a low pH. Following nitrification, ammonium sulfate nitrogen is available after two to three weeks. When spread around olive trees, it should be mixed into the soil to prevent evaporative losses of ammonia. Conversion of ammonium nitrogen to nitrate by microorganisms is slow and dependent on soil moisture and pH; it is more rapid in alkaline soils.

Ammonium nitrate.   Ammonium nitrate, containing around 34% nitrogen, is available in a very water-soluble granular form, ideally suited to situations where water is limited. Olive roots readily take up the nitrate component, but excessive amounts can leach into groundwater creating a potential environmental problem. When spread around the tree ammonium nitrate nitrogen is available within a few weeks after application. Like ammonium sulfate, ammonium nitrate can acidify soils. Nitrate nitrogen is more susceptible to leaching than ammonium nitrogen.

Calcium ammonium nitrate.   Calcium ammonium nitrate, containing around 25% nitrogen, is available as a granulated form. It is ideally suited to olive orchards with more acidic soils, as it has a neutral reaction. A further advantage is that it also provides calcium.

Urea.   Urea is a water-soluble organic nitrogen compound releasing its nitrogen as ammonia. Approximately 2 kg of urea delivers 1 kg of nitrogen. Applications have an acid reaction so its use should be avoided with acid soil. After applying urea it should be covered to reduce evaporative losses of ammonia, particularly with calcareous soils.

Potassium fertilisers

Potassium sulfate.   Potassium sulfate is available in several forms: powder, granular and soluble. It contains around 40% potassium and 17% sulfur. It is the preferred form of potassium for alkaline soils.

Potassium nitrate.   Potassium nitrate, available in soluble and granulated forms, contains around 37% potassium and 13% nitrogen. It is the preferred form of potassium for use in irrigation systems and as a foliar spray.

Phosphate fertilisers

Superphosphate.   Superphosphate contains 8–9% phosphorus and 11–12% sulfur and other minerals such as calcium. It is ideal for top-dressing grass-legume crops.

Soluble phosphate salts.   Soluble salts, such as mono-ammonium phosphate, containing both nitrogen and phosphate (see Table 3.10) are used for application by fertigation. Both these phosphate salts can come in forms that are not 100% pure. Commercial mono-ammonium phosphate for use as a fertiliser can have a profile with the following composition: mono-ammonium phosphate (47–78%), iron ammonium phosphate (3–6%), aluminium ammonium phosphates (2–13%), ammonium sulfate (3–5%), magnesium ammonium phosphates (2–5%) as well as calcium and fluoride. Di-ammonium phosphate can have a profile with the following composition: di-ammonium phosphate (60–85%), aluminium ammonium phosphates (4–10%), iron ammonium phosphates (3–5%), ammonium sulfate (4–5%) and magnesium ammonium phosphates (2–5%), as well as calcium and fluoride.

Olive orchard nutrition specifics

In practice, most of the nutrients needed by olive trees can be supplied through the soil. Supplying additional nutrients to olive trees, mostly nitrogen, potassium and trace elements such as boron, in commercial orchards should only be made on the basis of leaf and soil analyses. Nutrients translocate between the leaves, branches and fruit, and when leaves senesce and fall most of their nutrient has been reabsorbed for reuse – an important conservation mechanism.

Nitrogen nutrition. Nitrogen is required by olive trees for the synthesis of many essential compounds including proteins, nucleic acids and chlorophyll. Nitrogen is essential to the olive tree for vegetative growth and fruit production. It increases leaf chlorophyll levels and photosynthesis, hence promoting shoot growth, flowering and fruit production. Providing olive trees with additional supplies of nitrogen well before flowering and fruit set has proved beneficial, increasing growth and yield. It also increases the ability of the olive tree to utilise other nutrients.

As only the most fertile soils contain sufficient nitrogen, olives benefit when extra nitrogen is applied. Note that nitrogen is lost when olives are harvested and when the olive trees are pruned. Nitrogen can also be lost from the soil through leaching by rain and/or irrigation, by run-off and by denitrification: the conversion of nitrite/nitrates to gaseous nitrogen and nitrous oxide. Mulched olive prunings can return some of the original nitrogen back to the soil. As nitrogen is required during flowering and all events to stone hardening during late spring to summer, nitrogen applications should be made in autumn and early spring. Providing a balanced NPK fertiliser is useful in most situations.

Under most circumstances nitrogen in the soil has poor bioavailability because it is organically bound. It is slowly released through the action of microorganisms in the soil. Through their actions, water-soluble compounds such as nitrate ion, ammonium ion and urea are released. As both nitrate ions and urea are relatively more mobile than ammonium ions they are readily absorbed by the roots. Uptake of ammonium is slower because it attaches to soil particles due to its positive charge. The olive tree can assimilate nitrogen derived from a number of different sources: by nitrogen fixation, from dead and decaying organic matter, and through the application of fertilisers. Another potential source of nitrogen is groundwater. To be absorbed from soil by the roots, nitrogen compounds need to be in a biologically available form, such as NO_3^- (nitrate) – the most common form – NH_4^+ (ammonium), urea or amino acids.

Nitrogen fixation occurs when gaseous nitrogen is chemically reduced then incorporated into nitrogenous compounds such as NH_4^+ (ammonium) and NO_3^- (nitrate). In nature this process can occur by several mechanisms:

- the action of nitrogen fixing microorganisms in the soil;
- by symbiotic nitrogen-fixing bacteria associated with leguminous plants (these can fix up to 350 kg of nitrogen per hectare per year);
- during thunderstorms (nitrogen compounds formed in the air by the interaction of lightning and gaseous nitrogen are carried to the soil with rain); and
- by photochemical fixation in the atmosphere.

Another important process involving bacteria and nitrogen dynamics is nitrification. This involves the oxidation of ammonia (NH_3) and/or nitrite (NO_2^-) to nitrate (NO_3^-) in the soil. Nitrate ions are then taken up by the roots to be used for the synthesis of nitrogen containing compounds by the olive tree.

Nitrogen deficiency in olives. Nitrogen deficiency results in the following characteristics:

Tree. Restricted spindly growth with sparse foliage and little shoot growth.

Leaves. Generalised chlorosis (Plate 15) especially of older leaves; small pale green/yellow leaves without necrosis are also features of nitrogen deficiency. Younger leaves remain greener longer because they receive soluble forms of nitrogen transported from older leaves. Initially older leaves then younger leaves are affected.

Fruit. Size and oil content is reduced.

Care must be taken when using straw (predominantly comprised of carbon compounds) as a mulching agent for olive trees because when straw is broken down by microorganisms, these also consume nitrogen. Adding animal manures or nitrogen fertilisers to the straw can prevent this problem.

Excessive nitrogen applications before fruit set can result in small fruit due to high fruit load, accentuate the alternate bearing phenomenon and favour vegetative growth over flowering. In contrast, nitrogen application after fruit set promotes new wood production and a high yield the following year.

Nitrogen levels are adequate in the olive tree if the nitrogen concentration in dried olive leaves is >1.5–2.0% w/w but deficient if the concentration is <1.4% w/w.

Nitrogen can be applied as granular fertilisers, by fertigation or foliar spray. Applications of 0.5–1.5 kg nitrogen per tree annually up to 150–200 kg nitrogen per hectare are recommended as long as the dried leaf nitrogen content is >1.5% w/w nitrogen. The level and timing of nitrogen applications should be determined on the basis of annual rainfall, soil quality and available soil moisture, and guided by leaf analysis measurements.

When using urea, care must be taken not to exacerbate losses through volatilisation. If applied on the surface of the orchard floor, up to 20% can be lost through conversion to gaseous ammonia. Losses can be prevented by covering applied urea with soil and by watering it in so that the ammonia released is taken up by the soil. With most soils, urea is completely converted to the ammonium form of nitrogen within one week. Urea is not suitable for newly planted olive trees. The indicative amounts of nitrogen to be applied to olive orchards under various rainfall conditions are given in Table 3.11.

Table 3.11. Indicative nitrogen applications for olive trees relative to available water

Available water	Farming technology	Nitrogen application/tree	Nitrogen application/hectare
Less than 400 mm annual rainfall	Dryland	100–150 g/100 mm rainfall	10–15 kg/100 mm rainfall
400 mm to 700 mm annual rainfall	Dryland	Increase proportionally up to 1.5–2.0 kg	Increase proportionally up to 150–200 kg
Greater than 700 mm rainfall	Dryland	Add up to 1.5–2.0 kg depending on soil fertility	Add up to 150–200 kg depending on soil fertility
Irrigation water	Complete irrigation	Add up to 1.5–2.0 kg depending on soil fertility	Add up to 150–200 kg depending on soil fertility

Fertigation with nitrogen. As plants absorb and use most of the nitrogen in the nitrate or ammonium ion forms, the main types of nitrogen used in fertigation are ammonium nitrate, ammonium sulfate and urea or mixtures of these. Urea is often preferred because of its solubility and high nitrogen level. When using urea, the following points should be noted: as urea leaches easily it should be applied at the end of the irrigation cycle, it induces alkalinity of the water and soil until it breaks down, and daily urea fertigation is not recommended. Research in the Mediterranean region has shown that additions of around 600 g of nitrogen per tree per year as a urea solution applied by fertigation resulted in good growth and olive fruit production. Combinations of nitrogen as nitrate and ammonium have been suggested as providing the best balance for absorption by olive trees. When providing nitrogen to olive trees by fertigation its application should be evened out among all irrigations throughout the irrigation season. This strategy also takes into account the relatively high mobility of nitrogen in soil and reduces the effects of nitrification and leaching.

Foliar applications of nitrogen. Nitrogen can also be applied as 4–6% urea foliar spray, particularly in dryland farming olive orchards where nitrogen absorption by roots is restricted by a lack of water. Leaves, inflorescences and fruits rapidly assimilate the urea when applied using this method, and repeated applications during the growing period result in improved fruit set and retention. Researchers in California have noted increases in leaf nitrogen three days after foliar application of urea. There is some evidence that foliar urea treatment reduces fruit drop selectively for fruit at the base of the branch; a similar effect as occurs with apex removal. Although foliar application of urea is more effective in increasing yield than nitrogen applied to soil, it has no additive effect or commercial value if nitrogen has already been added to the soil as a fertiliser.

Phosphorus nutrition. The olive tree requires phosphorus to promote root growth and flower bud formation. At the cellular level phosphorus is needed for many biochemical processes including the following: cell division (phosphorus is present in nucleic acids (DNA, RNA)), development of meristematic tissue (new growth), photosynthesis linked carbon fixation from carbon dioxide, intermediary metabolism (the Tricarboxylic Acid Cycle involved in energy exchange), and the utilisation of sugars and starch.

The need for applying phosphorus to olive trees is questionable and should be guided by laboratory tests. Phosphorus deficiencies have been identified in some Mediterranean olive growing regions, but not in California. As olive is grown over a scattered but wide range of sites in Australia, further research is required to identify problem areas. It should be noted that many sites turned over to olive growing had broad acre and pastoral activity in the past and superphosphate had been applied on a regular basis.

Depending on the soil phosphorus levels, bearing olive trees may require no additional phosphorus by fertigation, or amounts up to 50–70 kg phosphorus per hectare (0.2–0.3 kg/tree).

Phosphorus levels are adequate in the olive tree if the phosphorus concentration in dried olive leaves is >0.1–0.3% w/w but deficient if the concentration is <0.05% w/w.

Phosphorus deficiency in olives. If phosphorus is deficient, the tree may exhibit restricted spindly growth, similar to that seen in nitrogen deficiency. The leaves are usually a dull, dark bluish green colour with a purpling of petioles and veins on the underside of young leaves. Otherwise, nitrogen and phosphorus deficiencies have similar signs.

Fertigation with phosphorus. Phosphorus for fertigation can be supplied as phosphoric acid, mono-ammonium phosphate or di-ammonium phosphate. Ammonium phosphates provide additional nitrogen but have the disadvantage that with irrigation water high in calcium or magnesium, insoluble phosphates form that can clog outlets and pipelines. Phosphoric acid has the advantage that it is easy to use and, providing the pH is kept low, will not form insoluble phosphate. As phosphorus has similar mobility characteristics to potassium, it is preferable to make its application by fertigation in a few irrigations.

Nitrogen, magnesium and calcium deficiencies can also occur under severe phosphate deficiency whereas boron deficiency can occur at both very low and high phosphorus levels. The requirement of phosphorus by olive trees is generally lower than for other crops and application rates should be checked against leaf analysis data. Regular use of superphosphate or complete NPK fertilisers helps maintain soil fertility, probably supporting ground cover (particularly clovers) rather than the olive trees. Other forms of phosphate that can be applied to olive trees include ground phosphate rock, or organic manures (blood and bone, chicken manure). Applying excessive amounts of phosphate to olive trees does not translate to increased yield. Furthermore, overuse of phosphorus rich fertilisers, particularly in sandy soils, can inhibit the growth of beneficial mycorrhizal fungi in the soil that would otherwise extend the trees' capability to absorb phosphorus.

Foliar application of phosphorus. Empirical application of foliar phosphorus at around 30–40 mg phosphorus per mature tree is not an uncommon practice undertaken by olive growers.

Potassium nutrition. Olive trees require potassium, particularly when they are bearing fruit. It is the dominant cation in plants.

Potassium has a number of important physiological and biochemical functions including: photosynthesis, respiratory processes, carbohydrate transport, synthesis of nitrogen compounds and carbohydrates, movement of water, and balancing nitrogen fertilisers.

Potassium is not readily available in the soil, particularly in the absence of moisture. Low ground temperatures reduce water absorption, hence potassium absorption. Much of the absorbed potassium (60%) ends up in the fruit and so is lost by harvesting. High yields have been found to correlate when leaf potassium levels are high. Excessive use of potassium leads to magnesium deficiency. Of course potassium applications should be guided by leaf analysis data.

Under normal conditions, a bearing olive tree needs around 300–400 g of potassium per year up to 0.6–0.8 kg/tree (160–210 kg/ha). As potassium is able to fix to soil, it is less susceptible to leaching and is less mobile than nitrogen. Potassium applications are not usually required in newly planted and non-bearing olive trees.

Potassium levels are adequate in the olive tree if the potassium concentration in dried olive leaves is >0.8% w/w but deficient if the concentration is <0.4% w/w.

Potassium deficiency. In cases of gross deficiency there is substantial defoliation (leaf loss) and reduction of internode spaces and shoot growth, even though the total node number is normal and the resistance of the olive tree to cold and dry conditions and disease is reduced. Olive trees may be in a potassium deficient state long before these leaf problems occur.

Interveinal chlorosis of older leaves, followed by dehydration and necrosis (scorching) (see Plate 14) and progressive discolouration of the leaves from tip to base, occurs, particularly with basal leaves.

Potassium can be applied to the soil as potassium sulfate, potassium chloride, NPK or by fertigation from spring to autumn. Potassium chloride has the disadvantage that the extra chloride can add to pre-existing salinity conditions. If only potassium is required, then digging in potassium sulfate (4–8 kg/tree) at the drip line of the tree or through irrigation will last for a number of years. Although composts and organic fertilisers contain only low levels of potassium, regular application should suffice.

Fertigation with potassium. Potassium for fertigation is available as potassium nitrate, potassium chloride and potassium sulfate. Although expensive, potassium nitrate is preferred because of its solubility and nitrogen content. Potassium sulfate is less soluble and potassium chloride can increase the chloride load if salt is a problem. When given by fertigation it should be as a small number of concentrated applications.

Foliar application of potassium. Foliar applications of potassium nitrate, 5% for deficient states and 2–3% for maintenance applications, can also be used. Foliar applications of potassium to olive trees with adequate potassium levels show no further beneficial response.

Calcium. Calcium has a number of important physiological and biochemical functions including: strengthening cell walls, stabilising polysaccharides by forming intermolecular complexes with pectins, controlling numerous metabolic reactions, and supporting the growth of young roots.

As cations, calcium as well as magnesium occupy a substantial proportion of cation exchange sites on soil particles. With very acid soils, calcium and magnesium leach out and are replaced by ions such as hydrogen, manganese and aluminium. Amending the soils with lime or dolomite reverses this situation by supplying calcium. Hence, an adequate amount of calcium in the soil reduces the risk of aluminium and manganese toxicities.

Olive trees are very sensitive to calcium deficiency. Calcium mobility is low in plants, flowing towards growing points rather than the fruit. Soft Nose in olive fruit, which renders the fruit unsuitable for table olive processing, is possibly caused by calcium and boron deficiencies, particularly under conditions of water stress. As fungal infections of the fruit have similar symptoms, then this cause must be excluded or confirmed. As calcium shortages manifest in the fruit rather than the rest of the tree, for an immediate effect foliar applications of calcium are more successful than if it is broadcast around the tree.

Calcium levels are adequate in the olive tree if the calcium concentration in dried olive leaves is >1% w/w but deficient if the concentration is <0.3% w/w.

Calcium deficiency in olives. Characteristics of calcium deficiency include: retarded tree growth, terminal dieback with increased lateral growth, chlorosis (yellowing) of the leaves and curling of terminal leaves with tip necrosis, reduced fruit production and possibly Soft Nose.

Calcium deficiency in the olive can be corrected by applying 3–5 kg of calcium oxide per tree. However, initial site preparation with crushed lime or dolomite should reduce the risk of calcium deficiency. Such applications to acid soils increases uptake of nitrogen, phosphate, potassium, sulfur, calcium, magnesium, boron, copper, zinc and molybdenum. Superphosphate also provides calcium. Soluble calcium salts can also be applied by fertigation or by several foliar applications.

Magnesium.   Magnesium has a number of important physiological and biochemical functions including: chlorophyll synthesis, participation in most reactions involving ADP and ATP, activation of enzymes for nucleic acid synthesis, activation of key enzymes in carbon dioxide fixation, and it has structural roles in cellular membranes.

Magnesium levels are adequate in the olive tree if the magnesium concentration in dried olive leaves is >0.1% w/w but deficient if the concentration is <0.08% w/w.

Deficiency is characterised by yellowing of basal leaves due to interveinal chlorosis followed by leaf drop. Fruit has a yellow (chlorotic) appearance although yield is reasonable. Magnesium deficiency in the olive is rare, and if present can be corrected by applying magnesium sulfate (Epsom salts) either to the soil (1 g/L; use 20 g/m^2 to dripline) or as a 0.7% foliar spray. If initial magnesium levels in the soil are low then dolomite should be applied prior to planting the olive orchard.

Sulfur.   Sulfur is needed for the synthesis of the essential amino acids cysteine and methionine. It is an essential component of membranes and of the biochemical reaction co-factors biotin and thiamine. Sulfur is available to the olive in its soluble sulfate form. In general, sulfur deficiency results in pale yellow leaves (similar to nitrogen deficiency), particularly with new vegetative growth. Deficiency is generally not a problem as sulfur is present in many fertilisers and manures. With low sulfate fertilisers, leaf analysis can be a guide to suitable additions.

Trace elements for olive trees

The olive tree requires small amounts of boron, iron, zinc, manganese, copper and molybdenum. A deficiency in any of these elements can reduce growth and fruiting in the olive. Deficiencies of trace elements are commonly associated with alkaline lime-rich (calcareous) soils where they are retained in an oxide form. Lowering soil pH by adding elemental sulfur, which is converted to an acid form by microorganisms, can overcome this problem. Ammonium sulfate applications have a similar effect.

Boron.   Boron is the least mobile of the elements in the olive tree and is well tolerated at high soil levels. Olive trees can also tolerate 1–2 ppm of boron in irrigation water. Depending on the soil pH, boron is present as either boric acid or the borate ion.

Boron levels are adequate in the olive tree if the boron concentration in dried olive leaves is >19–150 ppm, and deficient if the concentration is <14 ppm. It is toxic if levels are >185 ppm.

Boron deficiency is not uncommon in olive trees. Signs of boron deficiency are chlorotic (yellow) leaves and a delay in vegetative growth compared with that in healthy trees. In addition to leaf chlorosis other problems are: leaf tip necrosis followed by leaf drop, bud abortion, lack of flowering and abnormal fruit ('monkey face'). With 'monkey face' parts of the mesocarp lignify causing a distortion of the fruit. Tufts of growth, rather than normal vegetative growth, known as 'Witches broom' is a further consequence of severe boron deficiency.

In the case of boron deficiency, soil applications of 40 g boron per tree (200 g borax per tree) with an annual maintenance dose of 25 g of boron per tree (125 g of borax per tree) can be used. Foliar sprays of boron can also be used but are only effective in the current season.

Iron. Iron deficiency symptoms include yellowing of immature leaves with the midrib and veins greener than inter-vein areas (Plate 16). The fruit, which can also be affected, tend to be pale yellow rather than green-yellow in appearance. Iron chlorosis can occur even though the soil can have many times more iron than the olive tree needs because of high soil pH (and/or high pH irrigation water). Competition between ions such as manganese, zinc and potassium can also contribute to iron deficiency by displacing iron from chelating agents in the soil. Specifically, iron deficiency can be corrected by spraying the canopy with iron chelates, for example iron sodium ethylenediamine tetracetic acid. This contains 12% iron and should be applied at 50 g/L of water. Foliar sprays of iron are quick acting but are not long lasting. Other treatments are drenching the soil with iron sulfate. Iron sulfate contains 20% iron and should be applied at 20 g/m^2 in water, or injecting iron sulfate or iron citrate directly into the trunks of young trees.

Zinc. Zinc is thought to activate a number of enzymes. Signs of zinc deficiency are small pale green leaves with interveinal chlorosis. Otherwise signs are similar to iron and manganese deficiency. Specifically zinc deficiency can be corrected by foliar spraying with 0.1% zinc sulfate or zinc containing fungicides.

Zinc levels are adequate in the olive tree if the zinc concentration in dried olive leaves is >10 ppm.

Copper. Copper is important for its oxidation-reduction properties. Copper is essential to photosynthesis and the formation of a number of plant hormones. Sandy soils often lack copper and this problem is exacerbated if excessive amounts of phosphorus fertilisers are used. Signs of copper deficiency in the olive tree are stunted growth, distorted leaves, leaf rosettes and pale yellow-white leaves. Deficiencies can be treated with applications of copper sulfate (0.25–0.5 kg/tree) to soil or by foliar sprays (Bordeaux or copper sulfate 0.05%). However, overuse of copper can be detrimental to soil biota.

Copper levels are adequate in the olive tree if the copper concentration in dried olive leaves is >4 ppm. Excess applications of copper can be toxic to plants.

Manganese. Manganese deficiency starts with interveinal chlorotic mottling of immature leaves similar to that seen in iron deficiency. Flower buds often do not fully develop, turn yellow and abort. With severe deficiency, new growth is yellow in colour, but in contrast to iron deficiency, necrotic spots usually appear in the interveinal tissue. Manganese deficiency can be corrected by foliar spraying with 0.2% manganese sulfate or manganese containing fungicides. In severe cases, manganese supplements can be used. Acidifying fertilisers can increase the availability of some trace elements including manganese. Finally in waterlogged soils, where soil oxygen is depleted, high concentrations of soluble ferrous and manganous ions are released, which are toxic to the roots if their concentrations are excessive.

Manganese levels are adequate in the olive tree if the manganese concentration in dried olive leaves is >20 ppm.

Molybdenum. Deficiencies are rare but more likely in acid soils due to low bioavailability. Symptoms often consist of interveinal chlorosis in older leaves. Young leaves may be severely twisted.

Fertiliser requirements and monitoring nutritional status

The only reliable method for determining the fertiliser requirements for an olive orchard is leaf analysis. Leaf analysis results are of real value when they fall at either end of the scale for each single nutrient. For the home grower or hobby farmer, a rough guide to nutrient requirements can be determined with knowledge of the soil type and pH, and observation of leaf symptoms. Solid fertilisers and manures should be applied in autumn (half) and early spring (half). With sandy soils it is more effective to make three to four applications of fertiliser over the year.

A protocol for annual fertilisation of mature bearing olive trees extrapolated from data presented in the *World Fertilizer Use Manual* is given below as a guide. A per tree application (based on 250 trees/ha) is given in brackets (see also Table 3.12).

Nitrogen 200–250 kg/ha (0.8–1.0 kg/tree)
Phosphorus 50–70 kg/ha (0.2–0.3 kg/tree)
Potassium 160–210 kg/ha (0.6–0.8 kg/tree)

Horticultural suppliers can provide specific nutrient combinations for individual olive orchards or parts of orchards based on laboratory tests. Trace elements can be applied individually or as commercially available trace element mixtures such as a mixture of sulfur (11.5%), iron (12%), zinc (1.1%), calcium (3.5%), copper (0.5%), manganese (3.1%), boron (0.1%), magnesium (2%) and molybdenum (0.04%) used at a rate of 10 g/m^2. Applying mixtures can be a hit and miss approach risking under- or over-application of some of the nutrients.

For young non-productive olive trees, if nitrogen (N) is required the following data, together with leaf analyses, can be used as a guide. A sensible approach is to review the nutritional status of the olive trees on a yearly basis and correct nutritional problems according to the trends.

Year 1. 50 kg/ha N (200 g/tree)
Year 2. 70 kg/ha N (280 g/tree)
Year 3. 100 kg/ha N (400 g/tree)
Year 4. 150 kg/ha N (600 g/tree)
Year 5. 150 kg/ha N* (600 g/tree)

*If trees are bearing, use amounts for mature bearing trees as above; that is 0.8–1.0 kg/tree/year.

Diagnosis of nutritional problems. Common problems that occur in the olive due to nutritional deficiencies include: lack of vigour, leaf discolouration, excess vegetative growth, poor fruit size and productivity, and dieback.

Chlorosis (yellowing of leaves) is a common symptom of poor nutrition, but could be misinterpreted if the olive is diseased (Plate 15). Drought can also cause 'apparent' nutritional problems, the basis for which is a lack of water, hence nutrient availability.

Unfavourable soil pH can also reduce nutrient availability. When confronted with these problems the cause needs to be sought by a process of observation and elimination. Visual signs and leaf, soil and water analyses are important in identifying the cause of the problem. Checking the soil pH is essential. Causes other than nutritional deficiencies, such as problems in the immediate growing environment (soil, water availability, waterlogging and solar radiation), climatic effects, toxic chemicals and disease should also be considered.

Olive leaf analysis. Whenever a nutritional problem is suspected in the olive grove, a leaf analysis can be revealing. Analysis can be undertaken on a single tree, a section of the orchard or the whole orchard. The latter should be standard practice each year. With leaf analysis several aspects need to be noted: how and when to take a representative sample, and how to interpret the results.

Time for olive leaf sampling. The best time to take olive leaf samples for leaf analysis is in summer; namely, January–February, because the elements in question are at their most stable levels at this time. Outside of this period nutrient levels are more labile because of active translocation. As nutrient uptake is also a function of soil pH, this should also be assessed.

Taking a representative olive leaf sample. Eighty to 100 olive leaves are required for analysis. These leaves should be collected from as many trees as possible with three to four leaves taken from each tree to give a representative sample. When selecting the trees to be sampled, consideration needs to be given to tree age, soil type and level of irrigation. Trees with specific problems should be sampled separately. Olive leaves should be two to five months old, fully expanded, and picked halfway between the tip and base of non-bearing shoots.

Interpretation of leaf sample analysis results. Table 3.12 summarises the levels of key elements in olive leaves that can be used as a guide for interpreting leaf analysis reports.

Table 3.12. Critical nutrient levels in olive leaves (% w/w dry weight)
% w/w, g/100 g dry matter; ppm, mg/kg dry matter.

Element	Deficiency state	Target range	Toxic range
Nitrogen (N)	<1.40%	1.50–2.00%	
Phosphorus (P)	<0.05%	0.10–0.30%	
Potassium (K)	<0.40%	>0.80%	
Magnesium (Mg)	<0.08%	>0.10%	
Calcium (Ca)	<0.30%	>1.00%	
Manganese (Mn)		>20 ppm	
Zinc (Zn)		>10 ppm	
Copper (Cu)		>4 ppm	
Boron (B)	14 ppm	19–150 ppm	185 ppm
Chlorine (Cl)			>0.50%
Sodium			>0.20%

Soil analysis. Soil analysis is useful in assessing soil texture, soil water holding capacity, pH and soil toxicities. These are used in assessing soils on new orchard sites as well as the ongoing management of the olive orchard. They cannot be used on their own to evaluate nutrient deficiencies but are very useful when used in conjunction with visible symptoms and leaf analysis.

Because soils can vary across an orchard, ideally representative samples should be taken at 5–10 metre intervals. Depending on the soil type, three to four samples are taken at each point: surface (0–10 cm), mid-level (10–50 cm) and deep (50–100 cm). Soils collected at each depth are analysed separately.

Textured sandy soils have the lowest water holding capacity but as the loam and clay content increase, so does the water holding capacity (WHC) and cation exchange capacity (CEC). High CEC values are an indication of better soil fertility. Soils with pH values lower than 5.5 require amendment to at least pH 6.5. Soils with pH values above 8.5 are harmful to the olive.

Irrigation requirements for olive production

Two factors make up the water use/losses in an olive orchard. Losses include water lost from the soil around the olive trees by the process of evaporation (E) and by run-off and the water vapour discharged through the process of transpiration (T) from leaves. The sum of E and T is termed the crop water use or evapotranspiration (E + T = ET). The rate of evaporation from soil is a function of environmental temperature, humidity levels, wind intensity and solar radiation (direct sun or shade). Evaporation of water from the soil is only of significance when the soil surface is wet, such as after rainfall or irrigation. Cover crops increase the ET of the olive orchard, particularly in orchards with young trees. Total leaf area, reflected in the canopy size and planting density, also influences crop water use.

Evapotranspiration (ET) can be determined in a number of ways. Two common methods are Epan and ETo:

Epan. Epan is defined as the amount of water loss by evaporation from a standard size pan placed in irrigated close-cut grass.

ETo. ETo is defined as the estimated water use for close-cut grass derived from composite climatic real-time data including environmental temperature, humidity levels, wind direction and intensity, and solar radiation. ETo is more closely linked to the physiology of plant transpiration than Epan.

Long-term averages of monthly evaporation data for specific sites is also available from meteorological stations. These can be extrapolated into approximate daily evaporation rates (Table 3.13). These values can be used as a guide for Australian olive growers.

Table 3.13. Approximate daily evaporation rates (mm/day) based on long-term average monthly evaporation rates for a number of Australian centres based on national weather data

	Hobart	Melbourne	Sydney	Canberra	Brisbane	Adelaide	Perth
January	5	7	6	8	6	8	9
February	4	6	5	7	5	8	9
March	3	4	5	6	5	6	7
April	2	3	4	4	4	4	4
May	1	2	3	2	3	3	3
June	1	1	2	2	2	2	2
July	1	1	2	2	2	2	2
August	1	2	3	3	3	3	3
September	2	3	4	4	4	4	4
October	3	4	5	5	5	5	6
November	4	5	6	7	6	7	7
December	5	6	6	8	6	8	9

Both Epan and ETo are estimates of water use by close-cut grass and not by specific crops such as the olive. The water use by a specific crop is termed the ETc (the evapotranspiration rate for that crop), and can be correlated with ETo to produce a ratio ETc:ETo that is defined as the crop factor Kc. The Kc varies over the year and by using the specific Kc and ETo for a certain part of the growing period then the actual ETc for the olive can be estimated.

As water is an ever-diminishing resource, it is important that it is used effectively. Knowing when to apply water to olive trees and for how long is important. Traditionally, assessing the soil moisture by feel has been very effective in determining when to irrigate. With large olive orchards the 'feel' method is impractical, so instruments such as tensiometers, gypsum blocks, neutron probes and reflectometry sensors are used to detect the moisture levels in soils. Irrigation water is then applied as required. Another approach is to use a 'water budget' method in which the water requirements are estimated and the amount of irrigation to achieve this is applied.

Adequate water is required for new vegetative growth and optimal flowering. A lack of water results in reduced new growth and, in severe cases, defoliation. Lack of water also retards flowering and increases the percentage of imperfect flowers, thus decreasing yield. Adequate water is also required from stone hardening and during fruit growth. Water application (or rainfall) during the fruiting period increases fruit size and the Flesh:Stone ratio; both are favourable factors for table olive production. A potential problem of regular water applications is prolonging the ripening period and delaying fruit pigmentation.

Olive orchards in regions with little natural precipitation in winter will require water application prior to flowering otherwise production falls because of increased flower and fruit drop. Table 3.14 gives an indication as to when olive trees need irrigating. For oil production, irrigation water must be stopped early enough to give a dry period during ripening that facilitates oil extraction. For table olive production irrigation is required until there is sufficient natural precipitation.

Table 3.14. Suggested irrigation protocols for table olives

With sufficient winter rain
• Irrigation water is required during and after stone hardening.
• Irrigation water needs to be continued to maximise fruit size or olives are harvested after significant rainfall.
With no winter rain (summer rain, arid areas)
• Irrigation water is required in early spring to bud differentiation.
• Irrigation water is required in late spring, two to three weeks prior to anthesis (flowering).
• Rainfall or irrigation water is required in summer, during and after stone hardening.
• Rainfall or irrigation water is required in summer/autumn, during fruit growth.
• Rainfall or irrigation water needs to be continued to maximise fruit size.

Over-watering olive trees can cause excessive vegetative growth resulting in dense canopies and reduced yield. Over-watering can also raise humidity levels within the canopy as well as increasing the risk of erosion, waterlogging and disease. Olive tree roots are very sensitive to submersion in water and a short period of excess water can reduce photosynthetic activity. Good soil drainage also helps prevent toxicity problems if contaminated or salty water is applied. Olive trees can tolerate water with high salt levels, for example 4 g/l.

Irrigation requirements. Mature trees require a seasonal average of 500–800 L/tree/week. Irrigation of olive trees can increase yields five-fold compared with unirrigated trees by increasing the number of fruit and their size. Less relevant to table olive production than to olive oil production, the oil content in olives drops with irrigation, but increases on total oil production per unit area. As a lack of water before and during flowering will reduce fruit set, the soil in irrigated orchards should be kept near field capacity for at least until after fruit set.

Irrigation applications are delivered at times based on the amount of water in the soil profile, the uptake rate of moisture by the trees and on climatic conditions. Irrigation rates need to be consistent with soil infiltration rates; that is, the rate at which water passes into the soil. If irrigation water is applied at greater rates than the soil can absorb, then water is lost through run-off. Infiltration rates for some 'bare' common soil types are as follows:

Soil type	Infiltration rate
Sand	20–25 mm/hr
Sandy loam	15–20 mm/hr
Loam	10–15 mm/hr
Clay loam	5–10 mm/hr
Clay	1–5 mm/hr

Water infiltrates quicker into soil when it is dry than when wet. Hence, unless the soil is non-wetting, such as some sandy soils, or the soil surface has a crust, water infiltrates more easily and rates slow down as the soil becomes wetter. Coarse-grained soils, for example sands and sandy loams, are able to absorb water at a much faster rate than clays that have a much finer grain. Rates are doubled if the soils are vegetated. Infiltration rates are also reduced in situations of surface crusting, compaction and if soils are saline. With duplex soils, water infiltration rates depend on the nature of the soil layers.

As olive weight is related to water availability, targeted irrigation will ensure that table olives of a good size with favourable Flesh:Stone ratios are produced. Californian research indicates that with increased irrigation levels fruit size increases, which correlates with increased financial returns. Where water is readily available, 2–5 ML of water per hectare per year needs to be distributed at the appropriate times, particularly during flowering, fruit setting time and prolonged dry periods. For mature olive trees at a planting density of 250 trees/ha, annual yields should range from around 10–20 t/ha depending on water availability. If the fruit are 'thinned' to produce large fruit, such yields may not be achieved. The actual amount of water applied will depend on the soil characteristics, climatic conditions, competition from ground covers and the efficiency of the irrigation system.

For commercial production, irrigation or alternative methods of applying water must be commenced once the olive trees are planted. This ensures that trees come into production earlier. Ridged groves need more frequent irrigation because of their lower water holding capacity. Irrigated olive trees react poorly if irrigation is stopped and natural precipitation is inadequate.

Soil moisture is poorly available to olive roots in saline soils and as salinity levels increase the olive crop yield can fall as much as 50%. Salt accumulating around the olive roots should be flushed with good quality irrigation water. Leaf analysis can be used to evaluate the level of the salinity problem.

Regulated deficit irrigation (RDI) and table olives. Research in California has shown that deliberate under-irrigation of mature olive trees for three to four months after fruit set can conserve water by 25–40% without affecting the oil yield per tree. However, for table olives 20% more water is required to achieve maximum sized olives, indicating RDI is not really suitable for table olive production.

Calculating irrigation schedules. A sample calculation (Table 3.15) for determining the irrigation schedule for olive trees growing on sandy soil is given below. The assumption is made that the profile at the end of each irrigation is at field capacity; that is, the maximum amount of water that the soil can hold without it draining away under gravity, and during the interval between irrigations the Readily Available Water (RAW) is taken up by the olive trees.

For olive trees in home gardens, monthly deep watering in dry periods is normally adequate.

Table 3.15. Calculating irrigation schedules

Irrigation schedule using RAW (Readily Available Water). Using a root depth of 0.7 m for the olive tree and a RAW of 65 mm/m (sandy loam), an irrigation efficiency of 80% and an evapotranspiration rate of 6 mm, the following calculation will give the amount of water to be applied.

RAW within the root zone	= depth of roots × RAW
	= 0.7 metres × 65 mm/m
	= 46 mm
Time between irrigations	= RAW ÷ evapotranspiration rate
	= 46 mm/6 mm
	= 7–8 days
Theoretically the irrigation schedule is	= the equivalent of 46 mm of rainfall per 7–8 days
As the efficiency of the irrigation system is 80% the irrigation schedule would be	= the equivalent of 57.5 mm of rainfall per 7–8 days
As 100 mm of rainfall = 1 megalitre of water the amount of water	= 0.575 megalitre of water per 7–8 days
Using a Kc of 0.75, the corrected irrigation schedule (in the absence of rainfall)	= approximately 0.40 megalitre of water per 7–8 days

Olive productivity

Productivity. Because olive trees are often planted as 1.0 to 1.5-year-old trees, most varieties will commence production within two to three years after planting. The time from planting to first harvest is dependent on variety and management techniques. Most olive varieties will take at least four to five years to bear commercially useful crops. Even if trees have only a few olives, these should be removed within the season. If not cared for properly the olive trees may take longer to be productive. (See also Chapter 2.)

Producing large olives. The size of olive fruit depends on a number of factors, including: variety, growing conditions, irrigation levels and crop load.

This is in contrast to simply sorting out the larger olives from a typical crop using a mechanical sorting device. Maximising the first three factors, then thinning out the crop will contribute to the olives' ultimate size. Such strategies are used for other crops, such as apples and pears. With the latter fruits, espalier techniques for commercial application have been developed. Further research is needed to determine whether this technique is also useful for commercial olive production. It should, however, be recognised that small crops of over-large olives may not be commercially viable. Three ways by which olive crops can be thinned are: removal of some of the fruit by hand, removal of some productive wood, and using the chemical loosening agent Naphthalene Acetic Acid (NAA).

Chemical thinning of olive crops for table olives is an economically favourable procedure increasing the value of the crop. The trees are sprayed between 12 and 18 days after full bloom using 10–15 L/tree of a dilute spray of NAA 10 ppm for each day after full bloom; that is, 12 days = 120 ppm and 18 days = 180 ppm. Full bloom is when 80% of the flowers are open. In areas where the weather is abnormally cool or hot, the spraying time is based on fruit size. In this case, trees are sprayed with a 150 ppm NAA solution plus a wetting agent. The NAA is absorbed through the leaves and within two weeks of spraying an abscission zone develops on the olive stalk, causing some immature olives to fall. As significant leaf loss will occur in stressed trees after NAA treatment, only healthy trees should be treated. Thinning can be undertaken annually. Furthermore, sufficient olives should be removed to ensure new shoot growth as well as reduce biennial bearing. For more detailed information the reader is referred to the *Olive Production Manual* (Sibbett et al. 2005).

Maintenance of the olive tree

Pruning olive trees for production. The aim of pruning is to control vegetative growth, maximise productivity and facilitate harvesting. A feature of the olive tree, different to other fruit trees, is that its branches allow permanent, continuous renewal of fruit bearing buds. Correct training and pruning both shapes the tree for easier harvest and regulates production. Although the olive is a slow-growing evergreen, it can grow extensively making harvesting difficult and dangerous for workers particularly if they need to use ladders or mechanical aids such as cherry pickers. Because olive trees can tolerate radical pruning, it is relatively easy to maintain them at the desired height. For a single trunk, suckers and low branches (1.0–1.5 m above the ground) should be removed. Where a multi-trunk bush/tree is desired, three to four low branches or basal suckers should be selected and staked, removing other unwanted growth. Pruning can be undertaken using manual, electric (12V battery) or pneumatic secateurs (with or without extension poles), hand saws or pneumatic chain saws.

In contrast to some practices, long-term pruning studies in Spain found that pruning:

- every two years is excessive;
- should be practised at longer intervals or be less intense;
- every three to four years gives larger crops than every two years; and
- every 10 years is not viable.

Note: As accidents and injury are a possibility during pruning operations, a simple safety plan should be in place that includes first aid and evacuation if necessary. Safety pruning gloves will reduce the risk of hand injury and safety glasses will reduce the risk of eye injury. Persons using mechanised pruning equipment should be trained, particularly in ensuring the safety of fellow workers.

For olive trees planted in home gardens and as street trees, since fallen olive fruit can stain paths and patios, pruning in summer when the olives are green and small can reduce this problem. Alternatively, the Australian non-fruiting *Swan Hill* variety can be grown. In home gardens, and where space is limited, olives can be grown in an espalier style.

Pruning the olive tree keeps the fruiting sections close to the roots, hence avoiding embolism and transport limitations that can occur in unpruned trees. Ideally olive trees for table olive production should be trained and managed so that all operations can be undertaken at ground level. Initially pruning is kept to a minimum to encourage early bearing. Heavy pruning and training of olive trees will delay initial bearing and productivity. Branches crossing into the canopy can be bent out rather than removed by pruning. If pruning and training follow the natural growth habit of the olive tree the resulting canopy shape will be a bush, bush-vase or an open vase. Some individual olive varieties, such as *Barnea*, have a natural upright growth tendency.

The olive tree should be shaped to allow air circulation and light penetration through the canopy without increasing the risk of radiation damage to limbs and fruit. Effective airflow through the canopy reduces humidity and the risk of some diseases. A correctly trained tree should show light penetration through the canopy so that maximum-sized olives are produced. As olive fruit develops on one-year-old wood, pruning operations should be directed to new growth. All dead and non-productive wood should be removed.

During the productive phase of the olive, selective light pruning should be undertaken yearly after harvest (autumn to early spring) but severe pruning less frequently. Dead, weak and dense branches should be removed from the fruiting zone to facilitate new vegetative growth, ventilation within the canopy and the entry of sunlight. Where olives are grown with limited amounts of water (via rainfall or irrigation) and/or infertile soil, more severe pruning is required to reduce the canopy size to allow water and nutrients to be directed to new vegetative growth. Heading cuts, which involve making numerous well-distributed cuts along thin branches rather than removing medium-thickness branches, stimulates more vegetative buds to grow, resulting in new wood. The physiological reason for this is that the stimulus for bud growth is distributed over the whole canopy.

Older olive trees can be rejuvenated by either cutting the trunk below the branching points of the trunk and limbs or by removing individual scaffold branches. Either way new shoots develop that can be trained to be productive three to four years after the amputation.

Mechanised pruning.   With large-scale olive groves some form of mechanical pruning is essential to be cost effective. Attempts have been made to mechanically prune olive trees with scything bars or circular saws with varying levels of success. Topping olive trees is not difficult, but, unlike grapevines, productivity after mechanical pruning does not reach similar levels to manual pruning. As mechanical intervention reduces the production of vegetative buds, some olive growers mechanically prune one side of their mature olive trees each year over a four-year period every 8–10 years. This means that trees maintain some level of productivity. Another innovation is to rejuvenate overgrown olive trees

(10% of the orchard each year over 10 years) by cutting them back to a stump from which two trunks are encouraged to develop. Yields of such rejuvenated olive trees after 9–10 years is of the same order as those managed under normal cultural practices, despite a 30–40% period of non-productivity. More research is needed into the mechanised pruning of olive trees.

Tree protection.   Olive trees need to be managed to reduce the risk of damage by environmental factors (radiation, wind and frost), agricultural chemicals, pests and diseases. Preventive measures include using tree guards, painting the trunk with latex paint or placing protective paper around the trunk to protect young trees from sunburn or herbicide damage. Pests and diseases can affect tree vigour, productivity and fruit quality. Significant olive pests and diseases, their effects and management are presented below.

Pruning tips for olive trees

Heading cuts.   This involves removing the top part of a shoot or branch. Use when invigorating a weakly growing branch, to reduce the tree's height or width. Heading cuts promote the development of new shoots and if too vigorous will cause shading.

Thinning cuts.   The entire limb or shoot is removed to reduce shoot density, facilitating light entry into the canopy and directing energy into the remaining shoots. Thinning cuts are also used to maintain olive tree shape.

Watershoots.   These should be removed unless a new branch is required.

All diseased and dead limbs or shoots should be removed and disposed of. With young developing olive trees, bending limbs is preferred to pruning when shaping the tree. Older trees should be pruned harder at the top than at the bottom. Thin and long pendulous branches should be removed. Finally lateral shoots should be thinned and any shots growing towards the centre of the tree or vertically should be removed.

Stress factors and diseases that can affect olive trees

Health problems in olive trees are associated with a number of abiotic and biotic factors. Problems can be localised (leaf, root) or systemic, on one tree or more trees, acute or chronic. As prevention is better than cure, there are a number of control options to ensure stress- and disease-free olive trees.

Cultural control.   Cultural control measures include:

- proper site selection;
- planting resistant or tolerant varieties;
- meeting water requirements;
- providing appropriate nutrients;
- controlling weeds;
- avoiding mechanical injuries;
- planting and pruning at the correct times;
- providing adequate spacing between the trees;
- improving tilth (physical condition of the soil); and
- optimising the pH of the soil.

Sanitation control. This involves planting stock sourced from nurseries that can provide vigorous, healthy and disease-free olive trees, removing and disposing disease-affected parts of olive trees and disinfecting all tools and pruning equipment.

Biological and chemical controls. Biological control involves the use of biological agents, such as specific or non-specific parasites and companion planting. Chemical control involves the use of chemicals (pesticides, fungicides and herbicides). A concern when using chemical agents is the accumulation of the parent chemical agent and its metabolites in olive fruit and olive oil. To reduce toxicity problems, the proper selection of chemicals, the timing of application and observing withholding periods is of utmost importance. Of equal importance is that the chemicals are handled and used correctly so as not to injure oneself, other orchard workers or visitors, and not to contaminate adjacent properties or environment. Workers using chemicals should be trained in their safe handling and how to manage chemical spillage and emergencies.

Pests and disease management

Some pests and diseases affect olive trees. In the Mediterranean region, Olive Fruit Fly and Medfly (Mediterranean Fruit Fly) are major pests. Pests and diseases can affect the growth, vigour and productivity of the olive tree as well as the quality of the fruit.

Australia has been considered relatively free of olive pests and diseases; however, since the resurgence of the Australian olive industry, several unexpected problems have emerged including: Olive-scale, Peacock Spot, Olive Knot and Olive Lace Bug, Curculio-weevil, bird and animal attacks and soil pathogens such as Phytophthora Root Rot, nematodes and *Verticillium*. In drier areas, such as in Western Australia, Rutherglen Bug and grasshoppers have attacked young trees leading to dieback. Some growers have reported olive fruit damage by insects and Soft Nose, caused possibly by the fungus Anthracnose, stress or a nutritional deficiency. To date, Olive Fly and Olive Moth have not been found in Australian olive groves. Some indigenous insects, however, do attack young trees and olive fruit. A number of common problems affecting the olive in Australia are presented below.

Diseased and damaged olives should not be used for table olive production. Preventive measures such as paying attention to the following will reduce the risk of pests and diseases:

- hygiene and biosecurity;
- minimisation of mechanical damage to tree trunk and limbs;
- canopy management;
- adequate moisture and nutrition;
- copper sprays; and
- insect traps.

Where pest and disease problems occur, samples should be collected and examined by a plant pathologist for accurate identification and management, particularly if chemical treatments are to be used. With each specimen (affected plant part/pest), the locality, date of collection, collector details and additional information such as the type of damage should be included. For a more extensive discussion on olive pests the reader is directed to the IOOC publication *Olive Pest and Disease Management* (Lopez-Villalta 1999); the *Olive Production Manual* (Sibbett *et al.* 2005); and a recent RIRDC Report: *Sustainable pest and disease management in Australian olive production* (Spooner-Hart 2005).

Black Scale. Black Scale (*Saissetia oleae*) is the principal scale insect pest of olives in Australia (Plate 17). Hosts include many species such as citrus, fig, stone fruit and apple. Adult females of about 3–5 mm diameter attach themselves to leaves, stems and possibly fruit, where they lay up to 2000 eggs. The female scale is dark brown to black in colour with distinct, hard, H-shaped ridges on their backs. The eggs hatch over two to three weeks into red/brown crawlers that can be seen from summer to late autumn. The scale produces honeydew, visible as a shiny film on the leaves, which attracts ants and supports a black Sooty Mould (a complex of *Spilocaea oleagina*, *Capnodium* sp., *Cladosporium* sp. and *Alternaria* sp. has been observed in some infestations), which, if severe, inhibits photosynthesis and respiration. The ants feed on the honeydew, deter scale predators and move crawlers from one olive tree to another. Young scales are yellow to orange in colour and are visible on leaves and stems.

With severe infestations substantial leaf drop occurs. Therefore, scale infestations reduce vigour and productivity of the olive tree. Affected trees generally have a dense canopy, so part of their management involves cutting out severely affected parts and opening the canopy. Standard treatment is spraying crawlers with an oil spray commonly called white oil, mineral spray oil or vegetable oil sprays. The tree should receive two applications two to four weeks apart. Other types of scale infestations are also treated with mineral spray oil. Reducing the ant population also helps control Black Scale. Natural parasites and predators also control Black Scale. For severe scale infestations, methidathion can be used, as can the insect growth regulator buprofezin. Ants, which should not be allowed to increase in numbers, can be controlled with a pyrethrum spray. Recent research in Australia using a commercially available vegetable oil spray found that two to three sprays with an interval of three to four weeks between sprays fully controlled Black Scale infestations on olive trees at the same level as methidathion without reducing yield. Biological control with Black Scale predators is another possibility. Several introduced species of parasitic wasp (*Metaphycus* sp. and *Scutellista caerulea*) are very effective in controlling Black Scale. Ladybirds (*Cryptolaemus montrouzieri*) and lacewing larvae are also significant predators. Some 683 parasitoids, with five species predominating, that have been reared from Black Scale specimens collected in South Australia could provide useful Black Scale parasites in the future. A potential problem, however, is that where insecticides are used, biological predators also succumb.

Olive Scale. Olive Scale (*Parlatoria oleae*) occurs worldwide and has a wide distribution. It is oval in shape and smaller (2.5 mm long) than Black Scale. This scale appears grey in colour because of a waxy covering, with males a little longer than females and having a distinct black spot at one end. The scale feed on branches, leaves and fruit. Early in the season newly hatched olive scale feed on young fruit and leaves causing deformities. Newly hatched scale later in the season cause distinct purple spotting of green olives making them unsuitable for table olives. Severe cases of Olive Scale reduce the productivity of the olive tree. With mild cases treatment is not required and any affected olives can be culled. In California, biological control with *Aphytis maculicornis* and *Coccophagoides utilis* has proved successful against Olive Scale, reducing the need to use chemical pesticides. With severe cases treatments are more effective when crawlers are present. Post-harvest treatments are also effective.

Californian Red Scale. Californian Red Scale (*Aonidiella aurantii*) is a significant pest in citrus and can also be found on grapevines. It has similar characteristics in size and shape to Olive Scale except it has an orange/reddish appearance. All parts of the olive tree can be affected, but in contrast to Olive Scale, it does not cause purple spotting of the olives. Nevertheless, affected fruit cannot be used for table olives. With severe cases, pesticide treatments are more effective when first and second generation crawlers are present (in summer).

Oleander Scale. Oleander Scale (*Aspidiotis nerii*), which can attack oleander and a wide range of fruit species, has similar size and shape characteristics to Olive Scale except it has a waxy white covering with a yellow/tan coloured spot in the centre. This scale is more likely to be found on leaves and branches in the lower part of the canopy. When the fruit is affected, a green spot is visible on naturally black-ripe olives because the scale inhibits maturation at the spot of contact. Such spotting reduces the commercial value of the olives. Heavy infestations of Oleander Scale cause fruit deformity, and significantly reduces the oil content of the flesh. If the scale injects toxins into the tree, severe damage occurs. With severe cases treatments are more effective when crawlers are present (late autumn/early summer).

Olive Lace Bug. Olive Lace Bug (*Frogattia olivinia*), native to New South Wales and southern Queensland, has been detected in most olive growing areas in Australia, particularly eastern and southern regions. It is the second-most common pest (Plate 18). Olive lace bugs are 3 mm long, mottled dark brown/cream in colour with adults having wings with a lace-like texture and long black-tipped antennae. Female lace bugs can lay three to four generations of eggs per year. Eggs are laid along the leaf mid-vein then protected by a cover of black excreta. Eggs hatch in five to six weeks into oval-shaped wingless nymphs that after several nymphal stages (instars) develop characteristic wings. The lace bugs are often found on the underside of the leaves where they feed. The result of such activity is seen on the upper parts of the olive leaves as stippled greenish-yellow rusty spots or patches. Severe infestations can cause leaf desiccation, thereby reducing tree vigour and productivity. Thrips and mites can cause similar leaf damage, except with olive lace bug the spots when observed under magnification are seen as incomplete rings. Current treatment involves spraying the olive tree canopy, especially the underside of the leaves, with soap sprays and/or organophosphorus insecticides (fenthion, dimethoate), ensuring the withholding period is observed in productive trees. Soap sprays are less effective in treating late nymphal stages, whereas pesticides have little effect on unhatched eggs; pesticides are most effective a week or so after the eggs have hatched. In severe infestations a follow-up spray with an alternative pesticide two weeks after the first application can be used.

Olive Bud Mite. The Olive Bud Mite (*Oxycenus maxwelli*), originating from California, has a predilection for the olive tree and is widely distributed in Australian olive growing regions. The mite, which is difficult to see by eye, is pear-shaped, yellow to dark brown in colour and slow moving. Generally it is not a major problem. The mites feed on young stems, buds and upper leaf surfaces. In severe cases, vegetative and reproductive structures are affected, leading to reduced shoot growth and deformed sickle-shaped

Figure 3.11 Typical Olive Bud Mite damage to vegetative growth. Note smaller sickle-shaped leaves. Sickling of leaves may also occur with nutritional deficiencies.

leaves (Fig. 3.11), loss of flower buds and inflorescences leading to reduced productivity. Leaves that are sickle-shaped can also signify a nutritional problem such as boron deficiency.

Another problem attributed to Olive Bud Mite is reduced internodal length leading to the appearance of 'Witches broom' (tufts of stunted vegetative growth). A similar problem noted in other fruit species has been attributed to a fungal or viral agent or nutritional deficiency. With respect to management if there is no history of past problems then no treatment is required. If olive productivity is reduced over a number of years for no apparent reason, particularly if dried brown flowers are present, spring flower buds should be examined for the presence of large numbers of mites and treatment instigated if needed. In some centres, sulfur is used to control the mite.

Fruit flies. Several types of fruit fly can attack olives. Medfly, *Ceratitis capitata*, is prevalent in Western Australia, Queensland Fruit Fly (*Bactrocera tryoni*) in eastern Australia, and Olive Fruit Fly (*Bactrocera oleae*) has not yet been found in Australia. Although Medfly and Queensland Fruit Fly can attack the fruit of a number of species, their effect on olive fruit is of minor importance and in many cases is detectable only due to 'tell tale' pricks in the olive skin with little fruit damage. In severe cases damaged fruit fall or ripen prematurely. The insecticides dimethoate and fenthion can be used for fruit fly control. For other fruit fly species, yeast autolysate/insecticide baits have been used with some success in regions of high fruit fly activity.

Olive Fruit Fly. Olive Fruit Fly is a devastating problem for the olive industry in a large number of olive growing countries, and although not yet found in Australia is still worthy of discussion. Olive Fruit Fly infestations are generally restricted to traditional olive growing countries in the Mediterranean region, with a history of infestation of at least 2000 years. In 1998, it appeared in California, where it is now considered to be a serious threat to the local olive industry. An additional problem is the presence of the Olive Fruit Fly in abandoned olive orchards, in olive trees used as street trees, and those planted for landscaping purposes and in home gardens. Growers and owners need to pay attention to such trees, as they are likely to harbour other pests and diseases as well.

The female Olive Fruit Fly pierces the skin of immature olive fruit and deposits one fertile egg/fruit and an inoculum of bacteria. As the olive fruit matures, the bacteria breakdown the olive flesh, which facilitates the feeding of 'maggot hatches'. When the 'maggot hatches' feed upon the flesh, galleries are created and the fruit fall off the tree (Plates 19 and 20). Affected fruit cannot be used for table olive or olive oil production. Olive oil made from infested fruit is high in free fatty acid (6% or more) and has a grubby taste. Fallen fruit allows Olive Fruit Fly to spend the winter near the olive trees and so the cycle continues. Pesticides mixed with baits are used in commercial olive orchards to control Olive Fruit Fly. Plant regulators sprayed at bloom time in spring have been used by some. An additional strategy involves disposing of affected fruit from trees and the ground to reduce the number of Olive Fruit Flies.

Research in Greece has shown that in isolated groves and where Olive Fruit Fly is in low to medium population densities, one killing device – a tree baited with ammonium carbonate and pheromone – has the potential to keep Olive Fruit Fly populations and fruit infestation low. Results from this study were similar to conventional treatments of at least three ground bait sprays of protein-hydrolysate-dimethoate. In non-isolated olive groves and regions where the Olive Fruit Fly develops high population densities, one killing device per tree gave inadequate protection and at least one treatment of ground bait spray is required to keep the fly population and the fruit infestation at low level. In the long term, researchers believe such control methods could progressively replace the use of insecticides for the control of the Olive Fruit Fly.

Curculio Beetle.   Curculio Beetle, also known as Apple Weevil (*Otiorrhynchus cribricollis*), has been found to attack olive leaves in inland Western Australia, South Australia and New South Wales. The Garden Weevil, *Phlyctinus callosus*, has also been reported as an olive pest in Western Australia. With both species the adult insects are nocturnal, emerge from the orchard floor during summer, climb the olive tree and chew the outer margins of the leaves. The leaves take on a ragged appearance (Fig. 3.12). Heavy infestations, particularly in young trees, can damage growing tips and even causing defoliation. Butt spraying non-bearing olive trees with alpha-cypermethrin provides some protection. Strategies for controlling this pest include placing free-range poultry in the olive orchard or placing sticky or fibrous barriers on tree trunks to trap the weevils.

Rutherglen Bug.   Rutherglen Bug (*Nysius vinitor*), which is polyphagous, attacks olive tree twigs and leaves when the orchard floor and surrounding vegetation dries off. With severe infestations, particularly with young olive trees, dieback and tree deaths occur.

Peacock Spot.   Peacock Spot or Olive Leaf Spot (*Spilocea oleaginea*, or *Cycloconium oleaginum*) is a fungal disease associated with rain and high humidity during autumn/early winter, particularly in warmer regions. It is less of a problem in drier regions. It is

Figure 3.12 Olive leaves with characteristic notched appearance after attack by *Otiorrhynchus cribricollis* (Curculio Beetle).

widespread in most parts of Australia, excluding Western Australia. Both leaves and fruit can be affected. Infectious conidia (spores) survive throughout the year and germinate in the presence of water. First signs of the infection in early winter are small sooty patches that develop into characteristic yellow/green/black spots by spring. Some degree of defoliation occurs. Affected fruit develop brown spots and do not mature at the same rate as uninfected fruit. Management involves thinning the canopy by pruning and applying copper spray (Bordeaux, copper hydroxide or copper oxychloride) annually after harvest. Pruned material should be burnt. Varieties relevant to table olives that are susceptible to Peacock Spot are *Manzanilla*, *Frantoio* and *Arbequina* (Plate 21).

Olive Knot. Olive Knot is a condition in olive trees caused by the bacteria *Pseudomonas syringae* pv. *savastanoi*. It is not uncommon in traditional olive growing regions where its presence was known in the Old World. Olive Knot has been detected in South Australia and Victoria on *Barnea* variety olives. If Olive Knot appears in unaffected areas, it is likely that it has been introduced on olive cuttings or young olive trees. However, there are many susceptible varieties including ones currently being grown in Australia. The condition is not lethal; however, severely affected trees have reduced productivity because of defoliation as well as loss of branches. Other signs of the disease are galls or swellings 1–5 cm in diameter on the trunk, scaffold and branches. Within infected olive trees the bacterium produces the chemical indoleacetic acid that causes cells to proliferate, resulting in characteristic galls (tumours). The bacteria survive in the galls and can be released at any time of the year and spread by water (natural precipitation or irrigation) and picked up through wounds created during pruning and harvesting. Prevention is the best form of management. Firstly, disease-free planting stock should be sourced. Where the problem exists, sanitary protocols for pruning should be followed; affected parts should be removed and disposed of by fire and harvesting should be avoided in wet weather. Sodium hypochlorite (1.2% available chlorine) can be used to disinfect equipment. Copper sprays applied to wounds, pruning cuts and leaf scars give some protection if applied in spring and autumn. Where possible and practical, olive trees should be protected against radiation and frost damage that would provide entry points for the bacteria.

Anthracnose. Anthracnose is a fruit rot of olives caused by a strain of *Colletotrichum acutatum* resulting in an orange slimy mass of spores on the fruit surface (Fig. 3.13). Other organisms have also been implicated. The fungus can penetrate intact healthy skin although olives damaged through stress, sunburn, over-ripening, poor harvesting and post-harvesting practices may facilitate its penetration. Infection usually occurs at the tip of green olive fruit, where a drop of water may sit after dew, rainfall or irrigation. The fungal spores can survive in mummified fruit for one year in low temperatures. In severe cases the vegetative parts are affected causing defoliation, shoot death and loss of vigour in affected trees.

Figure 3.13 Fruit rot typical of Anthracnose or adverse environmental conditions.

Anthracnose can affect olives at all stages of maturation with symptoms appearing as the olive ripens. Management of Anthracnose involves disposing of infected fruit, preferably by burning, followed by spraying the trees with a copper spray (Bordeaux, copper hydroxide or copper oxychloride).

Soft Nose. Fruit rot caused by environmental conditions rather than an infective agent can be mistaken for Anthracnose. The tip of the olive softens and rots, hence the name 'Soft Nose'. This problem has been linked to unexpected changes in temperature and humidity that partially dehydrate the flesh. Water stress, overzealous application of nitrogen fertilisers, as well as deficiencies in calcium and boron have been suggested as triggering 'Soft Nose'. Prevention involves ensuring the moisture and nutritional needs of the tree are met. Unless there is a secondary fungal infection, fungicide applications have no role in managing 'Soft Nose'.

Phytophthora. Phytophthora (or dieback) is a condition where soil fungi of *Phytophthora* species cause root/collar rot in the olive tree, particularly in poorly drained waterlogged soils. Reducing water applications can help by reducing this risk. Affected trees experience reduced growth and leaf dieback resulting in drastic thinning of the canopy. Eventually most trees die. Prevention is the best strategy; do not plant olives in suspect soils, or if a problem is anticipated undertake soil amendment and mounding prior to planting. To prevent infection being transferred from propagating facilities, potting soils can be treated with Metalaxyl-M. Phosphonic acid can be used for field control.

Note: As dieback can also result from overuse of herbicide sprays (Fig. 3.14), soil fungi, nematodes, lack of water and insect attack, a full laboratory assessment needs to be undertaken to determine the cause and formulate a management plan.

Verticillium Wilt. Verticillium Wilt is a condition caused by the soil-borne fungus *Verticillium dahliae*, which attacks the roots and vascular tissue to the extent that distinct segments of the olive tree die. It is widespread in many olive growing regions. With persistent infection over the years, whole trees eventually succumb. Verticillium Wilt is not uncommon in olive growing areas that experience cool conditions in spring and summer. As the fungus is spread by nematodes, land previously used for growing crops such as potatoes, eggplant, peppers, tomatoes, stone fruit, brassicas and cotton should be avoided. Nematodes are microscopic unsegmented worms that invade plant roots, with some (Root Knot nematodes)

Figure 3.14 Dieback (top of the tree) due to overuse of herbicide spray. Note regrowth from the base of the tree.

forming gall-like lesions that restrict water and nutrient uptake. Preventive measures include planting disease-free olive trees, soil fumigation/solarisation and paying attention to sanitation so that infected soil is not carried into unaffected areas and orchard implements are kept clean and relatively 'sterile'. Soil solarisation, a method used to kill soil pathogens, should be undertaken during the warm season. One method is that after old roots and root debris have been removed, the soil is moistened then worked to a fine tilth to a depth of at least 25 cm. The soil is then covered with a thin, transparent polyethylene sheet, with its edges buried to a depth of 25 cm or more, for at least four weeks. Following this procedure most pathogens die, while normal soil microbes survive in sufficient numbers to multiply, preventing the establishment of pathogens. Solarisation also kills weed seeds and insect pests. Maintaining a weed-free orchard floor also helps. Affected plant parts should be removed and disposed of by fire.

Note: Several species of nematodes that feed on olive roots can also cause root damage that can also lead to 'dieback'.

Integrated pest management

The use of synthetic pesticides and herbicides in agriculture has the potential to cause significant environmental problems. Concern regarding the indiscriminate use or overuse of such chemicals has lead to taking a more integrated approach including non-chemical preventive and maintenance measures. Low levels of damage due to pests and diseases should be tolerated, rather than treated with chemicals, which increases the risk of secondary pest infection. Several approaches have been suggested by international commentators and scientists regarding the overuse of chemicals in horticulture/agriculture. Approaches include:

- completely eliminating synthetic agricultural chemicals;
- permitting the use pesticides derived from plants or microorganisms;
- using targeted applications of synthetic agricultural chemicals, together with non-chemical measures; and
- reducing the use of agricultural chemicals.

The following strategies have been suggested specifically for olives:

Black Scale.   Shortening the periods between pruning and thinning the canopy, and the use of predators.

Olive Fly.   Protein baits with chemical pesticides, mass trapping by combinations of attractants and pesticides, and the use of parasites.

Olive Moth.   Using the predator *Bacillus thuringiensis*.

Very few chemicals should be needed for successful olive cultivation, particularly in dry inland areas. Most problems that do occur can be controlled but they should be positively identified and expert advice on management sought to minimise indiscriminate spraying of broad-spectrum insecticides – which will also kill beneficial insects – and fungicides. Correct pruning to allow adequate airflow through the canopy will help keep many problems under control. Copper sprays applied to the tree canopy after harvest and pruning can be used as a general antifungal treatment. Integrated pest management strategies (IPM) using cultural techniques and safe sprays, such as *Bacillus thuringiensis*, should be adopted.

Table 3.16. Selected chemicals used in olive growing and their withholding periods

Chemical application	Use	Withholding period and/or comment
Insecticides		
Natrasoap (potassium oleate)	Lace Bug	None Application 1–2 l/100 l
Fenthion	Lace Bug, Green Vegetable Bug, Rutherglen Bug, Queensland and Mediterranean fruit flies	Do not harvest olives for 14 days after last application
Dimethoate	Lace Bug, Green Vegetable Bug, Rutherglen Bug, Queensland and Mediterranean fruit flies	Do not harvest olives for seven days after last application
Natural pyrethrum concentrate	Lace Bug	Do not harvest olives for one day after last application
Chlorpyrifos	Ants, African Black Beetle, Light Brown Apple Moth	Not required when used as directed
Methidathion	Scale insects: Black Scale, Olive Scale, Californian Red Scale	Do not harvest olives for 90 days after last application
Mineral spray oil (petroleum oil)	Scale insects: Black Scale, Olive Scale, Californian Red Scale	Do not harvest olives for one day after last application For green olives do not apply six weeks before harvest to prevent spotting
Buprofezin	Scale insects: Black Olive Scale, Olive Scale, Californian Red Scale	Do not harvest for 28 days after last application
Alpha-cypermethrin	Curculio Beetle control on non-bearing olive trees	Newly planted non-bearing orchard for the first one to three years
Fungicides		
Copper hydroxide	Fungal leaf spots including Peacock Spot and some fruit rots	Do not harvest olives for one day after last application
Copper (cupric) oxychloride	Fungal leaf spots including Peacock Spot and some fruit rots	Do not harvest olives for one day after last application
Metalaxyl	Phytophthora Root Rot and Crown Rot in potted trees only	Mainly for nursery use
L-alpha-cypermethrin	Rutherglen Bug control on non-bearing olive trees	Newly planted non-bearing orchard for the first one to three years
Herbicides		
Glufosinate-ammonium	Numerous weeds and grasses	Do not harvest olives for 21 days after application Do not graze or cut treated areas for stockfeed for eight weeks after application
Fluazifop-P	Numerous weeds and grasses	Not required if used in accordance with directions Grazing animals: hold for seven days after grazing on treated areas before slaughter
Pendimethalin	Numerous weeds and grasses common in many orchards	–

A number of pesticides and fungicides are available for use on olive trees and these are presented in Table 3.16. There are other products registered for orchard use that can also be used in olive orchards, provided label rates and critical comments are followed. Information provided in the table is based on *Interim permits for chemicals used in olives* from the publication by The Australian Pesticides and Veterinary Medicines Authority (2005) and should only be used as a guide. Regular updates should be sought as to the currency of the information as permits are temporary and their extension will depend on the availability of chemical residue data.

In summary: mineral oil spray is used for scale insect pests; Natrasoap for Lace Bug; copper hydroxide or copper oxychloride for various leaf spots and fruit rots in olives; granular metalaxyl is used for Phytophthora Root Rot root and crown rot in potted nursery trees; glufosinate-ammonium and fluazipop-p-butyl for weed control; chloropyriphos for ants (around the tree butt), African Black Beetle (as a drench around the tree base) and Light Brown Apple Moth (foliar spray on non-bearing trees); methidathion for scale insects; dimethoate for Lace Bug, Green Vegetable Bug and Rutherglen Bug; fenthion for Lace Bug, Green Vegetable Bug, Queensland Fruit Fly and Mediterranean Fruit Fly; and Alpha-cypermethrin as a tree butt drench for Curculio Beetle and cutworms.

Note: It is most important that chemicals are used correctly, taking into account orchard workers' health and that withholding periods are observed. Such information should be documented as part of the GHP in the olive orchard.

Residues from some chemical sprays have been found in raw olives. Although procedures involving lye treatment of olives reduce chemical residue levels in final products, more research needs to be undertaken to determine residue levels in table olives processed by all methods. It must be recognised that chemical contaminants may come from sources other than chemicals used in the olive orchard, for example water, inputs and containers.

Maturation states for table olive production

The three maturation states for raw olives (Plate 1) relevant to processing table olives are green-ripe, turning colour and naturally black (ripe olives).

The exact time of harvesting will depend on the variety, region, crop load and growing conditions. Generally green-ripe olives can be picked in late summer to early autumn and naturally black-ripe olives from late autumn to late winter. Turning colour olives are picked between these periods. A point to note is that olives on heavily laden trees are smaller and take significantly longer to reach the naturally black-ripe stage than those with a lesser load. Features of olives at different ripening stages are given in Table 3.17.

Changes in firmness and texture occur as the olive fruit proceeds from green-ripe to the black ripe stage. Olives at different maturation stages can have a difference of 30–40% in fruit firmness.

Green-ripe olives. Olives at the green-ripe stage are most suitable for processing as green table olives by any method or style. When olives change from a leafy green to a yellow-green to straw colour (green-ripe) they have generally reached their maximum size and are ready for picking. When squeezed between the fingers the olives should release a creamy white juice that has an oily feel and the characteristic fruity aroma of olive flesh. Using olives at the unripe leafy green stage for processing poses technical problems such as lower fermentation rates because of poor skin permeability and a lower Flesh:Stone ratio. When processed, these olives have a 'rubbery' texture and poor organoleptic characteristics. Green-ripe olives, such as *Manzanilla*, can also be salt-dried.

Table 3.17. Features of olives at different ripening stages

Maturation stage	Description
Green-ripe	Olives at this stage are normal sized, green to yellow in colour, with firm flesh resistant to pressure within the fingers and without marks other than the natural pigmentation.
Turning colour/ semi-ripe	Olives at this stage are still firm and have started to accumulate purple pigments in the skin and appear multicoloured, rose, brown or purple in colour. The flesh lacks pigment or is partially pigmented close to the skin.
Naturally black-ripe	Olives at this stage are close to full ripeness with near total purple pigmentation of the flesh. Oil content has also reached maximum levels.

Olives turning colour. Olives that have started to change to a light rose to red-brown colour are termed 'turning colour'. At this maturation stage the flesh is still creamy white in colour but is softer than that of olives at green-ripe stage. Turning colour olives can be processed in brine, by natural fermentation, or in lye followed by oxidation to produce Californian/Spanish-style black-ripe olives. Note that trade products called Black-ripe Olives are not produced from naturally black-ripe olives. Turning colour olives, such as *Manzanilla*, can also be salt-dried.

Naturally black-ripe olives. Olives that are allowed to fully ripen on the tree are called naturally black-ripe olives. Good quality naturally black olives with adequate flesh pigmentation should be hand-harvested before the first frosts to ensure the best final product. Harvesting with hand rakes or long poles results in damaged fruit; these process to a mushy product with poor organoleptic qualities. Damage may be more difficult to see with black olives than green olives. Fruit from most olive varieties ripens to a deep red to black colour. The skin changes colour due to the accumulation of red/purple anthocyanin pigments. The olive is termed naturally black-ripe when the pigments diffuse and accumulate in the flesh right down to the stone. Olives with black skins and white flesh do not meet the requirements of naturally black-ripe olives and so will not give the desired product. For example, raw *Kalamata* olives with only black pigmented skin process to a light brown colour. The best stage to harvest for naturally black olives in brine so that a firm product results is when the pigment is half to three-quarters through the flesh. When over-ripe naturally black olives are processed they lose much of their texture, resulting in soft products, and have an increased likelihood of off-flavours. Over-ripe black olives are preferred for some types of dried olives.

Harvesting for table olive production

The two most important factors in harvesting for table olive production are picking the olives at the correct maturation stage for the processing method or style, and ensuring that only quality fruit is picked.

Olives should be harvested by hand in preference to mechanical harvesting to reduce the risk of damaged fruit, which when processed result in inferior products. Trees may need to be worked over at least three times to obtain fruit at the correct maturation stages. It may be sound economically to pick the whole crop at once, but the olives will have to be sorted later. The development of sorting machines that detect different skin colours will be very useful for sorting green-ripe olives from olives at other maturation states.

Harvesting techniques for table olive processing.   Hand picked olives should be collected into padded baskets to prevent damage, then placed in 20–25 kg slotted crates (Fig. 3.15). Olives are picked using a milking action to improve efficiency. Inexperienced pickers should be given on-site training before picking is commenced, especially in the prevention of tree and olive damage and which olives to pick. While in the orchard, picked olives should be stored in a cool place out of the sun until transported to the packing shed or processing facility (Fig. 3.16). As olive varieties process at different rates, individual varieties should be kept in separate crates.

Figure 3.15 Olives in crates ready for transporting to the processing facility.

Figure 3.16 Sunburnt green-ripe olives. Olives were left exposed to direct sunlight for a few hours.

Harvesting olives with hand-held or mechanised rakes, tree shakers or overhead harvesters increases the risk of bruising, leading to gas pocket spoilage and softening when the olives are processed. More serious damage occurs with black-ripe olives than green-ripe olives of a particular variety.

Mechanical harvesting with shaking or vibration devices has limited application for olives destined to be processed as quality table olives. As yet, machine harvesting has not been a real option, even though some growers are using this technology, because of bruising and marking of the fruit. More research needs to be undertaken in this area such as identifying varieties that are not sensitive to the action of the mechanical harvesters. Olives are damaged when they come in contact with branches during shaking or vibrating and when they fall to the ground. Damaged fruit when processed reveals surface scars such as brown spots and the fruit is more likely to form blisters or gas pockets. There is possibly some scope to mechanically harvest green-ripe olives for bruised or cracked olives and naturally black-ripe olives for drying by heat or salt.

Ethylene releasing compounds, which weaken the olive fruit attachment, have also been used to facilitate mechanical harvesting so that less force is required, but their application is limited because with indiscriminate use substantial leaf loss occurs. This is detrimental to future productivity. Although some developers of mechanical harvesters have made claims that their equipment is suitable for table olives, further studies comparing processed hand picked olives with processed machine harvested olives need to be undertaken.

Olives show appreciable injury even at the green-ripe stage when mechanically harvested (Fig. 3.17). This becomes more obvious during processing. Immersing machine harvested green-ripe olives into weak lye solutions within 20 minutes of harvest limits bruising. Although these olives are generally as good as those harvested by hand, this procedure is not widely practised. For green-ripe olives produced under irrigation a sodium hydroxide solution of 0.2–0.3% w/v in potable water is used, but higher concentrations (0.6–0.8% w/v) are used for unirrigated olives. The olives are transported in this solution from the field to the processing facility to continue processing. Immersion time in this transporting solution should be short and limited to 1.0–1.5 hours. Some varieties may require longer periods to remove browning damage, and if this is the case, lower concentrations of sodium hydroxide are used to avoid skin sloughing. Nevertheless, green-ripe olives treated in this way can be used for producing Spanish-style olives but are less suitable for producing table olives by natural fermentation methods.

Figure 3.17 Machine harvested olives two to three days after harvesting.

With the large olive crops expected in the future as the large number of olive trees in Australia come into production, serious consideration must be given to varieties that can resist damage by mechanical harvesting. Costs for hand harvesting olives are currently A$1.50–$2.00/kg depending on the variety, tree shape and height, weather, availability of labour and distance from major community facilities. In contrast, the costs of machine harvesting olives are estimated at around 30c/kg, and if utilised would radically reduce table olive production costs. Such an advantage would be lost if culling of damaged olives is necessary. Hand harvesting is more economical when olive trees are trained for easy picking by hand.

Harvest date. Harvest date is influenced by the final product to be made; disposal of crop, for example contracts and processing capacity; growing conditions and by the state of the weather.

In most parts of Australia, olives are harvested during reasonable weather conditions. Harvesting should be undertaken before early frosts as extreme cold can severely damage the fruit tissue. In areas prone to rain during harvest, olives are likely to get wet and dirty. Furthermore, in orchards with heavy soils it becomes more difficult for vehicle movement and transporting the picked olives within the grove under wet conditions. In California, researchers have developed a harvesting protocol based on the proportions of different sized olives. When 50% of olives picked randomly from three olive trees are within predetermined sizes and their percentage is increasing at a rate of 3–5%, picking is commenced within one to two weeks. This method is mostly suitable for green-ripe olives.

Post-harvest handling of raw olives

Deterioration of olives must be prevented during the post-harvest handling, storage and transport stages. Green-ripe, turning colour and naturally black-ripe olives are all subject to damage if handled badly or stored harshly. Discolouration is obvious if green olives are injured. Although damage is less obvious with black olives, they still need careful handling because injury may lead to the development of soft olives or gas pocket formation with fissures during processing.

Transporting raw olives. Olives should be transported carefully to the processing facility and processing commenced as soon as possible. To prevent post-harvest deterioration olives must be packed and transported (if over long distances preferably at night) in shallow ventilated crates that allow air circulation, but never in closed crates, bags or sacks. Crates should be cleaned and sanitised before use. The transporting vehicle must be in a clean and hygienic state, and not be carrying at the same time other chemicals, petrol in cans or animals. With large-scale table olive production, olives are packed and transported in 250–500 kg crates with perforated sides. Olives are more likely to deteriorate during transport if packed in large, rather than smaller sized crates.

Storage of raw olives. All olives should be processed as quickly as possible after picking, preferably within one to two days, to avoid deterioration. Mould can be a problem for the olive fruit between picking and processing. During poor storage and handling of olives undesirable breakdown products are formed, for example methanol, ethanol, 3-methyl-butanol, 2-methyl-propanol, acetic acid and ethyl acetate as well as others that give 'fusty' and 'musty' malodours and flavours. Depending on the maturation stage, olives stored for longer periods under ambient conditions can, within seven days, develop fusty and musty defects through heating and sweating. Fresh olives stored at ambient temperatures will deteriorate quickly through the action of enzymes within the olive flesh and by the action of bacteria, yeasts and moulds on the fruit and from the external environment. Processing such olives yields soft products with poor organoleptic qualities. Black-ripe olives deteriorate more quickly than green-ripe olives.

Storing unprocessed olives at low temperatures prevents their deterioration by decreasing their respiration rate and by retarding the action of spoilage organisms. As raw olives are a living entity, they continue to respire and lose moisture during storage, causing weight loss and shrivelling. Such weight loss can be significant. Storing for prolonged periods leads to nail-head markings on the skin (Fig. 3.18) and injury thought to be of bacterial origin.

Figure 3.18 Nail-head defect due to prolonged storage of olives before processing.

Research in California has shown that green *Manzanilla* olives stored between 5°C and 10°C will keep from four to eight weeks respectively before deterioration is significant. Storage temperatures below 5°C caused browning reactions in the olive flesh radiating from the stone to the surface. Green olives can be stored for longer periods under controlled atmospheric conditions, for example 5.0–7.7°C with 2% oxygen and a relative humidity of 90–95%. More research is required in this area. Green-ripe olives generally store better than naturally black-ripe olives. Images of cold damaged green-ripe olives stored at 2°C for seven days are presented in Fig. 3.19.

Figure 3.19 Olives stored under refrigeration showing cold induced damage. (Photo: Australian Mediterranean Olive Research Institute.)

Grading raw olives. Sorted and graded olives are more desirable for processors than delivering the olives 'as is' off the orchard. With sorting and grading the expectations are that the batches will be of a single variety, be of uniform size, and have no leaves or small, defective or damaged olives.

Sorting and grading can be undertaken by hand or by machine. Olives can be damaged during any vigorous grading procedures so appropriate precautions must be taken to prevent this. Injuries such as bruising when the olives pass through mechanical sorters and graders can lead to the formation of brown spots, gas pockets or blisters during processing. Systems of grading based on differences in the specific gravity of the olives at the different maturation stages have been developed, and proponents have reported that the quality of raw olives is better than compared with other grading systems. All sorting and grading equipment should be cleaned and sanitised before use.

Undesirable qualities of raw olives for table olives

As mentioned earlier, quality raw olives are essential for quality table olive production. Undesirable qualities for raw olives and actions to improve olive quality are listed in Table 3.18.

Skin blemishes can be due to bruising of fruit by machinery during harvesting, or during post-harvest handling and grading. Flesh damage can also occur during harvesting and poor storage, particularly at low temperatures. Shrivelled olives can result if trees are stressed, particularly if irrigation is unavailable at times of excessively high temperatures. Olives soften and shrivel on storage.

Changes in olive colour may be due to chemical sprays and some infestations. Deformed fruit occurs through nutritional problems such as boron deficiency and superficial insect damage, whereas moulds can lead to changes in skin colour and partial dehydration. Insects such as fruit fly can mark the skin, whereas Olive Fruit Fly causes substantial damage to the flesh. Severe scale infestations can render olives unsuitable for table olive processing. If pests or diseases are present, growers need to check their pest management program. Picked olives should be without stems unless required for specialty products. When olives are smaller than expected, the authenticity of the variety should be checked.

Note: Raw olives can be contaminated with pathogenic and spoilage organisms throughout the pre-harvest and post-harvest periods. Wild and domestic animals, soils, faeces, irrigation water and orchard workers are potential pre-harvest sources, while harvesting tools, crates, bins, rinse water, processing equipment and unhygienic processing operatives can cause post-harvest contamination.

Table 3.18. Undesirable qualities in raw olives for table olive processing

Defect	Description	Operations needing checking
Skin blemish	Superficial marks affecting the skin, such as bruises, blows and stains induced by brushing against branches, but do not penetrate into the flesh and are not the consequences of disease.	Cultural and agricultural practices Harvesting Post-harvest handling, storage and transport Sooty Mould
Flesh damage	Imperfection or damage to the flesh which may or may not be associated with superficial marks.	Cultural and agricultural practices Harvesting Post-harvest handling, storage and transport
Shrivel	Olives are so wrinkled, the appearance of the fruit is materially affected. An exception is when shrivelled olives are used for preparing dried olives.	Cultural and agricultural practices Post-harvest handling, storage and transport
Softness	Olives are excessively soft or abnormally flabby.	Harvesting Post-harvest handling, storage and transport
Fibrous or woody	Olives that are excessively or abnormally fibrous or woody.	Variety Harvesting time
Abnormal colour	Olive colour is distinctly different from the characteristic colours of green-ripe, turning colour and naturally black-ripe.	Cultural and agricultural practices Post-harvest handling, storage and transport
Damage due to abnormal growing practices	Skin of the fruit has been accidentally burnt.	Cultural and agricultural practices
Deformed fruit	Fruit deformed through factors including nutritional deficiencies and environmental damage.	Cultural and agricultural practices
Cryptogamic and mould damage	Includes lustreless fruits and those with scattered, more or less dark stains caused by the mycelium of certain fungi (e.g. *Macrophoma* and *Gleosporum*) growing either within the olive (leading to dehydration of the tissues) or on the skin and affecting fruit colour.	Cultural and agricultural practices
Insect damage	Deformed fruits and those with abnormal stains, or flesh has an abnormal aspect. Exit holes are often present.	Cultural and agricultural practices Scale infestation
Stems	Significant number of stems still attached to olives (unless as a specialty item).	Harvesting practice
Small olives	Inappropriate size.	Check variety authenticity If small, thin out the crop

Concluding remarks

In setting up an olive orchard for the production of olives for the table olive market, a number of important strategies need to be taken into account. The first is the average rainfall and the availability of irrigation water for the growing of olives. This may be from dams on site or from scheme water. The amount of rainfall (or equivalent by irrigation) for successful commercial olive growing should be between 600–1000 mm, and effective. The regional temperature patterns also need to be taken into consideration, including frost factors, wind strength and finishing temperatures. The soil structure and the lay of the land is important as this will determine the layout of the olive orchard so that the trees are able to receive maximum light, be protected from hot dry winds, and the soil does not become waterlogged. The consideration of olive varieties is one of personal selection, but the most popular are *Kalamata* and *Manzanilla*. It is recommended, however, that growers have a number of table olive varieties in their orchards. Pollinator trees are important and these should be distributed around the orchard. A number of agronomic parameters need to be considered: planting density, plant nutrition, training of the trees, pruning, tree thinning, tree protection, and pests and disease management.

Care during harvesting olives for table olive production is important, as the fruit should have no blemishes. Currently the best way to harvest is by hand; however, this is expensive. Advances in mechanical harvesting are currently being investigated and new developments should improve harvesting efficiency and lower the cost. The transport and storage of the olives is an important factor at harvest time. The recommended time after harvesting for processing the olives is one to two days.

4

Table olive processing: general aspects

There are different processing procedures for the production of table olives, and in setting up a production facility there are a number of considerations that need to be taken into account. Key areas in the processing facility need to be separated, such as raw olive storage and grading areas, chemical storage (both for production and cleaning) and staff facilities. The materials to be used in the production of table olives must be of food grade quality and potable water must be used. The equipment must be made from food grade stainless steel, plastic or fibreglass and the facility must comply with statutory health and environmental regulations. The types of table olive processes used for black table olives, shrivelled black olives, turning colour olives, green-ripe olives, bruised olives and split olives are explained. The chapter also discusses secondary processing and the finished products. The authors have also examined the storage of raw olives prior to processing and have suggested a general protocol for the processing of table olives from receipt of the raw olives to the finished product. The environmental needs of the processing area are also considered with the types of processing to be undertaken. Records of all the processes used in the processing facility need to be kept so that any problems with the final product can be traced back to a specific point in the production line.

Introduction

Table olive processing involves the transformation of bitter inedible olives into an edible foodstuff. Processing methods also preserve the olives from natural deterioration so that they can be stored for significantly long periods and consumed as required. As well as being palatable, when processed the transformed olives must be safe to eat and have retained their nutritional qualities. Additional preservation techniques such as pasteurisation, heat sterilisation and preservatives are also used for some packaged olive products.

Processed table olives, a manufactured food, need to be produced by acceptable technologies under safe conditions for consumers and table olive operatives. The olives to be processed need to be grown, harvested, stored and transported by methods that minimise physical damage as well as prevent chemical and/or microbiological contamination.

Raw olives at all stages of maturation are bitter and mostly inedible and require processing before being suitable for consumption. As with other types of food processing, table olives should be processed under *Good Hygienic and Manufacturing Practices* using quality assured ingredients. Raw and processed olives must meet quality standards including varietal authenticity as well as physical, chemical and microbiological standards to ensure safe and nutritious products.

Table olive processing should be approached systematically. Processing facilities must be appropriate for the production and storage of table olive products and meet food, environmental, and occupational health and safety requirements. All processing must be undertaken in ways that minimise environmental impacts, such as fewer washing steps, and, where possible, reusing fermentation brines as packing solutions.

Requirements for the production of high quality table olives are:

- potable water (water that meets drinking standards);
- quality raw olives;
- food grade chemicals (salt, lye, food acids);
- food grade herbs, spices and condiments; and
- food grade cleaners and sanitisers.

All material used in table olive processing, such as water, salt, lye (caustic soda, sodium hydroxide), acetic acid, lactic acid, citric acid, hydrochloric acid, dextrose, sorbic acid, ferrous gluconate or lactate, herbs and spices, must meet food grade standards. Equally important in table olive processing is the water used for all operations in the plant, including washing and processing, as it can be a source of contaminants. Only potable water, which meets prescribed microbiological, chemical and physical standards set by national/regional authorities, must be used to reduce the risk of food poisoning and product spoilage.

Unlike olive oil production, where the technology is well defined and the processing automated, table olive production needs careful planning with regard to the size of the operation, the processing methods to be used and the products to be made. Safety issues are a high priority and if processing is not controlled effectively, spoilage of the olives and possibly food poisoning are both potential negative outcomes.

Table olive processing methods and varieties

During the planning phase, decisions need to be made regarding the types of olives to be produced and which varieties to use. Such information is given in Table 4.1. Even if other varieties are used, the processor must ensure that they have sufficient quantities of raw olives for commercial table olive production.

Table 4.1. Olive processing methods and suggested varieties

Processing method/style	Suggested varieties
Green	
Untreated in brine	*Conservolea, Chalchidikis, Manzanilla, Hojiblanca, Barouni, Picual, Verdale*
Spanish-style green (lye treated)	*Manzanilla, Hojiblanca, Sevillana*
Turning colour	
Ligurian olives	*Taggiasca* or *Frantoio*
Greek-style in brine	*Jumbo Kalamata, Verdale*
Black	
Naturally ripe in brine (untreated in brine)	*Conservolea, Frantoio, Leccino, Ascolana, Picual*
Kalamata-style	*Kalamata, Hojiblanca, Leccino*
Californian/Spanish-style (lye treated)	*Manzanilla, Californian Mission, Hojiblanca*
Salt-dried	*Thrubolea, Manzanilla, Frantoio, Kalamata, Leccino*
Heat-dried	*Manzanilla, UC13A6, Kalamata, Leccino*
Specialty products	
Marinated	Processed destoned green/black olives of any variety
Destoned – stuffed	*Manzanilla, Sevillana*
Pastes and tapenades	Destoned processed green or black olives of any variety

Good Manufacturing Practice (GMP) and table olives

Processors need to develop procedures that ensure the olives to be processed are of the correct variety, at the same maturation state, not damaged or defective and of a size relevant to market standards. When packed for sale, olives should be of uniform size, except for specialty mixes where several varieties are used. Sizing and grading is undertaken manually or by machines. The Codex Alimentarius (1987)/IOOC (2004) Table Olive Standards have listed specific size ranges (see Table 4.2) for processed table olives. Exporters will also need to check with individual countries for any local requirements.

Table 4.2. Summary of the size scale for grading table olives as set by the Codex Alimentarius (1987)/IOOC (2004) Table Olive Standards

Size range olives/kilogram	Interval within a size range
60/70–111/120	10
121/140–181/200	20
201/230–381/410	30
Above 411	50

Processing olives should be commenced as soon as possible, preferably within one to two days after harvesting. Olives must be washed in potable water prior to entering the processing line. Process control (microbiological, chemical and physical) is essential during table olive production. Procedures must be developed and documented according to Hazard Analysis Critical Control Point (HACCP). Product profiles must also be developed for each olive type being produced, for example *Kalamata*, Greek-style black or green, Spanish-style green, Californian black, Ligurian, Thrumbes or dried. All inputs, such as olives, water, herbs and spices, chemicals and cleaning agents, must have product profiles and be under inventory control. A system allowing for product recall, an essential element in food production, must also be in place.

Water requirements for table olive production

Water is required for all washing procedures as well as preparing brine and lye solutions. As water can contain a number of contaminants (physical, chemical or microbiological), all water used for table olive processing must meet drinking water standards (potable water). The general characteristics of potable water are:

- physical: there should be minimal particulate matter;
- chemical: salts, iron, heavy metals and organic chemicals must be within health and safety limits; and
- microbiological: no harmful organisms are present.

Scheme water (water from state or municipal authorities) is generally of potable water quality and so can be used as is. However, this water source should be tested at the point of entry into the processing plant and thereafter should be tested annually. Groundwater and tank water can contain a number of contaminants, so water from these sources should be tested and treated before use. Groundwater, tank water and water from dams, lakes and rivers can typically have one or more of the following types of contaminants:

- physical: particulate matter, mineral, organic matter;
- chemical: herbicides, pesticides, industrial chemicals, naturally occurring salts and metals; and
- microbiological: microorganisms of faecal and/or plant origins.

Water of unknown quality needs testing by an accredited laboratory and must be treated before use. If contaminated with microorganisms the water must be sanitised, for example by UV irradiation, in the following cases: (1) if coliform/thermotolerant coliform organisms are present, or (2) if there are no coliforms/thermotolerant coliform organisms present but the Heterotrophic Plate Count is >500 CFU/ml.

Typical results of tank water analyses presented in Table 4.3 indicate that both tanks tested had higher than normal Heterotrophic Plate Counts and coliforms. The Heterotrophic Plate Counts, although high, are typical of rainwater tank water. The coliform count is unacceptably high. Successfully sanitised water should have a Heterotrophic Plate Count of 500 CFU/ml or less.

Table 4.3. Typical microbiological analyses of untreated rainwater tank water undertaken by the authors CFU/ml, colony forming units per millilitre.

Parameter	Results		
	Potable water	Tank water 1	Tank water 2
Appearance	Clear, colourless	Clear, pale, straw coloured	Clear, colourless with debris
Heterotrophic plate count			
At 21°C CFU/ml	<500	>1000	>1000
At 31°C CFU/ml	<500	>1000	>1000
Coliforms CFU/100 mL	0	23	33
Thermotolerant coliforms/100 mL	0	<2	<2

UV light disinfection of water. Equipment for UV light irradiation of water is easy to install and the process is effective against most microorganisms. Filters should be fitted before the UV system for the removal of particulate and organic matter by filtration and/or reverse osmosis. This is critical to effective UV disinfection. Using reverse osmosis is expensive and the system requires regular maintenance, so its use in table olive processing is optional. Variable flow rates can also reduce the effectiveness of the process.

Water source → Filtration/reverse osmosis → UV sanitiser → Storage tank → End use

End uses of treated water relevant to table olive processing include: washing water for olives, preparing lye solutions and fermentation brines, and for packing solutions. An alternative to UV sanitising of water is to use a continuous heat exchange pasteuriser where incoming water is treated at 72°C for 15 seconds. Other disinfection methods for water are available but some that use chemical sanitisers are less suitable as they may interfere with table olive processing, particularly where naturally processed olives by fermentation are being made. Where treated water is retained in water storage tanks, such tanks should be covered and the water utilised as quickly as possible. Tanks not being used should be emptied, cleaned and kept clean and dry.

Planning table olive processing facilities

Table olive activities can be divided into four broad categories: producing raw olives (see Chapter 3), primary processing of raw olives, secondary processing of primary processed olives, and packing and marketing olive products.

The Australian table olive industry is in a position to implement the latest practices in table olive processing. Much of the recent international research effort has been directed to the chemical, microbiological and organoleptic control of table olive production. Emphasis is placed on the prevention of spoilage due to poor quality fruit, contaminated inputs (such as water), lack of documented procedures, use of unhygienic premises and equipment, and unhygienic staff practices as well as poor packaging and storage of final products.

Processing facilities require careful planning with respect to processing methods and capacity. As indicated earlier, facilities and processing procedures must meet occupational health, safety and environmental standards. All equipment must be constructed so that it can be easily cleaned and sanitised. Processing equipment should be constructed from food grade material, such as plastic, fibreglass or stainless steel, and be suitable for the function required. Processing tanks and containers vary in size, with some exceeding 15 tonnes. Boutique table olive processors often use plastic barrels to process olives. Larger scale enterprises use processing tanks of substantial sizes (see Figs 4.1, 4.3 and 4.4).

Excluding the cost of buildings and land, a small table olive processing plant with a capacity of up to 20 tonnes of olives can cost as little as A$50 000–A$100 000 to establish. Large-scale facilities of 500 tonnes or more will cost between one and two million dollars (Australian) depending on the level of sophistication. Ancillary equipment and facilities for large-scale processing such as waste disposal, pumps, sorting tables, graders, destoners, bottling lines and a testing laboratory can account for at least a further A$300 000. Australian developed machines for slicing and destoning olives are expected to cost approximately A$30 000–A$40 000. The cost of similar machinery imported from overseas is of the same order. When purchasing equipment, processors should ensure that efficient backup for repairs and maintenance is available.

Key considerations for establishing a table olive processing facility include the following:

- understand the industry, products and markets;
- decide on the size of operation: boutique, small, or large scale;
- determine the environmental issues, such as disposal of washing water, brines and lye solutions;
- determine sources of quality fresh olives: from your own olive crop or olives sourced from others;
- consider how olives will be delivered to the processor. Use quality systems;
- establish sources for acquiring quality water and food grade ingredients. Use recognised suppliers;
- plan functional and hygienic processing plants. Check building, food, and environmental codes;
- employ trained operatives: formally or in-house trained;
- establish documented processing procedures with controls. Implement HACCP;
- adhere to occupational health and safety requirements. Check appropriate codes;
- establish procedures to maintain product profiles. Implement laboratory testing; and
- establish product recall procedures. Check the FSANZ or other statutory food codes.

Functional table olive processing facility

Processing facilities must be appropriate for the production and storage of table olive products. They must meet statutory food regulations and good manufacturing practice as well as occupational health and safety and environmental requirements. The plant design should ensure that all operations can be undertaken effectively without the risk of contamination or cross contamination. All areas and equipment must be kept clean, hygienic and free from insects, rodents and other pests. One such processing plant is depicted in Fig. 4.1.

Figure 4.1 Small-scale Australian table olive processing enterprise at Casino, NSW. (Photo: Newton and Merryl Ellaby.)

Pest control. Pests, such as rodents and insects (especially cockroaches), can carry pathogens like *E. coli* and *Salmonella*. Doors and windows should be placed so they prevent the entry of animals, birds and insects. Keeping areas clean and tidy and removing unwanted gear that could harbour vermin are important strategies. Where infestations are present, baits and traps should be used rather than pesticides, which could increase the risk of contamination.

Producing Table Olives | 137

Plate 1. *Kalamata (Kalamon)* variety at different maturation stages: green-ripe, turning colour and naturally black-ripe olives. Note natural bloom on the black olives.

Plate 2. *Arbequina* olives.

Plate 3. *Barnea* olives.

Plate 4. *Frantoio* olives.

Plate 5. *Kalamata* olives.

Plate 6. *Leccino* olives.

Plate 7. *Manzanilla* olives.

Plate 8. *Picholine* olives.

Producing Table Olives | 141

Plate 9. *Sevillana* olives.

Plate 10. Four large-fruited olive varieties. (From left to right: *Barouni*, *UC13A6*, *Hardy's Mammoth* and *Nab Tamri*.)

Plate 11. *Jumbo Kalamata* olives and stones. Note the large white patches (lenticels) on the surface of this variety. (See also Fig. 3.4.)

Plate 12. *UC13A6* (Californian Queen) olives.

Producing Table Olives | 143

Plate 13. *Verdale* olives.

Plate 14. Burnt olive leaf tips typical of excess salinity or potassium deficiency.

Plate 15. Chlorosis of olive leaves typical of nutritional deficiencies.

Plate 16. Signs of iron deficiency in olive leaves. Note green venation.

Plate 17. Black Scale with typical Sooty Mould on stems and leaves of an infested olive tree. (Photo: Australian Mediterranean Olive Research Institute.)

Plate 18. Olive trees infested with Olive Lace Bug. Note the sparse grey foliage.

Plate 19. Olives infested with Olive Fruit Fly. Note entry marks. (Photo: Paul Vossen, California.)

Plate 20. Single olive with grub consuming the flesh. (Photo: Paul Vossen, California.)

Plate 21. Olive leaves with characteristic lesions due to Peacock Spot, *Spilocea oleaginea*. (Photo: Paul Vossen, California.)

Plate 22. Evaporation pond with heavy plastic liner for the disposal of table olive wastewater. (Photo: Newton Ellaby and Merryl Ellaby, Australia.)

Plate 23. Loose olives for sale at the Victoria Markets, Melbourne, Australia.

Plate 24. Cracked turning colour *Manzanilla* olives prior to processing.

Plate 25. Heat-dried olives with olive oil, crushed chilli and rosemary.

Plate 26. Turning colour *Manzanilla* olives processed in brine with crushed dried chilli and olive oil.

Plate 27. Provencale-style black *Leccino* olives, processed in brine, with crushed dried chilli, sliced garlic, rosemary and olive oil.

Plate 28. Destoned green-ripe *Ascolana Tenera* olives processed in brine, with added crushed chilli, chopped garlic, fennel seeds and olive oil (Sicilian style).

Plate 29. Various tapenades prepared with processed green-ripe, turning colour or naturally black-ripe olives.

Plate 30. Lye treated *Sevillana* olives with and without fermentation. Note yellow coloured olives are the ones that have undergone fermentation after lye treatment, whereas the green olives have undergone lye treatment only.

Plate 31. Oxidative moulds growing on the brine surface during processing of olives in a poorly controlled barrel.

Plate 32. A number of problems that can occur during processing. Notice the black gas pockets and fissures on the brown *Kalamata* olives, white yeast spots on the *Kalamata* and green *Sevillana* olives, and pressure marks on the *Sevillana* olives.

General facilities and equipment required

Facilities should be designed and built for easy cleaning and maintenance and to minimise the entry of contaminants. Makeshift facilities should be avoided, as in the long term it may be more practical and cost effective to construct new purpose-built facilities rather than trying to alter existing structures.

The main equipment required includes washing machines, sorters, graders, tanks (stainless steel or food grade fibreglass), pumps and packing equipment. Large-scale commercial olive processing tanks can hold up to 10 tonnes of olives and 5000 litres of brine (approximately 15 tonnes capacity). They are cylindrical or spherical in shape and made of polyester and fibreglass. Tanks can be placed underground, above ground, in large sheds or in covered areas. Washing machines for table olives are similar to those used for washing olives for olive oil production. Storage areas, chemical stores, wash-up and toilet facilities should be segregated and away from receiving and production areas. All chemical materials should be of food grade quality and under inventory control to prevent errors, mix-ups or contamination. All equipment must meet food grade standards and be of a design that allows for easy cleaning, maintenance and safe use by workers. The use of equipment made of galvanised iron or aluminium should be avoided because of the risk of corrosion and contamination.

The table olive processing facility should be divided into a number of physically or functionally delineated areas:

- management and service area;
- laboratory for records, testing;
- receiving and storing chemicals;
- receiving raw materials;
- raw material storage area;
- pre-processing operations;
- primary processing operations;
- bulk storage of processed olives;
- secondary processing operations;
- packaging operations;
- pasteurisation and/or sterilisation facility;
- storage of packed products; and
- loading area for finished products.

Buildings and work sites should meet the standards required for food processing, including adequate lighting and ventilation. Walls, ceilings and floor surfaces should be made of approved material for cleaning and drainage. Equipment used for table olive processing will depend on the scale of the operation and the products to be produced. The list below is indicative of the types of equipment that can be used:

- sound and functional building and work site;
- conveyor system for olives entering the facility;
- washing machines to rinse olives before processing;
- sorting machine for size-grading raw olives;
- sorting tables with moveable belt to facilitate removal of defective olives;

- slitting machines to slit olives to facilitate processing (Fig. 4.2);
- bruising or cracking olive machines to facilitate processing;
- tanks for solution preparation, fermentation procedures (Figs. 4.3–4.4) and rinsing;
- brine pumps to transfer and circulate brines;
- olive pumps to transfer olives from tanks;
- destoning machines to destone olives for stuffing, slicing or making tapenade and olive pastes;
- tapenade and olive paste machines, purpose-built or commercial food processors;
- stuffing machines to stuff olives with food fillings;
- slicing machines to slice destoned olives;
- commercial ovens for heat drying olives;
- storage tanks and barrels;
- filling, packing and labelling equipment;
- pasteurisation or sterilising equipment; and
- laboratory for testing olives and brine.

Figure 4.2 Australian developed olive slitting machine. (Photo: The Olive Centre, Queensland.)

Figure 4.3 Medium-scale table olive processing tanks with brine pump. (Photo: Koorian Olives, Gingin, Western Australia.)

Figure 4.4 Large-scale table olive processing tanks. (Photo: Viva Olives, Loxton, South Australia.)

Plant maintenance and sanitation. Because table olive production involves food processing activities, sanitation is important to prevent consumer illness and spoilage of the olives. Cleaning procedures should be defined, particularly for the chemical agents used. Pest control is best left to a licensed third party that can provide records to ISO standards for inspection by health surveyors. Workers must conform to stringent hygiene protocols including the need to report to employers any illness that can impact on the safety of the table olives being produced. Cleaning procedures should ensure against microbiological, physical and chemical contamination. Records should be kept so that an effective maintenance program can be developed.

Cleaning procedures. The processing plant and all equipment should be kept in a clean state. Cleaning procedures are undertaken with food grade detergents (often based on sodium lauryl sulfate) and water that generally removes chemical contaminants, soil, organic matter and microorganisms from surfaces. It should be noted that detergents are not lethal to residual microorganisms. The processing plant and all equipment used should be kept in a clean state.

Sanitising procedures. Sanitising procedures significantly reduce the number of residual microorganisms after cleaning procedures have been undertaken, but do not kill all microorganisms. Common sanitising agents suitable for surfaces and equipment include: steam sprays, chlorine compounds, iodine compounds, quarternary ammonium chloride compounds, peroxy compounds and carboxylic acids.

Care must be taken to ensure that residual levels of sanitisers on equipment are within food safety limits.

Protocol for cleaning and sanitising equipment

The facility, fittings and equipment must be kept clean and in a good state of repair.

The item or equipment is rinsed with cold water, rinsed with hot water, rinsed with caustic rinse (1.0–1.2% w/v sodium hydroxide in water) if required, sanitised with peroxide-sanitiser or steam, and dried with compressed air or disposable paper towels.

Protocol for cleaning and sanitising food contact surfaces

Sanitising of contact surfaces can be achieved by using hot water (77°C or above), a food grade sanitiser and dilute bleach (sodium hypochlorite solution).

After treatment the surfaces are allowed to air dry or are dried with disposable paper towels.

Trained operatives. People working in the processing plant must have an understanding of all the activities within the facility, particularly as they relate to occupational health and safety. More specifically they should understand the importance of personal hygiene so the risk of contamination or cross contamination of the olive products is minimised. Those workers involved more directly in processing must understand the procedures and controls to produce olives that meet quality and safety standards. The table olive processing facility must have personnel trained to the level of the operations required, especially in the areas of hygiene, food handling and record keeping. Workers must be provided with information for the safe handling of toxic and microbiological materials and other possible hazards that may be encountered when using equipment and machinery. As processing involves the use of salt, lye and concentrated acids, workers must be trained to handle chemicals, prepare solutions and be able to handle spills and emergencies especially for personal and third party safety. Workers must be aware of Material Safety Data Sheets (MSDS) for all chemicals used in the processing of table olives. Health and safety implications and decontamination procedures need to be documented and brought to the attention of operatives. Processing procedures should include pre-calculated quantities of raw materials for safety and quality purposes.

Operator health and hygiene. Adverse medical conditions, such as food borne problems, should be reported to management and a clearance obtained before returning to food handling operations. Unsanitary practices such as smoking, eating and drinking should not be undertaken in production areas. Clothing must be clean and suitable for the task. Operatives should not wear cosmetics and jewellery during processing and packaging of table olives to prevent accidental loss or contamination. As long as personal hygiene procedures and requirements (hand washing, no open cuts or sores, no infectious disease) are adhered to, wearing gloves while undertaking table olive processing procedures is not necessary. Gloves should be worn when hand stuffing table olives, for example with cheese, nuts and fish. Workers should wear approved closed footwear and disposable head or beard covers. However, approved gloves and safety glasses should be worn when handling corrosive substances, preparing brines and where spillage or spoilage may occur.

Practicalities of operatives' health and hygiene. Clothing worn by operatives in food processing areas should be clean and changed daily. Eating, drinking and smoking should never occur inside the processing area or when olives are being unloaded and loaded. Hands should be washed and/or sanitised before starting work, after each toilet visit, after eating and smoking, after blowing the nose or coughing, after handling rubbish or undertaking maintenance procedures, and after any break in work from the processing area. Operatives with colds need to take extra precautions to prevent contamination, such as blowing the nose rather than sneezing and coughing over the olives. Increased handwashing and using disposable tissues can help. It may be best for affected persons to undertake non-food contact duties. With more serious health conditions where the risk of contamination with pathogenic microorganisms is likely, such as gastroenteritis, diarrhoea, vomiting, sore throat with fever, fever or jaundice, it is best for them to stay away from the table olive processing plant or, where practical, be given non-food contact duties.

Clothing and practicalities when working in a food production area. Olive processors should be aware of the desirable clothing and practices for operatives working in food production areas. Furthermore, these should not be considered as 'Rolls Royce' requirements but best practice that will reduce the risk of contamination and spoilage of food products, such as table olives.

Clothing. Clothing to be worn in food production areas to protect the worker from spillage and reduce the risk of contamination includes a two piece suit made up of a V-neck cotton top with no pockets and elastic-waisted cotton pants with an inside pocket. It is optional as to whether workers wear a clean T-shirt and/or light pants inside the suit or just their underclothes. Suits are changed daily and more often if they are soiled. Clothing is also changed when working in high-risk areas. Workers are not allowed to sit on the facility floor, on equipment or outside, for example on grass. Steel-cap safety work shoes (not rubber boots) supplied by the establishment are worn in the facility, but left in the facility at the end of the shift. Generally, gloves are not worn during processing procedures except when the foodstuff is closely handled. In this case disposable gloves are worn, with regular changes during the day. Circumstances relevant to table olive processing are destoning olives and stuffing olives with food fillings, such as cheese and peppers, by hand. Gloves are also worn when the worker has a hand injury, which should be protected with a coloured adhesive dressing. Workers should also wear protective gloves and safety glasses when handling corrosive liquids such as lye solutions and acids.

A typical procedure follows. The worker arrives at the facility and changes into their two piece working suit and puts on safety shoes, a hair net and, if appropriate, a beard mask before entering the processing area. The worker then walks in a shoe sanitiser bath containing chlorine or iodine based solutions. They should then dry-clean their vest, particularly at the shoulders, with a roller with a disposable sticky surface to remove hair and specks. Hands are then washed with food grade non-perfumed liquid soap and hot water at 55°C or above (no-touch soap dispenser, no touch taps or infra-red sensor panel to turn on water), rubbed with an alcohol sanitiser (optional), liquid or wipes (tissues impregnated with sanitiser), then with a disposable towel.

Note: Effective hand sanitisers, especially wipes, remove and absorb germ laden biofilm from hands preventing cross contamination when handling food. The sanitiser kills most pathogenic skin organisms within 20 seconds and is effective against *Pseudomonas aeruginosa*, *Staphylococcus aureus*, *Salmonella choleraesuis*, *E. coli* and *Streptococcus pyogenes*.

Recall procedures. Table olive processing facilities must have a recall system for unsafe or contaminated products. All olives, whether stored in bulk or final consumer containers, should be traceable through records and batch numbers. The recall process must be within guidelines of government agencies responsible for public health surveillance. Responsible persons such as processors, packers, wholesalers and government health agencies can initiate recalls. All products requiring recall should be reported immediately to the appropriate authorities.

Products are recalled when there is a real or perceived health risk (see Chapter 6 for more detail).

The first scenario is that a local health department receives notification or a complaint regarding a food poisoning incident. This matter is then investigated, a procedure that includes testing of the suspected foodstuff, for example to detect the presence of organisms such as *Escherichia coli*, coagulase positive Staphylococci, *Clostridium perfringens* and *Listeria monocytogenes*, and calling for information from the point of sale or use and from the manufacturer.

A second scenario is when the manufacturer has found a contaminant in their product, or some other problem that impacts on safety, for example stones in stuffed olives or pH outside the safe range.

A third scenario is that a food is produced in a way that the absence of harmful organisms such as *Clostridium botulinum* or its toxin cannot be guaranteed.

A fourth scenario occurs where an ingredient that may cause an allergic reaction in susceptible persons has been left off the ingredient list, for example nuts and milk protein.

In all four cases the suspect product needs to be recalled. These four scenarios illustrate the need to ensure that documented procedures with the appropriate controls are used along the processing chain to the consumer and that samples of the final packaged products are cleared by a registered laboratory.

Table olive processing

As defined by the Codex Alimentarius (1987)/IOOC (2004) Table Olive Standards, table olives are prepared from the sound fruits of suitable varieties of the cultivated olive tree (*Olea europaea* L.). When treated or processed the olives are ready for consumption subject to packaging requirements. Olives used for processing are harvested at the appropriate level of maturation and processed so that microbiologically safe and edible products are produced. For international trade in olives, as per the IOOC, individual countries are required to indicate the varieties considered suitable for processing.

The table olive processing method, together with variety and growing conditions, has a major impact on the taste of the final product. Olives are edible after primary processing, but they are often finished off (secondary processing) in a number of ways, for example by the addition of vinegar, extra virgin olive oil, herbs and spices.

Sources of inputs. Sources of olives should be secured well ahead of processing and contingency plans put into place in case there are shortfalls in supply. Processors must ensure that the olives used meet quality and safety criteria. They must also be certain that the source of potable water is secured and the amounts required for processing are available. Water sources other than from water authorities or agencies should be tested for physical, chemical and microbiological contaminants and treated to potable water standards before use. All chemicals, cleaners, disinfectants and ingredients used must meet food safety standards. Salt, sodium hydroxide and acids must be of food grade quality and only dried herbs, spices and condiments of satisfactory microbiological quality should be used for commercial products.

Herbs and spices. Herbs and spices should be suitably decontaminated for microorganisms before incorporating into table olive products. Traditionally, herbs and spices have undergone decontamination with chemical agents such as ethylene oxide and disinfestation with methyl bromide. Irradiation of herbs and spices is also used to destroy microorganisms, such as yeasts and moulds that can cause spoilage and may be harmful to health, as well as insects. This technique involves passing the food through a radiation field of gamma rays generated from a Cobalt–60 source or from an 'electron beam' generated from electricity. With irradiation the food remains cool, does not become radioactive, or lose flavour, aroma or nutritional value.

Salt use in table olive processing. Salt, chemically known as sodium chloride, is commonly used in table olive processing and packaging of table olive products. Coarse, dry salt is used for processing salt-dried olives, while coarse salt in water (salt brine) is used in fermentations and packaging brines.

Food grade salt with no additives must be used for all table olive operations. Non-conforming salt can cause the following problems and should be avoided in olive processing:

- anticaking agents (as in table salt) make brines cloudy;
- lime impurities can reduce the acidity of final products;
- iron can darken olive products;
- magnesium impurities can impart a bitter taste;
- carbonates can alter texture, causing softening; and
- iodised salt may darken olives and possibly give the olives a chemical taste.

Sodium chloride is soluble in water. One gram of salt will dissolve in 3 ml of water. This amount is not much lower than if the salt is dissolved in boiling water, so there is no advantage in using hot water to speed up its dissolution. A saturated solution of pure salt (100 degrees) contains 26.36 g/100 ml of solution.

To convert salt concentrations to approximate Salometer (Salinometer) degrees, multiply salt concentration by 3.8; for example, for a 10% w/v salt solution: 10 × 3.8 = 38 Salometer degrees.

Conversely, to convert Salometer degrees to approximate salt concentrations, divide Salometer degrees by 3.8; for example, for a 40 Salometer degree solution: 40 ÷ 3.8 = 10.5% w/v salt solution.

Alternatively, the graph in Fig. 4.5 can be used. After measuring the salt level in a particular brine, by extrapolation on the graph below, the % w/v salt concentration can be determined. If required, additional salt can be added and checked with the salometer or other salt measuring instrument.

The use of glass equipment such as salometers, where practicable, should be discouraged in table olive processing. Portable conductivity meters, salt refractometers and a titration method are a more suitable means by which salt content in brines can be assessed. If glassware is used, strategies need to be in place to handle situations such as breakages or if the device is accidentally dropped into brines.

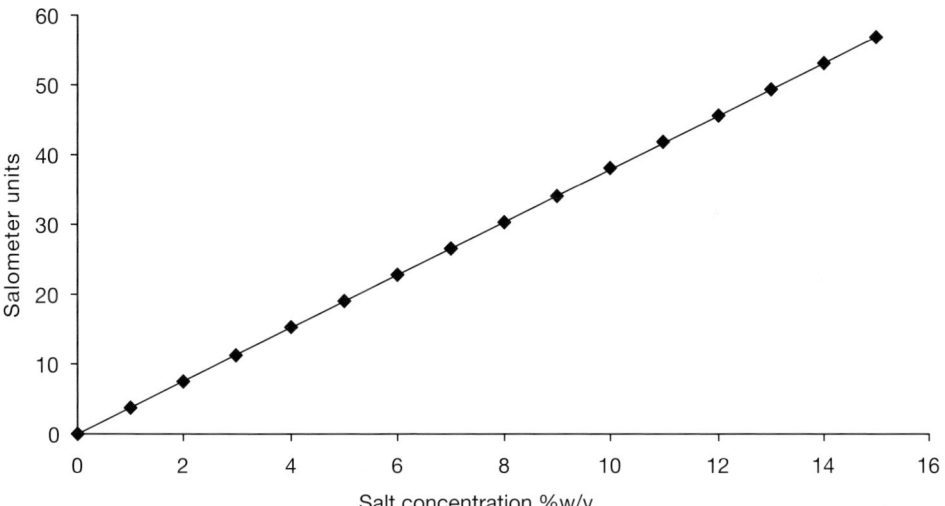

Figure 4.5 Graph for correlating salt concentration in % w/v with salometer units.

Care must be taken when preparing salt solutions. Following these procedures will reduce problems:

(a) double check calculations and weighed quantities;
(b) wear safety glasses to avoid salt solutions splashing into the eyes;
(c) add salt to water and ensure that all the salt has dissolved; solutions must be clear, not cloudy before adding to olives; and
(d) check salt concentration using a salometer, refractometer, conductivity meter or by titration.

Note: More research is required in developing and promoting low-salt table olive products. Salt, made up of sodium and chloride, is essential in the diet, but all too often intake is excessive. Health advice is to reduce salt intake, particularly in people with high blood pressure. As it has been estimated that 75% of salt intake comes from manufactured foods, low-salt table olive products, where technically and microbiologically possible, could help reduce salt loads.

Sodium hydroxide used to debitter olives in table olive processing. Sodium hydroxide solutions are used to debitter olives, for example Spanish-style green olives and Californian/Spanish-style black olives. Lye, caustic soda and 'alkali' are alternative names for sodium hydroxide. Potassium hydroxide has also been used as a debittering agent, as has wood ash. The use of wood ash, because of possible impurities and its variable quality, needs careful consideration before using in commercial table olive production.

Sodium hydroxide, which generally comes in the form of pellets, is very soluble in water (1.0 g/1.0 ml). It should be stored in sealed containers to prevent it reacting with carbon dioxide in the air forming sodium carbonate.

Sodium hydroxide is an extremely dangerous substance to handle and if it comes into contact with skin and eyes, severe burns can occur. Therefore, gloves, protective clothing, industrial footwear and safety glasses should be worn at all times when handling sodium hydroxide, preparing sodium hydroxide solutions and when transferring these solutions during processing operations. These comments also apply to the use of potassium hydroxide in table olive processing.

Care must be taken when preparing sodium hydroxide solutions because dissolving sodium hydroxide in water is an exothermic reaction generating a significant amount of heat. Using the following procedures will reduce problems:

(a) double check calculations and weighed quantities;
(b) wear safety clothing at all times;
(c) always add sodium hydroxide to water when making lye solutions rather than adding water to solid sodium hydroxide to prevent spattering and the risk of burns to skin and eyes; and
(d) ensure all the sodium hydroxide has dissolved before adding the solution to olives.

Table olive methods and styles

Numerous table olive processing methods are available. These depend on olive variety, degree of ripeness, processing technology, cultural and traditional factors. Using any processing method, one can process most olives, regardless of variety. However, from a commercial viewpoint, specific varieties are preferred because of superior technological and organoleptic factors and consumer preference. Examples are given below. *Manzanilla* is used for Spanish-style green olives, *Kalamon* (*Kalamata*) for Kalamata-style, Californian *Mission* for Californian green and black-ripe olives and *Conservolea* for Greek naturally black-ripe olives. (Also see Table 4.1.)

Kalamata-style: *Kalamon*
Greek-style black or green: *Conservolea*
Spanish-style green (Sevillian): *Manzanilla*
Spanish/Californian black: *Mission* (Californian), *Manzanilla* and *Hojiblanca*
Sicilian green: *Nocellara di Etnea*

Ligurian: *Taggiasca*
Dried olives: *Thrumbes*
Olive de Nimes: *Picholine*

Important international table olive trade products listed in the International Olive Oil Council's Codex Alimentarius (1987)/IOOC (2004) Table Olive Standards are provided in Section 4.11. All listed trade products are available in Australia, either loose or packed in containers, from fresh food markets, supermarkets, delicatessens and specialty food shops. Australian table olive processors are currently producing or have the ability to produce most of these trade products in competition with imported products, albeit in relatively small quantities. Australian growers/processors should be familiar with these, especially if they are to enter the international table olive trade or compete against imported table olive products.

Raw olives used for table olive production

Olives used for table olive production are classified in one of the following categories according to the degree of ripeness of the raw olives. Only firm, sound, unmarked olives are used for top quality table olives (Fig. 4.6).

Green olives. These are olives that are yellow-green in colour prior to pigmentation developing.

Olives turning colour. These are olives that are multicoloured or ones that have not reached complete ripeness. Turning colour olives are ones that have started to develop colours such as rose, wine-rose or brown before the olives are fully ripe.

Note: For practical purposes, turning colour olives have a partially pigmented flesh. When olives are nearly fully pigmented, they are used for preparing black olives.

Figure 4.6 *Manzanilla* olives at different stages of maturation (turning colour, black and green).

Black olives. These olives are harvested when fully ripe, or slightly before full ripeness. With the latter the flesh is not fully pigmented; the red/purple colour can be seen at least halfway between skin and the stone. Depending on the region 'black' olives can be reddish black, violet-black, deep violet, greenish-black or deep chestnut with both the skin and the flesh coloured.

Common trade preparations of table olives

A brief summary of common olive preparations based on the Codex Alimentarius (1987)/IOOC (2004) Table Olive Standards is given below.

Natural olives. Raw green, turning colour or black olives are placed directly in brine where they undergo partial or complete fermentation and may be preserved by the addition of acidifying agents. The resulting products are natural green olives, natural turning colour olives (Fig. 4.7) or natural black olives (Fig. 4.8).

Natural olives in brine are relatively firm and smooth with a glossy skin. The finished product retains some fruity and bitter flavours. During processing, raw black or turning colour olives can lose some of their black purple pigments resulting in pale to dark brown coloured olives. The colour of black olives can be partially restored by exposing them to air after processing where the phenolic compounds present in the skin and flesh oxidise.

Note: With incomplete fermentations or when new packing solutions are used, food acids are often added to ensure preservation. Fermentation brines from black olive processing take on a rich burgundy colour due to the presence of anthocyanins and are strong in aroma and flavour compounds, and where practical should be used to prepare packing brines. The latter also applies to green and turning colour olives where brine colours range from pale yellow to pale orange to pink. Some repackers of table olives add caramel to covering solutions when packing black olives, giving them a brown colour. Also the antioxidant, ascorbic acid, is often added to covering solutions or brines for green or turning colour olives to prevent discolouration (see Chapter 6).

Figure 4.7 Turning colour *Jumbo Kalamata* olives processed as natural olives. (Photo: Australian Mediterranean Olive Research Institute, Perth, Western Australia.)

Figure 4.8 Black-ripe *Conservolea* olives processed as natural olives. (Photo: Australian Mediterranean Olive Research Institute, Perth, Western Australia.)

Treated olives. Here, raw green, turning colour or black olives are debittered with lye then placed in brine in which they undergo partial or complete fermentation and may be preserved by the addition of acidifying agents. The resulting products are treated green olives in brine (Fig. 4.9), treated turning colour olives in brine or treated black olives in brine.

Curing olives in lye is fast and cost efficient. Because lye can leach out flavour (fruitiness), and leave a slight chemical taste, they are less popular with discerning consumers and by boutique table olive processors than naturally processed table olives. However, in traditional olive producing countries such as Spain, lye treated table olives (*Sevillana*, *Manzanilla* and *Hojiblanca*) are well accepted, as are the naturally processed *Arbequina* olives.

Note: With incomplete fermentations or where new packing solutions are produced, food acids are often added to ensure preservation. The term 'treated' is used when olives are debittered by immersing them in an aqueous solution of sodium hydroxide, lye or caustic soda. This produces irreversible physical and chemical changes in the skin and flesh of the olives causing the loss of water-soluble compounds (including reducing sugars, proteins, organic acids, salts and amino acids) and softening the flesh. The latter is less of a problem in Spanish/Californian black olives (see below), which are processed when green. Changes in the skin and flesh after lye treatment increases its permeability to soluble components from olive flesh into the brine as well as sodium chloride and flavour compounds into the olive.

Figure 4.9 Green-ripe *Manzanilla* olives processed as lye treated Spanish-style green olives.

Dehydrated and/or shrivelled olives. Raw green, turning colour or black olives that may have undergone a mild alkaline pre-treatment are preserved in brine, or partially dehydrated in dry salt and/or by heating or by any other technological process. The resulting products are dehydrated and/or shrivelled green olives, dehydrated and/or shrivelled olives turning colour, or dehydrated and/or shrivelled black olives.

Shrivelled black olives are popular with consumers. Variations in preparation include processing naturally black-ripe or nearly black-ripe olives in a low temperature oven, with or without lye treatment in brine or with dry salt (Figs 4.10–4.12). During processing the olive flesh dehydrates resulting in a soft, slightly moist, shrivelled product. With salt drying, processing is undertaken in containers or crates where the moisture with bitter substances drawn out of the olives by the salt is allowed to drain away. Green-ripe or turning colour olives when salt-dried are relatively firm, and when exposed to air turn a chocolate brown colour rather than black.

Note: After curing, salt-dried olives can be plunged briefly into boiling water to remove the excess salt, allowed to dry, then stored in extra virgin olive oil. Home processors can also add olive oil, herbs and spices. Salt-dried olives can be vacuum packed to aid in their preservation.

Figure 4.10 Salt-dried naturally black-ripe *Kalamata* olives with olive oil and fennel.

Figure 4.11 Salt-dried green-ripe *Manzanilla* olives. (Photo: Australian Mediterranean Olive Research Institute, Perth, Western Australia.)

Figure 4.12 Heat-dried naturally black olives. (Photo: Australian Mediterranean Olive Research Institute, Perth, Western Australia.)

Figure 4.13 *Hojiblanca* olives processed as Spanish-style black olives.

Olives darkened by oxidation. Green or turning colour olives preserved in brine that may have been fermented are darkened by oxidation in an alkaline medium and preserved in hermetically sealed (airtight) containers subjected to heat sterilisation. These olives, often called Californian- or Spanish-style black olives, are characteristically pitch black in colour (Fig. 4.13).

Note: In some centres, raw green-ripe olives are used, while at others, the olives are placed in brine as a storage procedure or naturally fermented olives are purchased from others. Olives stored in brine for sufficiently long periods undergo natural fermentation.

Specialty table olive products

A number of specialty products can be prepared from green, turning colour or black processed table olives. Such specialties retain the term 'olive' as long as the fruit used complies with expected standards, allowing the consumer to understand that the products are prepared from olives. These products include green and black olive pastes, tapenades, stuffed olives, olives in marinades and olives with pickled vegetables.

After primary processing, large olives of any style can be destoned then stuffed, segmented, sliced or crushed into a paste. Large-sized olives called 'Queens' are most suited to pitting and stuffing (Fig. 4.14).

Figure 4.14 Commercially available green 'Queen' size olives of the *Sevillana* variety, whole, destoned and stuffed with pimento.

Table olive styles

The Codex Alimentarius (1987)/IOOC (2004) Table Olive Standards indicate a number of styles for processed table olives.

Whole olives. Whole olives (Fig. 4.15) are those that have retained their natural shape, with or without stems, and their stone is intact. Examples include primary processed (by any method) olives. Whole olives are commonly packed randomly (Fig. 4.14) or, less commonly, in an orderly arrangement (Fig. 4.16) into containers.

Figure 4.15 Whole green-ripe *Chalchidikis* olives processed with lye.

Cracked/bruised olives. These are olives that have undergone a procedure whereby the olive flesh is opened while leaving the olive stone intact and still attached to the flesh. Clingstone olive varieties are more suitable for cracked olives than freestone varieties (see Chapter 2). Bruised olives are mostly prepared from raw green-ripe or turning colour olives or green olives that have already undergone primary processing. Naturally processed black olives, because of their soft characteristics, are less suitable. When preparing cracked/bruised olives, the olives are struck with a blunt object or passed through a device so that the flesh is exposed without removing or breaking the stone. If raw, the olives are processed with or without lye treatment. Raw cracked olives are placed into brine where they may undergo a partial or complete natural fermentation. Where lye is used the olives are washed before placement into brine as for Spanish-style green olives. Once processed, the olives are placed in brine into which herbs, spices, vinegar and olive oil are added to extend aroma and taste.

Figure 4.16 Orderly packed green *Oliva di Cerignola* olives processed and packed in Italy.

Split/slit olives. Split/slit olives are prepared from raw olives: naturally black-ripe, green-ripe or turning colour olives, or olives that have already undergone primary processing. The olives are slit longitudinally with a knife or by some other device (such as a mechanical slitter) so that the skin is breached and the flesh is penetrated. Raw olives are processed with or without treatment with lye. The olives are placed into brine where they may undergo partial or complete natural fermentation. Where lye is used the olives are washed before placement into brine. As for bruised olives, once processed the olives are placed in brine into which herbs, spices, vinegar and olive oil are added to extend aroma and taste. A variation of the method, often used with Kalamata-style olives, is to split or slit the intact olives after primary processing, so that when embellishments are added, for example garlic, wine vinegar and lemon flavours, they penetrate into the flesh.

Stoned/pitted olives. Olives of this style have their stones removed while retaining their original shape. Removing the stone can be undertaken before or after processing. For this style, olives should be medium to large in size. *Manzanilla*, *Sevillana* and *Kalamata* olives are commonly available in the destoned form. Of these three varieties, *Kalamata* loses most of its original shape but this is not a major problem as olives of this variety are mostly used on pizzas or in prepared food.

Destoned olives are also sold in the following divided forms for use in cooking, salads and pizzas.

Halved olives. Stoned or stuffed olives sliced transversely into two approximately equal parts (Fig. 4.17).

Quartered olives. Stoned olives sliced lengthwise and transversely into four approximately equal parts.

Divided olives. Stoned olives cut lengthwise into more than four equal parts.

Sliced olives. Stoned olives or stuffed olives sliced transversely into segments of fairly uniform thickness. Sliced Spanish-style green and black olives are commonly available in this form (Fig. 4.18).

Pitted and stuffed. Stoned olives with various fillings, such as pimento (sweet chilli), anchovy, onion, garlic, almonds, celery, orange or lemon peel, hazelnuts or capers.

Chopped (minced) olives. Small pieces of destoned olives without a definite shape.

Broken olives. Olives accidentally broken when being destoned and stuffed.

Stuffed olives. Stoned olives stuffed with natural fillings or pastes, for example pimento, onion, almond, celery, anchovy, olive, citrus peel, hazelnuts, capers and cheeses.

Figure 4.17 Halved cracked Sicilian-style *Manzanilla* olives suitable for adding to salads and cooked food. Note the freestone has been ejected during the cracking procedure.

Figure 4.18 Sliced Spanish-style black olives suitable for adding to salads, cooked food and pizzas.

Salad olives. Whole broken olives (or broken and stoned) with or without capers or any filling product.

Olives with capers. Whole or stoned olives, generally of small size, packed with capers and peppers.

Note: With substantial destoning and stuffing operations, large quantities of stones are produced. In Spain, olive stones are used for the extraction of seed oil, but they are also used as a fuel. In California, major olive processing plants have burnt olive stones to produce steam for use in sterilising canned olives. Other uses are as BBQ fuel, fuel bricks, starting material for plastics and for the production of high quality activated carbon.

Common table olive methods and products

Common table olive styles and methods are summarised in Table 4.4. They are the types of products imported into Australia and eaten by most Australians. Australian table olive producers have developed similar products except for Californian/Spanish-style black olives. Products are sold in their primary processed form (loose or packed in brine) or as specialty products (secondary processing). Information in this table can direct Australian growers/processors to which table olive products to produce.

Plant and process control. Controlling table olive processing by using documented procedures and relevant tests ensures the quality of the final products is maximised. Microbiological, pH and salt tests are used to monitor fermentation during processing as well as for quality management and safety purposes. Microbiological tests are used to test for the presence and absence of pathogens and spoilage organisms. Orderly records should be maintained and in a form others can follow in the absence of key workers. Records should be kept for at least three years and archived if appropriate.

Primary processing specifications for table olives

The concepts of *primary* and *secondary* processing have been developed by the authors to simplify the understanding of processing table olives. After discussions with a large number of potential and existing table olive processors and home table olive processors around Australia, it became obvious that they were having difficulty grappling with basic principles.

Primary processing involves any process used to debitter and preserve the olive, such as fermentation and drying. The product may or may not be suitable for immediate consumption at this point; that is, packed in consumer size containers, or have not been embellished.

Specifications. Specifications must be drawn up prior to starting the manufacture of table olives.

Process specifications must be clearly documented for primary processing and should include:

- product specifications: intermediate and final;
- amount and type of olives to be processed;
- processing steps to follow;
- quantities of chemicals and additions to be used;
- information for operators on the safe handling of chemicals;
- additional preservation steps: pasteurisation, sterilisation, preservatives; and
- testing procedures.

Note: Workers in the olive processing plant should be in-serviced and made aware of the requirements of the standards to ensure production of quality table olives.

Table 4.4. Commonly available commercial table olive products

Olive styles	Processing method
Black olives (primary processing)	
Naturally black-ripe olives in brine (Greek-style)	Whole or slit naturally black-ripe olives are processed by spontaneous fermentation in 8–10% salt solution for three to six months. Exposing processed olives to air returns some of the original black colour.
Kalamata-style	Whole or slit naturally black olives, usually *Kalamata*, are either debittered in water and brine or subjected to spontaneous fermentation in brine followed by the addition of wine vinegar and olive oil.
Heat-dried naturally black olives	Whole naturally ripe black olives are: • Sun-dried until bitterness has reached an acceptable level • Ripe fruit is blanched then oven-dried at low temperatures (50°C) for a few days until bitterness disappears.
Spanish/Californian-style black olives (olives darkened by oxidation) Note: starting material is not black-ripe olives	Whole green turning colour olives treated with several lye solutions of different strengths to remove bitterness, washed, transformed to a black colour by oxidation in an alkaline medium with air and then packed in brine. Processed olives are heat sterilised in their final containers.
Treated black olives (primary processing)	Whole black olives are given a short treatment with lye solution followed by natural fermentation in brine until debittered.
Untreated naturally black olives in dry salt (Thrumba-style)	Whole naturally full ripe black olives, fresh or partially dried, are packed in alternating layers of dry salt until debittered.
Green olives (primary processing)	
Untreated green olives in brine	Whole, slit or bruised green-ripe olives are processed by a natural fermentation in an 8–10% salt solution for 3–12 months until debittered.
Spanish-style green olives (Sevillean-style)	Whole green-ripe olives are treated for a short period with 1–2% lye solution, washed and then partially or completely fermented (lactic) in brine.
Turning colour olives (primary processing)	
Untreated turning colour olives in brine	Whole, slit or bruised turning colour olives are processed by natural fermentation in an 8–10% salt solution for 3–12 months.
Treated turning colour olives in brine	Whole olives are treated with lye, then preserved by natural fermentation in brine or heat treatment.
Specialty products (secondary processing)	
Marinated green or black olives	Marinades added to processed olives: untreated green, turning colour or naturally black-ripe olives in brine, Spanish-style green or Kalamata-style olives.
Destoned olives	Green or black olives destoned by hand or machine.
Stuffed olives	Processed green or black olives destoned then stuffed with garlic, pimento, onion, almonds, celery, anchovy, citrus peel, hazelnuts and capers.
Olive pastes and tapenades	Destoned processed green or black olives crushed to a paste with or without the addition of other foodstuffs (e.g. capers, anchovies, olive oil, garlic, sun-dried tomato).

Secondary table olive processing specifications

Secondary processing involves procedures for marinating and stuffing olives, making olive pastes, tapenades and olive pickles. It can be undertaken by the grower or processor, and can involve the preparation of the packing solution and additions such as oil (vegetable, olive, sunflower or canola) wine vinegar, herbs, spices and aromatics.

Specifications. Processing specifications must be clearly documented and should include:

- product specifications: intermediate and final;
- amount and type of olives to be processed;
- processing steps to follow;
- quantities of packing brines, olives and additives (olive oil, herbs, spices and aromatics);
- information for operators on the safe handling of chemicals, additives, adjuvants, herbs, spices and aromatics;
- alerts on health and safety matters during secondary processing; and
- testing procedures.

Note: Workers should be in-serviced on these specifications.

Finished table olive product specifications

To make primary or secondary processed table olives ready for consumption they may require an additional preservation method such as the addition of a preservative, pasteurisation or sterilisation.

Specifications should include:

- packing solution details;
- additives;
- additional preservation method;
- packaging information; and
- labelling information.

Generic processing protocol for table olives

The generic processing protocol in Fig. 4.19 can be used as the template for all processing methods. Specific procedures are added for the different olive methods/styles.

There are a number of procedures common to all processing methods/styles; however, specific details relate to the actual method/style. Furthermore, most people confuse the nature of the final product with the method/style, particularly where the olives are embellished with herbs, spices, aromatics, fillings and marinades. Olive recipes are often requested. It is best to separate the actual processing procedure as

- Accept only quality raw olives
- Store raw olives correctly
- Wash raw olives
- Size grade and sort raw olives
- Use processing method or style with HACCP controls
- Size grade and sort processed olives
- Pack processed olives in final packing solution
- Pasteurise/sterilise the packed processed olives if required
- Label packed processed olives in accordance with food standards
- Implement a safety recall system for faulty packed processed olives

Figure 4.19 Generic processing protocol for table olives.

primary processing and recipes as secondary processing. The latter relates to the embellishments of olives that have undergone primary processing. Trying to process olives when the embellishments are included at the beginning can result in variable or anomalous products and should not be practised for commercial table olive production.

Process control (microbiological, chemical and physical) is essential during table olive production. Procedures must be developed and documented consistent with the processing steps, including HACCP.

Acceptance of raw olives by processors

Each batch of olives should be examined for quality: olive size, shape, damaged olives and the presence of leaves. Processors should ensure that they are able to recognise and distinguish between varieties, for example *Manzanilla* v. *Sevillana*, *Verdale* v. *Conservolea*, *Leccino* v. *Frantoio* or *Kalamata* v. *Barnea*. Knowledge of the growing region and the growing technologies used, for example irrigated v. non-irrigated, can be related to skin and flesh properties, so processing conditions can be modified accordingly. Olives from irrigated trees, because of their higher water content, are more sensitive to salt damage than those from unirrigated olive trees. Processors buying olives from growers should ask for a chemical use diary for the delivered batch of olives. The chemical use diary should include chemical names, dates of application and the recommended withholding periods (see Chapter 3) to ensure these have been met. Processors should not accept olives where olives or orchards have been treated with non-approved chemicals. Random testing for chemical residues, although expensive, is useful in establishing supplier confidence. Growers and suppliers should understand the table olive processor's requirements before they enter into supply arrangements. The following protocol can be used as a guide.

Protocol for accepting olives. At the processing facility raw olives should be assessed for variety, ripeness and soundness before acceptance. With each olive batch or load, the following information should be provided by the grower and checked by the processor:

- supplier's name;
- receipt date;
- harvesting method and date;
- variety;
- maturation state;
- olive size;
- growing region;
- growing technologies;
- chemical use diary; and
- weight of olives delivered.

Batches of olives with high proportions of undersized or defective olives are uneconomical for table olive production and should be rejected.

Correct storage of raw olives at the processing facility

Careful post-harvest handling of olives is essential to achieve quality table olive products. Raw olives, particularly naturally black-ripe olives, are sensitive to damage during handling and storage. These must be processed as soon as possible after harvesting and certainly within 24 hours of delivery to avoid deterioration and poor table olive products. Bruised or marked raw olives fetch low prices compared with good quality olives (see Chapter 3).

Protocol for general storage of raw olives. To minimise the risk of contamination or damage, olives should be stored at temperatures between 5–10°C for no more than 24 hours in shallow ventilated crates under clean and hygienic conditions. Green olives can be stored for longer periods.

Protocol for storing raw olives in brine. Quantities required to prepare salt brines are given in Table 4.5. Commonly used brine solutions for pre-process storage of olives contain 8–10% w/v sodium chloride. The pre-storage of olives in brine is used when the capacity of the processing plant is exceeded and the processor wishes to 'buy' time, particularly where processing involves lye treatment.

Table 4.5. Quantities of sodium chloride required to prepare salt brines

Sodium chloride % w/v in potable water	Brine volumes to be prepared		
	100 litres	500 litres	1000 litres
5	5 kg	25 kg	50 kg
6	6 kg	30 kg	60 kg
7	7 kg	35 kg	70 kg
8	8 kg	40 kg	80 kg
9	9 kg	45 kg	90 kg
10	10 kg	50 kg	100 kg
Potable water to	100 litres	500 litres	1000 litres

Such a procedure means that storage in air, with the risk of damage to the olives, is avoided. Of course, placing olives in brine is the exact procedure for processing natural untreated olives in brine by spontaneous fermentation.

Storage tanks are partially filled with the salt solution before the olives are introduced to prevent bruising or pressure damage. Depending on the storage time, a partial fermentation takes place. The salt concentration in the brine should be checked with a suitable instrument before adding the olives (see Chapter 6).

To reduce the risk of the olives shrivelling, a problem which is often variety dependant, an initial solution starting at 5.0–7.5% w/v food grade sodium chloride in potable water should be used.

The container should be filled to the brim with olives and brine so there is minimal air space above the brine, then tightly sealed to ensure anaerobic conditions. Keeping the olives submerged with a grating made of food grade material prevents them discolouring, but does not necessarily prevent the growth of fungi on the brine surface. After a few days, more food grade salt is added to increase the strength of the brine to 8–10% w/v salt. Brines should be mixed either manually or mechanically with pumps, initially every three days, then weekly to ensure homogeneous mixing and prevent uneven salt levels within the tank. If experience shows that shrivelling is not a problem with particular varieties and maturation states, the higher salt solutions can be used initially.

If anaerobic conditions are not achieved, oxidative yeasts and moulds develop on the surface of the brine, releasing enzymes that attack the fibrous structural components of the olives, causing them to soften. This can be avoided by carefully controlling the brine strength, having well-filled tanks with minimal air space between the lid and brine surface, and by keeping tanks well sealed. Prevention is better than cure but if yeasts or moulds start to develop on the surface of brines they should be skimmed off regularly. Allowing extensive fungal mats to develop on the surface of brines increases the risk of spoilage and more rarely the likelihood of these fungal contaminants producing cancer-inducing agents such as mycotoxins. In the case of large tanks, having only small openings at the top of the tank for loading olives can markedly reduce the surface area of the brine, thus reducing the problem of moulds.

Note: If there are heavy mould growths or bad smelling olives, the olives should be destroyed.

As indicated earlier, with prolonged storage a weak spontaneous fermentation occurs typical of processing olives in brine without initial lye treatment or the addition of starter cultures. Under these conditions the pH of the brine falls and the free acid levels should reach the equivalent of 0.40–0.45% lactic acid. The brine conditions need to be controlled for pH and salt levels to prevent spoilage such as olive softening and gas pocket formation by Gram negative bacteria. Ensuring the salt concentration is at least 8% w/v and ensuring a pH of around 4 can prevent spoilage. If a rapid lowering of pH is required, a food grade acid (lactic acid or acetic acid) is added to give tank concentrations of 0.5% w/v lactic acid, or 0.25% w/v acetic acid.

Protocol for storing raw olives in salt-free solutions. Salt-free storage solutions have been developed at some centres for storing olives prior to processing with lye. This method confers an environmental advantage; using such solutions avoids the need to dispose of large amounts of salt storage solution. Researchers have found salt-free storage solutions do not support fermentation and there is no deterioration in organoleptic qualities when olives are processed. A typical salt-free solution is presented in Table 4.6. It includes both acidulants and preservatives.

Table 4.6. Quantities of chemicals for preparing salt-free olive storage solutions

Chemical component	Action	%	100 litres	500 litres	1000 litres
Lactic acid	Acidulant	0.67	0.67 kg	3.35 kg	6.70 kg
Acetic acid	Acidulant	1.00	1.00 kg	5.00 kg	10.00 kg
Sodium benzoate	Preservative	0.30	0.30 kg	1.50 kg	3.00 kg
Potassium sorbate	Preservative	0.30	0.30 kg	1.50 kg	3.00 kg
Potable water to		100	100 litres	500 litres	1000 litres

Raw olives enter the table olive processing line

Special care must be taken when unloading and handling olives. With small enterprises most operations are undertaken manually, whereas in large-scale operations these procedures are mechanised, using belts and conveyors to move olives from collection areas to the sorting and washing facilities. Individual varieties should be processed separately to avoid variable final products.

Protocol for washing raw olives. Olives should be washed with potable quality water before entering the processing line. Spray washing is more effective than static washing. The olives are washed with spray rinsing machines using potable water to remove contaminants such as leaves, orchard dust, dirt, chemicals and soil microbes, such as *Clostridia, Bacilli* and coliforms, thus reducing the risk of spoilage during processing and harm to consumers. In the case of boutique/small-scale table olive producers, spray washing olives packed in slotted crates with a spray hose should suffice. Washing does not remove microorganisms on the skin that are required for natural fermentation processing procedures. If water of unknown quality from sources such as rainwater tanks and bores is used, it should be checked and treated so that it meets potable water standards. A diagrammatic representation of a typical line for washing, sorting and size grading olives for table olive production is presented in Fig. 4.20.

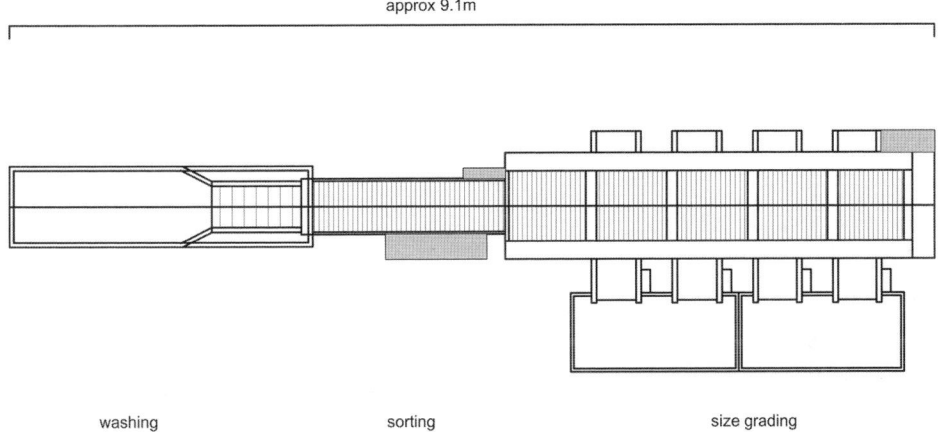

Figure 4.20 Diagrammatic representation of a raw olive washing machine with sorting table. (Image: Australia Olive Oil Supplies, Victoria, Australia.)

In some enterprises, where olives are processed with lye (caustic soda), the initial washing step is omitted as multiple washes with water are used during processing. In this case, the olives are generally washed in the same tanks as the lye treatment step.

Note: Lye solutions should be prepared in separate tanks and then fed into the processing tanks.

Protocol for preliminary size grading of raw olives and removal of damaged olives. Before the olives are placed in the processing tanks or containers they should be size graded into three to four sizes to remove undersized, misshapen and damaged fruit using a grading machine and sorting table respectively. Either the grower and/or the processor can undertake these operations. If there are sufficient culled olives they can be used for olive oil production.

Preliminary grading and sorting of raw olives has a number of advantages.

- Similar sized olives process at the same rates.
- Increased efficiency because reject olives are not processed.

- Reduced risk of contamination if large numbers of defective olives are present.
- Facilitates final sorting and packaging operations.

Note: All equipment should be in a clean and sanitary condition before commencing preliminary size grading of olives.

Placement of table olives into processing tanks

Processors should have a selection of different sized tanks made of food grade material (food grade plastic, fibreglass or stainless steel) – 250 kg barrels to 10–15 tonne fermentation tanks – and develop a processing plan for the season based on quantities, varieties and styles. Attention needs to be paid to loading and unloading olives to reduce any damage and contamination. Tanks with sloping bases and large valved outlets at the bottom facilitate unloading of the olives, otherwise pumps need to be used (Fig. 4.21).

Processing media for table olives include potable water, brine (salt 8–10% w/v in potable water) or lye (caustic soda) depending on the method used. Tanks are partly filled with water or appropriate processing medium then filled with olives. This procedure prevents bruising and pressure on the olives. The tank filling procedure should be undertaken as quickly as possible especially with green-ripe or turning coloured olives, certainly in less than 20–30 minutes, so that olives are not damaged by pressure or by exposure to air, resulting in discolouration. When filled with olives, the tanks are topped up with water or processing medium.

Figure 4.21 Stainless steel pump for pumping olives in brine.

During processing, particularly with brine, the liquid should be well mixed by circulating brine with a pump, stirring or barrel rotators to allow an even reaction in the tank and prevent the formation of salt gradients. In some centres, rotating stainless steel processing tanks are used. These tanks are often fitted with plumbing and valves to allow the addition and removal of processing solutions as a batch or continuous process (Fig. 4.22).

Tanks should be filled to capacity with olives only of the same variety, maturation stage and size before processing commences. This favours even processing of olives. Once processing has commenced, further additions of raw olives should not be made. Processing temperatures should be maintained between 20°C and 25°C during

Figure 4.22 Rotating stainless steel olive processing tanks. (Photo: Buffet Olives, Paarl, South Africa.)

fermentation. If temperatures are too low, processing is stalled or processing time is prolonged. To overcome this problem, in-line heat exchangers can be used or air conditioners installed to maintain brine temperatures. Tank temperatures higher than 30°C should be avoided as high temperatures can lead to the growth of anomalous organisms resulting in spoilage and possibly food poisoning in the consumer.

General methods for processing table olives

Olives are processed using one of the following procedures to remove the bitter principles and in their preservation (more details on olive processing procedures are presented in Chapter 5):

- repeated soaking in water followed by placement in brine;
- fermentation in brine;
- lye treatment followed by fermentation in brine;
- lye treatment without fermentation, in brine; or
- drying with salt or heat.

The first three methods involve a fermentation process, albeit weak in some cases.

When raw olives are debittered through multiple soakings in water (where the water is changed daily) and then placed in brine, a weak fermentation may proceed. With olives placed directly in brine a spontaneous anaerobic fermentation bacteria and/or yeast (depending on the brine pH and salt concentration) is initiated by native organisms on the fruit. With lye treated olives, for example Spanish-style green olives, the olives are placed in brine after washing out the excess lye and undergo a bacterial lactic fermentation. In the latter case, the addition of starter cultures of lactic acid bacteria, for example *Lactobacillus pentosus* (formerly *Lactobacillus plantarum*), to the fermentation tanks is often required as lye treatment destroys the natural flora on the olives that would have normally supported fermentation. In well-established olive processing facilities in olive producing countries of the Mediterranean, addition of starter cultures is not standard practice.

Starter cultures. The effect of the addition of pure starter cultures on the organoleptic qualities of finished products is yet to be concluded, with different researchers having opposing views. Some believe that organoleptic characteristics of olives processed with added pure cultures lack the nuances of the traditional product, whereas others believe there is improvement. In Australia, the starter culture Vege-Start™ (Chr. Hansen Laboratories, Melbourne, Australia), comprised of *Lactobacillus pentosus* (formerly *Lactobacillus plantarum*) culture, is available for a range of pickling operations and can be used for producing Spanish-style green olives. When starter cultures are used, in principle, complete fermentation can be achieved in three to nine weeks. In practice a sufficient amount of starter culture is added to the fermentation brine so that after a few days, when the acidity of the brine falls to around pH 6, the number of fermentative organisms reaches about 10^6/ml of brine. The amount of starter culture used will depend on the size of the tank. For example, a 50 g sachet of Vege-Start™ is added to tanks with an approximate capacity of 10 tonnes holding 6000 kg of olives. Samples taken from active fermentation brines can also be used as starter cultures. The effectiveness of starter cultures as identified by Spanish and Italian researchers depends on the variety, competing microorganisms and brine characteristics such as:

- fermentative substrate levels;
- growth factor levels: additional nutrients, such as vitamins and amino acids;
- salt concentration;
- pH and acid levels;
- temperature; and
- inhibitor levels, for example polyphenols released from olives.

Vege-Start™. Vege-Start™ is a freeze-dried culture, in a lactose carrier, of the homofermentantive (produces only lactic acid) lactic acid bacteria *Lactobacillus pentosus* (formerly *Lactobacillus plantarum*), used for pickling of olives and cucumbers. The *Lactobacillus pentosus* in Vege-Start™ has been selected by the manufacturers to have a high salt tolerance and a wide pH and temperature range. The product specifications are as follows:

- cells per gram = minimum of 4.8×10^{10} CFU/g;
- viable temperature range 15°C to 40°C with an optimum temperature of 30°C;
- viable pH limits are minimum pH 3.7 and maximum pH 8.0;
- principle fermentation product is a 50/50 mixture of the D and L forms of lactic acid;
- shelf-life is 18 months at −18°C and three months at +5°C; and
- levels of other microorganisms (in 25g of Vege-Start™): *S. aureus* <100 CFU/g; *B. cereus* <100 CFU/g; *E. coli*/coliforms <100 CFU/g; Streptococci <100 CFU/g; yeasts/moulds <100 CFU/g, and *Salmonella*-negative.

Vege-Start™ is effective in carrying out fermentations at salt concentrations from 0 to 8% sodium chloride. However, as the salt levels in the brine increase, fermentation rates decrease with the final pH of brines all reaching around pH 3.5. Optimal salt levels are 4–8% w/v. At salt concentrations of 9% w/v Vege-Start™ survives and grows, but the lag phase (period of low growth and activity) is prolonged so undesirable organisms early in fermentation could dominate. If brine salt levels are less than 4% w/v the risk of undesirable salt sensitive organisms growing increases.

The optimal pH conditions for inoculation with Vege-Start™ are between pH 5–7 but it can be used up to pH 7.5 and as low as pH 4.5. If adjustment of brine pH is required, the manufacturers recommend hydrochloric acid be used rather than acetic acid, as the *Lactobacillus* in Vege-Start™ is sensitive to the latter. If brine pH is too high (after lye treatment) or too low (over-correction with acid), when an inoculum of Vege-Start™ is made, the starter culture is inhibited. This problem can be managed in part by reducing the brine salt levels to below 7% w/v.

Optimal fermentation with this starter culture occurs at 25–30°C with very good results at 20–35°C. Reasonable fermentation results can be achieved at 15–20°C. Lower temperatures slow growth of the starter culture and result in acidification of the brine.

The Vege-Start™ culture is added directly to olives and brine already placed in the fermentation tank. Ideally the brine should have an initial pH of 5–7 (adjusted with a food grade acid, preferably hydrochloric acid) and salt levels of 6–8% w/v. The freeze-dried powder is quickly dissolved in water or brine then added to the tank as soon as possible with gentle agitation to ensure its dispersion.

Note: Leaving cultures too long in brine is detrimental so it should be added to the fermentation tank immediately after dissolution. Also, for optimal results, once the sachet has been opened it should be used straight away and not stored.

Starter culture research. Processors and researchers in Spain and Italy, as well as Australia, are collaborating with microbiologists to develop specific starter cultures that are advantageous to table olive fermentations. Sought-after characteristics include the ability to:

- resist the action of polyphenolic inhibitors;
- grow at low levels of growth factors;
- compete with other microorganisms, including wild strains;
- tolerate the extreme chemical conditions in brine;
- speed up fermentation to reduce processing time;
- be freeze-dried without losing potency when stored; and
- have multiple functions such as beta-glucosidase activity to break down oleuropein as well as support fermentation that produces lactic and acetic acids.

Californian-style/Spanish-style black olives are produced from green-ripe olives (the skin can have some pigmentation) by chemical processing with lye. They are then oxidised and brined without fermentation. Although there is an expectation by consumers that blackness in an olive is indicative of ripeness, artificially turning green olives into black is a contradiction. Olives processed by this method require heat sterilisation. Sterilisation is unnecessary for olives when fermented in brine as long as the brine/salt concentration and pH are favourable for preservation.

A major disadvantage of spontaneous brine fermentation is that it can take from three months for black-ripe olives and 12 months for green-ripe olives, whereas lye treated olives can be in the market-place four weeks after processing.

The final flavour of primary processed olives depends on the variety, processing method and final packing procedure. After primary processing, adding vinegar, marinades, and herbs and spices extends the flavour of the olives. Addition of food acids such as vinegar (acetic acid), lactic acid or citric acid is used to optimise shelf-life and the flavour of the olives. Olive oil is often added to dried olives.

Microorganisms relevant in table olive processing

Bacteria and yeasts are the main microorganisms of importance in table olive processing, particularly with respect to fermentation. Moulds, another important group, play no role in fermentation but can cause spoilage. Generally with fermentations, the order of organism growth is: bacteria, yeasts then moulds. Smaller organisms, *Leuconostoc* and *Streptococcus* spp., grow and carry out fermentation more rapidly than other related bacteria and are, therefore, the first species to appear in the fermentation brine.

Bacteria. Bacteria are a large group of unicellular organisms lacking chlorophyll and a nucleus. They mostly multiply rapidly by simple fission: each bacterium growing and then splitting into two bacteria and so on.

Lactic acid bacteria. Relevant to the fermentation of table olives is a group of Gram positive lactic acid bacteria that utilise reducing sugars to produce acids (lactic acid and acetic acid) that are released into the fermentation brine. The net effect is an accumulation of acids and a lowering of brine pH. The main taxa involved are *Lactobacillus, Leuconostoc, Pediococcus* and *Streptococcus*. *Streptococcus* and *Leuconostoc* species produce the least acid. Heterofermentative *Lactobacillus* species produce intermediate amounts of acidic compounds, followed by *Pediococcus* and then homofermentative *Lactobacillus* species producing the most acid.

Heterofermentative bacteria, *Leconostoc* spp. and some *Lactobacillus* spp. produce 50% lactic acid + 25% acetic acid and ethanol + 25% carbon dioxide.

Homofermentative bacteria, *Streptococcus* spp. and some *Lactobacillus* spp. produce 100% lactic acid. *Lactobacillus pentosus* (formerly *Lactobacillus plantarum*), a homofermenter that plays a major role in olive fermentations, produces high levels of acidity.

Note: When all conditions are favourable, the amount of acid produced depends on the availability of fermentative substrates in the processing system. At low brine salt concentrations the pH levels of brine can fall to <4 with the total acid produced reaching 1% w/v or more when calculated as lactic acid. At high salt concentrations the pH levels of brine are between 4.0–4.5 with the total acid produced reaching around 0.5% w/v or more as lactic acid. During spoilage, if organisms consume acid the pH of the brine increases and the total acid levels fall, jeopardising the commercial value and safety of the table olive product.

Spoilage bacteria. Bacteria can cause olives to soften. Three malodorous fermentations due to spoilage bacteria – putrid, butyric and '*Zapateria*' – are associated with table olive processing (see Chapter 6).

Yeasts. Yeasts and yeast-like fungi are unicellular organisms that reproduce asexually by budding, or in some cases sexually by conjugation. They are present in the environment, in the air and in the soil, in orchards and the intestinal tracts of animals. They are active at temperatures between 0–50°C with an optimum range of 20–30°C, which is also the temperature range for natural fermentations. As yeasts are usually acid tolerant (that is, able to grow at pH 4.0–4.5), they can be associated with spoilage of acid foods. Commonly found yeast species in natural fermentations include *Candida boidinie, Debaryomyces hansenii, Saccharomyces cerevisiae* and *Torulopsis candida*. Some yeasts, such as *Saccharomyces cerevisiae*, have the ability to shift their metabolism from a fermentative to oxidative pathway depending on how much oxygen is available, hence the need to keep strict anaerobic conditions during table olive fermentation. As yeasts produce alcohol and carbon dioxide from sugars such as glucose and fructose, lower lactic acid levels and higher pH values result. One advantage of yeasts growing in fermentation brines is that the texture and fruitiness of the olives improves. Yeasts are not inhibited by polyphenols released from the olives into the brine during processing. In fact some yeasts, like *Candida veronae*, can split oleuropein into a less bitter compound and sugar which is then available to *Lactobacillus pentosus* (formerly *Lactobacillus plantarum*) for fermentation. It is possible that some strains of *Lactobacillus pentosus* (formerly *Lactobacillus plantarum*) may be able to carry out both debittering and fermentation.

Spoilage due to yeasts. Yeasts create food spoilage by producing brines with a slimy or cloudy appearance or by producing unwanted metabolic by-products that taint the olives. Yeasts have not been found to produce human toxins or cause food poisoning. Yeasts, in particular *Saccharomyces oleaginosus* and *Hansenula anomala*, are also involved in gas pocket formation.

Note: A chemical combination of alcohol and acetic acid results in the formation and accumulation of ethyl acetate, an ester with a characteristic 'nail polish' odour.

Moulds. Moulds are a group of multicellular fungal microorganisms, some of which can cause olive spoilage. Although plant-like, they do not have chlorophyll and must absorb organic nutrients from their surroundings. They are often found growing as a mycelial mat, for example *Penicillium*, *Aspergillus* and *Rhizopus* spp., with powdery growths (spores) on the surface of brines when the conditions (pH, salt concentration and degree of anaerobiosis) are poorly controlled. Moulds are aerobic (require oxygen for growth) and grow in a temperature range of 8–35°C. As moulds are usually acid tolerant and able to grow at pH 2.0 to 8.5, they are associated with spoilage of acid foods. Moulds release chemicals into brines that can taint the olives, for example by giving them a mouldy taste. Others release enzymes or can colonise and grow on olives, particularly those at the brine surface. Some like *Aspergillus* can produce carcinogenic toxins called 'aflatoxins'.

In summary, with fermentations, the order of organism growth is bacteria, yeasts, then moulds. Smaller organisms *Leuconostoc* and *Streptococcus* spp. grow and ferment more rapidly than other related bacteria and are, therefore, the first species to appear in the fermentation brine.

Food spoilage. Spoilt foods are those that have become unpalatable due to microbial growth. Such products can have undesirable odours, flavours, appearance or texture. Food poisoning occurs when harmful microorganisms in food cause human illness or death. In addition to bacteria, yeasts and moulds, food contaminated with viruses, protozoa or nematodes can also cause food-related illness. (See Chapter 6 for a more detailed review of table olive spoilage and safety.)

Manipulation of microbial activity

Controlling microorganisms is of the utmost importance in table olive processing.

Prevention of contamination. As olives are grown in open orchards, controlling microbial contamination in this setting is mostly impossible. Establishing reliable procedures that can eliminate some microbial diseases (particularly fungal) from the olive tree could possibly help, but needs further investigation. Using quality irrigation water and undertaking harvest and post-harvest operations in a clean hygienic manner helps reduce contamination of olives. During processing, measures must be in place so that contamination from unpurified water, contact with sewage and air containing microbial spore laden dust is minimised. Equipment should be cleaned and sanitised before use. Preventing damage to the olive skin during harvesting and transport of raw olives also protects against microbial damage. Worker hygiene, another important factor, has been discussed in an earlier section of this chapter.

Physical removal of contaminants. Filtration of water through various types of filters reduces physical, chemical and some microbiological contaminants. Washing olives prior to processing is a basic form of microbial control.

Controlling the following six major factors that can affect growth and survival of microorganisms associated with table olive processing is also important:

- moisture;
- oxygen levels;
- temperature;
- nutrients;
- acidity and pH; and
- inhibitors.

It is the interplay of these factors that ultimately determine the viability of specific microorganisms.

Moisture. All living organisms, including bacteria, yeasts and moulds, require water for survival. The quantity of water required by these organisms is termed water activity (A_w). Water activity is defined as the ERH (equilibrium relative humidity) divided by 100. Pure water has an $A_w = 1$. When A_w is less than 0.6 the growth of most microorganisms is inhibited. For the different categories of microorganisms (accepting there will be exceptions) broad critical values of A_w are given below:

- bacteria: inhibited below A_w of 0.9;
- yeasts: inhibited below A_w of 0.8;
- fungi: inhibited below A_w of 0.7.

Moisture or water in living organisms and foodstuffs exists in two forms, free and bound. Water activity is a measure of the free water calculated by dividing the water vapour pressure of the material by the water vapour pressure measured at the same temperature.

The 'effective' water activity of olives can be lowered by reducing moisture (heat-dried olives), adding salt (natural olives and salt-dried olives) or adding sugar (such as in jams and conserves). With natural olives, the salt binds free water, whereas in salt-dried olives salt binds water as well as reducing the amount of water in the flesh.

The A_w of a food may not be a fixed value; it may change over time, or may vary considerably between similar foods from different sources. An A_w value stated for a microorganism is generally the minimum A_w that supports growth. At the minimum A_w, growth is usually minimal, increasing as the A_w increases. At A_w values below the minimum for growth, microorganisms do not necessarily die, although some proportion of the population may do so. The microorganisms remain dormant but still retain the potential to grow. Most importantly, A_w is only one factor and other factors such as pH and temperature of the food must also be considered.

Lactic acid bacteria can tolerate the high salt concentrations used in fermentation brines. Such tolerance is advantageous for these lactic acid fermenters over less tolerant species of microorganisms. The acids produced by the lactic acid bacteria lower the pH of fermentation brines, further inhibiting the growth of undesirable organisms. *Leuconostoc* spp. initiate the majority of lactic acid fermentations because of their high salt tolerance.

Oxygen levels. All forms of life, including microorganisms, require oxygen in one form or another. On the basis of this, microorganisms can be divided into two main categories: (1) aerobic microorganisms use atmospheric oxygen, and (2) anaerobic microorganisms use oxygen bound to compounds such as carbohydrates.

Some microorganisms, classified as 'facultative organisms', are able to switch from one form to another depending on the available oxygen levels.

Most olive fermentations are anaerobic, although research has shown that aerobic fermentations also have a place. Key organisms in olive fermentation such as lactic acid bacteria utilise reducing sugars (glucose, fructose) from the olive flesh to produce lactic and other food acids that lower brine pH, a requirement to ensure consumer safety. Moulds do not grow well under the anaerobic conditions created during fermentation. However, surface mould will grow if enough air space is available above the brine surface.

Since many yeasts and all moulds are strict aerobes, removing air by applying a vacuum or replacing it with a gas (such as carbon dioxide or nitrogen) as in the principle of vacuum packing will prevent their growth. However, it is still possible for anaerobic organisms such as *Clostridium botulinum* to reproduce in the absence of oxygen.

Temperature. Temperature affects the growth and activity of microorganisms. The temperature range for optimal fermentation conditions is between 15°C and 30°C with 25°C being very favourable. Below 15°C fermentation is very slow and above 30°C growth of anomalous food spoilage organisms is more prevalent. Most lactic acid bacteria operate more effectively at temperatures of 18–22°C including the *Leuconostoc* species that initiate fermentation. Temperatures above 22°C favour the *Lactobacillus* species.

Very low temperatures (refrigeration) improve the storage life of table olives, whereas exposure to high temperatures, for example during pasteurisation or sterilisation, destroys microorganisms. High temperatures kill spores and vegetative cells and destroy some microbial toxins. The effectiveness of heat treatments depends upon the temperature and duration, type and number of microorganisms present, pH and salt levels and the stability of the olives. The higher the temperature the less time is needed. In general, most olives show physical changes after pasteurisation. Pasteurisation is not necessarily designed to kill all microorganisms, just pathogens, whereas when olives are canned and sterilised the goal is to eliminate all organisms, including spores.

Note: When checking the temperatures of fermentation brines electronic thermometers should be used rather than glass mercury thermometers.

Nutrients. All microorganisms require nutrients for growth and survival. In the case of olive fermentation, nutrients released from the olive include reducing sugars, amino acids, vitamins and minerals. When all the sugars have been used to produce food acids and other metabolites, microorganisms are inhibited and growth rates decline. If environmental temperatures increase or the temperature of the fermentation tanks is poorly controlled, then anomalous organisms can utilise the food acids produced, raising the brine pH and increasing the risk of spoilage.

Acidity and pH. pH is a measure of the hydrogen ion concentration in a solution, such as fermentation brine or packing solution. It is measured by using specific pH papers or pH meters. pH values, which are based on a logarithmic scale range from 1–14, where a value of one represents high levels of hydrogen ions (acidic) and a value of 14 represents a low concentration (alkaline). A value of seven is classified as neutral (see Chapter 6 for testing details). The actual amount of acid produced during fermentation or added to brine for adjustment is generally calculated as grams of lactic acid per 100 millilitres equivalent, and must also be considered in the preservation of table olives.

The optimum pH for most microorganisms is pH 7 (neutral) or slightly acid. Foods with a pH 4.5 or less are classed as high acid foods. The growth of bacterial spores associated with food poisoning will not occur at these levels of acidity. Foods with pH values greater than 4.5 are prone to spoilage due to the growth of bacterial spores. Effective fermentations of table olives achieve brine pH values between 4.3 and 4.5 or less. If these pH values are not achieved, food acids, for example lactic or acetic acids, can be used to lower brine pH making the olives safe and resistant to bacterial spoilage. *Lactobacillus* spp. and *Streptococcus* spp. are acid tolerant bacteria.

Yeasts can grow at pH ranges of 4.0–4.5, so in spontaneous fermentations yeasts are often present with fermentative bacteria. Moulds prefer acid environments, but they can grow over a wide range of pH values (pH 2.0–8.5).

Inhibitors and preservatives. All substances, synthetic or naturally occurring, that can interfere with microbial cell membranes, enzymatic action or genetic, or act in a way to negatively affect the microbial environment, can be considered to be inhibitors or preservatives. Acid build-up during fermentation inhibits many bacteria. Fermentative bacteria such as *Lactobacillus* spp. can also be inhibited in acid environments and if too much acid is added during correction procedures no fermentation occurs and the possibility exists of not debittering the olives. Furthermore, polyphenols in olive flesh can inhibit a number of useful as well as harmful microorganisms.

Chlorine (sodium hypochlorite), used as a sanitiser, injures cell membranes, inhibiting or killing harmful microorganisms. Chemical preservatives if added to olives should have minimal impact on the health of consumers and not impart off-flavours to the olives. Preservatives such as sorbic acid injure microbial cell membranes. Salts of sorbic acid can also dehydrate microbial cells as well as inhibit enzyme activity within the microbes. Potassium sorbate, commonly included in table olive products as a preservative, should be held at concentrations within the range of 0.02–0.05% w/w.

Fermentation and table olives

Fermentation is the process by which organic substrates, such as sugars (for example glucose and fructose), undergo biochemical changes by the action of microorganisms or enzymes to produce food acids, ethanol, carbon dioxide and other metabolites. Fermentation involving microorganisms is a natural process, hence its use in the production of 'natural' olives. Controlled fermentation is generally an efficient process requiring low energy inputs and increasing the safety (when consumed) and shelf-life of olives.

The fermentation process, in most cases, occurs in the brine. Fermentable substrates need to diffuse out of the olive flesh into the brine and fermentation products (lactic and acetic acids) and salt need to pass into the olives. When processed correctly, preservation is due to the combined effects of salt, pH and the organic acids, and the olives will not need heat treatment to ensure safety and stability. However, as a precaution many fermented table olive products are packed and pasteurised with or without preservatives such as salts of sorbic or benzoic acids.

It is essential with olive fermentation to ensure that only the desired bacteria or yeasts start to multiply and grow in the brine at the expense of undesirable pathogenic and spoilage microorganisms. Most spoilage microorganisms cannot survive the salt/acidic environments of table olive processing.

Fermentation of olives involves the action of lactic acid producing bacteria, for example *Lactobacillus* spp. and/or yeasts, on fermentable substrates, such as sugars, released from the olives during placement in water or brine. During fermentation, acids such as lactic and acetic acid are produced, which increase the acidity level of the brine and lower its pH. Alcohol is also produced during some types of fermentations. The combination of high salt and low pH greatly reduces the risk of microbial spoilage of the olives. Controls are essential to reduce the risk of overgrowth of undesirable or harmful microorganisms that can lead to product deterioration or food poisoning. Process control involves maintaining the salt and acid levels by targeted additions of sodium chloride and food acids respectively. Such processing, generally undertaken at temperatures between 20°C and 25°C, requires negligible energy input. However, where average daily temperatures are low, additional heating is required. Where processing tanks are not under cover and tank temperatures are expected to exceed 30°C, they should be made of materials with a reflective colour and cooled with water sprays on excessively hot days.

A simple debittering process for any olive, green-ripe, turning colour or black-ripe, is placing them in 8–10% w/v salt brine solution for a period of time. Fermentation then takes place in the processing tank, and flavour compounds are formed through the interaction of microorganisms. Textural changes in the flesh also occur. If the process is well controlled, safe nutritious olives are produced and the fermentation brine can be used to prepare the final packing solution for 'natural' and 'traditional' olives.

Continuous records should be kept and the process controlled, especially pH, salt levels, microbiology, organoleptic changes and spoilage. All operatives need to be trained in food processing methods, handling chemicals and processing olives. Total quality management and HACCP systems should be in place.

Anaerobic fermentation

Anaerobic fermentation is commonly used in table olive production either as the first step or after debittering with lye. Processing olives by anaerobic fermentation in brine involves a number of sequential stages. Aerobic fermentation methods for table olive processing are also used, but to a limited extent in the olive industry. An advantage of aerobic fermentation of olives is a reduction of gas pockets in the product.

The process of fermentation involves the splitting of organic compounds by microbial enzymes into simpler substances, for example sugars are converted to lactic acid, acetic acid and alcohol. Raw olives have a natural microflora: Gram negative bacteria; homofermentative and heterofermentative lactic acid bacteria and/or yeasts; oxidative yeasts and moulds; and *Clostridia*, *Propionibacteria* and *Bacillus* spp. Some of these microorganisms are integral to processing and fermentation whereas others, if not controlled, can eventually lead to soft and malodorous olives. The exact combination of microorganisms varies with the olive maturation stage, but the principles above apply.

Olives are generally fermented in brine (8–10% w/v salt brine). The initial brine has a pH of around 6.5–7.5 and possibly higher if the olives are pre-treated with lye (Fig. 4.23). At the beginning, Gram negative bacteria predominate even in the nutrient-poor brine. These bacteria produce copious amounts of carbon dioxide, as does the fruit that is still technically alive. The carbon dioxide released dissolves in the brine producing carbonic acid. Some natural acids released from raw olives can also contribute to the initial acidity. Oxygen is also consumed. The net result is a moderate increase in the brine acidity and a fall in pH to around 5, which helps establish anaerobic conditions in the brine for fermentation. This process generally takes three to four days. If the pH does not fall, then Gram negative bacteria persist and the olives can develop gaseous spoilage: gas pockets and soft olives. The addition of food acid, for example lactic acid, so that the brine pH falls to around 5 can avoid this problem. The salt levels in the brine should be maintained around 8% w/v during processing.

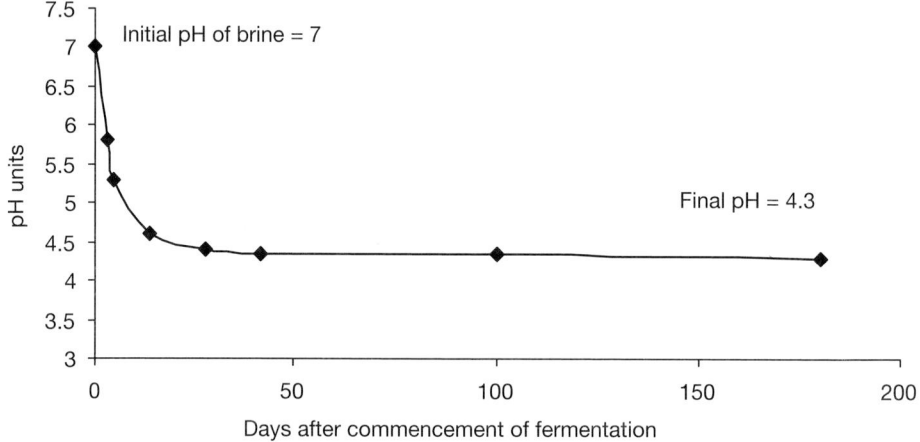

Figure 4.23 Typical changes in brine pH during spontaneous anaerobic fermentation of untreated raw olives.

Note: The fermentation containers (barrels, tanks) must be kept full of brine at all times. During the period of active fermentation (four to five days) when gas production causes excessive frothing and bubbling, care must be taken to replace all lost brine. When gas production subsides, the closures or lids should be tightened firmly to exclude air and keep oxidative yeast and mould growth at the surface to a minimum. If the olives being fermented are low in sugars, sugar (dextrose, sucrose or corn syrup) can be added three to four days after fermentation is underway.

Depending upon the final product, homofermentative and heterofermentative lactic acid bacteria and/or yeasts are able to proliferate under these anaerobic conditions and lowered brine pH. Normally, faster growing heterofermentative species dominate at this stage, utilising sugars and other fermentable substrates released into the brine from the olive flesh producing carbon dioxide, lactic acid, acetic acid and ethanol. A further lowering of brine pH occurs and anaerobic conditions are maintained, which prevents further proliferation of Gram negative bacteria. By ensuring that strict anaerobic conditions are maintained during processing, the growth of surface moulds and yeasts is inhibited. Otherwise these organisms would consume acids produced during fermentation, resulting in an increase in brine pH, hence reducing the stability of the olives and increasing the risk of spoilage.

As brine acidity increases, heterofermentative species are replaced by homofermentative lactic acid bacteria, for example *Lactobacillus pentosus* (formerly *Lactobacillus plantarum*), producing predominantly lactic acid. As indicated earlier, heterofermentative bacteria do not produce as much acid as the homofermentative bacteria. If the brine sugars are low or depleted, insufficient acid is produced and preservation problems can occur. During natural fermentations, yeasts are often present with the lactic acid bacteria; namely, mixed flora, and in some cases, for example Greek-style olives, yeasts predominate. If well controlled, the final products have desirable organoleptic qualities. If poorly controlled the olives soften, change colour and become gassy or 'fritzy'.

When fermentation is complete, the olives can be stored in the same brine for up to two years, particularly if the salt level is maintained at 10% w/v or more.

Environmental considerations with table olive processing

All processing must be undertaken in ways that minimise environmental impacts. Methods that use less energy and water and produce lower volumes of wastewater, such as natural fermentation in brine or the use of dry salt, are more favourable than methods involving lye treatments or heat. Processing brines can be filtered, pasteurised and reused, following suitable salt and pH adjustments. Also, heat or sun-dried olives do not produce any solid or liquid wastes. Salt-dried olives produce some salt/black water that requires disposal.

Olives treated with lye require multiple washes and can use up to five times the amount of potable water compared with natural methods. Furthermore, with lye treatments, energy requirements increase general costs and possibly labour costs. However, methods using lye treatments have a number of advantages, including shorter processing times than natural methods and specific organoleptic characteristics that are agreeable to many consumers. In table olive production, Spanish and Australian researchers have estimated the following likely levels of wastewater (per kilogram of olives) are produced during the processing stages:

- Water cured and Kalamata-style (fast method), 5–7 l/kg;
- Brine cured: natural, traditional green, turning colour, naturally black-ripe and Kalamata-style (slow method), 0.5 l/kg;
- Lye treated green olives: Spanish-style green, Californian-style green, 1.0–3.0 l/kg;
- Spanish/Californian-style black olives (lye treated), 1.5–6.0 l/kg;
- Heat/sun-dried olives, 0; and
- Salt-dried olives (depends on variety), 150 ml/kg.

Waste products, such as brines/black water containing salt, polyphenols, dissolved organic solids and cell debris, can be removed from the processing facility by contractors, but this can be impractical or expensive. It is more usual to place unwanted table olive liquid wastes into shallow evaporation tanks. In one such facility, presented in Plate 22, the wastewater is collected on a thick composite rubber liner where the water evaporates assisted by the wind tunnel created by the structure. It is important that liquids do not leak and contaminate the water table. Initial rinsing water and washing solutions, depending on pH and salt content, can be used to irrigate olive trees. Scientific studies have shown that wastewater from table olive processing, if high in salt content, is less favourable when applied to olive trees compared with controls. Physiological changes observed in the olive trees included decreases in leaf water potential, stomatal conductance to water, and rates of photosynthesis after two weeks of application. The observed reductions were greater after two months of irrigation. Furthermore, applications of the wastewater reduced nitrogen levels in the olive trees as well as yield.

Common methods of preservation for table olives

Common methods used in preserving foods such as table olives to prevent microbial spoilage and food poisoning are given below. (More than one preservation method may be used.)

- *Salting*, stops growth of most microbes by reducing the water activity of the flesh;
- *Acidifying*, stops growth of most bacteria;
- *Refrigeration*, inhibits microorganisms;
- *Preservatives*, as permitted by legislation;
- *Preserving atmosphere*, partial or total removal of air and its substitution with an inert gas;
- *Vacuum pack*, total air removal;
- *Drying*, stops growth of all microbes if the water activity (A_w) is less than 0.6;
- *Blanching*, at temperatures greater than 95°C vegetative bacteria and yeasts are killed;
- *Pasteurising*, olives undergo a thermic treatment at temperatures between 60°C and 80°C that kills non-sporing (vegetative forms) of pathogenic, non-pathogenic and spoilage bacteria and yeasts; and
- *Sterilising* (canning), olives undergo a thermic treatment, at temperatures greater than 100°C, which kills all pathogenic bacteria, making the product 'commercially sterile'.

Lowering the water activity of table olives by adding large amounts of salt or drying the olives increases the osmotic pressure in the cells of microorganisms and so impairs their ability to grow. When water activity is low, microorganisms must compete with other dissolved molecules (solutes) for free water molecules. Bacteria are poor competitors for water, except for *Staphylococcus aureus*, whereas moulds are excellent competitors. *Staphylococcus aureus*, a skin pathogen, if present in brines, can release life-threatening toxins during processing, increasing the risk of food poisoning.

Acidification of table olives by fermentation acids produced during processing or the addition of food acids impairs the activity of microorganisms by a direct pH effect and by the uptake of undissociated forms of acids, such as acetic and lactic, resulting in an increase in intracellular acidity and lowering of pH.

Heat application, for example blanching, pasteurisation and sterilisation, is commonly used to preserve foods. The exact effect of heat on microorganisms is not fully understood, but it most likely disrupts macromolecules in their cell membrane and cytoplasm as well as nucleic acids, for example DNA and RNA. Microorganisms can adapt to mild heat; they develop protective mechanisms such as modifications in their cell membranes, production of heat shock proteins and increased enzyme stability to heat. Some, such as *Clostridia* spp. and *Bacillus* spp., however, can develop heat resistant spores, hence the need for high temperature sterilisation in an autoclave for some olive products.

Novel methods for food preservation being developed by researchers may have future applications for table olive preservation. One such process involves the use of ultra high pressure; microorganisms in food are inactivated or destroyed when subjected to hydrostatic pressures of 130 mPa or more.

Packaging of table olive products

Once processed, table olives are packed for sale in containers made of tin, glass, plastic or other material (but not wood) that meets technical and health standards. Whichever material is used, it should ensure the preservation of the olives and not release harmful substances into the brine or olives. Containers should be well filled with the product (table olives and packing solution) and should occupy not less than 90% of the water capacity of the container when sealed. Using the IOOC Table Olive Standards (2004), the water capacity is determined by completely filling the container with distilled water at 20°C. For non-metallic rigid containers, such as glass jars, the basis of the calculation is the weight of distilled water at 20°C that the sealed container will hold, when completely filled less 20 ml. Containers that do not meet the requirements for minimum fill are classed as defective. However, some tolerance is allowed in product lots.

Drained weight. The net weight for a container with olives in brine is made up of the weight of the olives and the weight of the brine, for example 360 g. The drained weight is only the weight of the olives, for example 230 g. It is not unusual to see both net weight and drained weight on the label, particularly with table olives imported in original containers. The IOOC Table Olive Standards (2004) prescribes tolerance levels for drained weight. The tolerance concerning the net drained weight, as indicated on the container label, shall not exceed the following percentage scale, providing the sample's mean drained weight is equal to, or in excess of, the declared weight.

5% For containers with drained weight less than 200 g
4% For containers with drained weight between 200 g and 500 g
3% For containers with drained weight between 500 g and 1500 g
2% For containers with drained weight in excess of 1500 g

To achieve the correct drained weight, the required amount of olives is packed into the container then packing solution added to the desired fill volume. The container is tightly sealed ensuring there are no leaks or air above the olives.

Safety aspects. Packed table olives should comply with statutory/international microbiological criteria and be free of any objectionable matter. The olives and brine should also be free of any microbiological spoilage, especially putrid, butyric or 'zapatera' fermentation. When table olives are tested they should be free of pathogens or contaminating microorganisms likely to develop in the packaged product under normal storage conditions. Furthermore, the table olives should be free of any substances such as toxins in quantities that represent a health hazard when eaten. Olives preserved by heat sterilisation such as Spanish/Californian-style black olives (olives darkened by oxidation) should receive a processing treatment that ensures the destruction of *Clostridium botulinum* spores.

Bulk table olive products. Table olives in bulk, for example 200 L plastic barrels containing 150 kg of olives, are sold on the basis of their net weight, for example 150 kg. Bulk olives can also be transported in a food grade liner bag (Pallecon System, TNT, Australia). Fermented olives held in bulk in a covering liquid may contain fermentative organisms such as lactic acid bacteria and yeasts at levels of up to 10^9 CFUs/ml (Colony Forming Units) of brine, or per gram of flesh (using a selective culture medium) depending on the level of fermentation. Plastic barrels can be reused as long as they are in good condition and have been cleaned and sanitised in a manner that does not compromise the safety and organoleptic quality of table olives when refilled.

Consumer size table olive products. Table olives are mainly packed in glass or plastic jars. Where plastic jars are used they should be of food grade quality and capable of withstanding pasteurisation temperatures, such as polypropylene polymer. Table olives are also available in vacuum packs: plastic bags filled with olives and a small amount of brine and sealed after the air has been removed or replaced with an inert gas. Spanish/Californian-style black olives are packed and sterilised in specially lined cans or glass jars of various sizes.

Pasteurisation of table olives

The aim of pasteurisation is to reduce the numbers of pathogenic and spoilage organisms in the olives and brine. Although not all types of table olives need pasteurisation, its application is common for many types of olive products (see Chapter 5 for specific applications). Specifically for pasteurising olives, the maximum pH of the brine should be 4.3 for a salt content of 2% w/v or more and pH 4 if the salt content is less than 2% w/v. Brines with a total free acid value of between 0.6% w/v and 0.9% w/v, expressed as lactic acid, are consistent with the above pH values.

In practical terms, pasteurisation involves submerging the packed olives in a bath of boiling water for a sufficient time to bring the container and contents to a temperature of 70–80°C. Heating time depends on the container size and the type of olive product. During pasteurisation, air and liquid expand and escape past the lid. When the containers cool down, a partial vacuum is created. Note that deterioration in texture and flavour occurs if higher temperatures are used. Pasteurisation is unnecessary for bulk olive production with fermentation as long as brine salt, pH and acid levels are controlled.

Containers to be treated must be of a height dimension such that when submerged in the water bath their tops are 5 cm below the water level. An additional 5 cm headspace is also required if the water in the bath is allowed to boil.

An indicative pasteurisation procedure typically involves the following steps:

(a) Ensure that the containers, especially glass or plastic jars, and lids are food grade and will withstand prolonged heating.
(b) Check jars for cracks or chips.
(c) Clean/wash jars with potable water before use to ensure any factory contaminants are removed.

(d) Pack the required weight of olives into containers.
(e) Add the packing solution, preheated to 70°C, to the containers leaving a small headspace to allow for any expansion of the contents.
(f) Secure the lids onto the containers.
(g) Place the containers in the water bath preheated to around 45°C to prevent breakage or damage to the container.
(h) Raise the temperature of the water bath and maintain this at 70–80°C for a period of time so that the olives and brine reach this temperature for the prescribed period (with the contents at 70–80°C, small containers require five to six minutes, whereas larger ones require up to 20 minutes).
(i) Representative samples of the packaged table olives should be cleared for safety by a microbiology laboratory before sale.

Note: Over-heating may change the organoleptic qualities of the olives such as colour and texture. Also, if plastic containers are used, they should be heat resistant.

Codex Alimentarius (1987)/IOOC (2004) Table Olive Standards – Characteristics for Thermal Pasteurisation.   From a standards point of view the Codex Alimentarius (1987)/IOOC (2004) Table Olive Standards provide guidelines for pasteurising and sterilising olive products. For lye treated olives, natural olives (untreated), dehydrated and/or shrivelled olives a minimum of 15 pasteurisation units are required, where pasteurisation units are defined as the cumulative lethal rate during heat processes performed at temperatures below 100°C. The reference microorganism for these types of table olive products is *Propionibacteria* for which a reference temperature (Rt) of 62.4°C and a z curve of 5.25 defines the equation of the thermal death time (TDT). Rt is the temperature corresponding to a decimal reduction time which, together with the z curve, defines the logarithmic representation of the thermal death time curve of a given microorganism. The z curve plots the logarithmic representation of the thermal death times according to temperature (TDT curve); it is equivalent to the number of degrees necessary for the curve to transverse one log cycle. The TDT is the heating time, at a specific temperature and in specific conditions, required to reduce the initial microbiological population by a factor of 10^{12}. In practical terms this is achieved by heating the table olives to 62.4°C for 15 minutes. Note that if higher temperatures are used, the exposure time can be reduced. All pasteurisation procedures should be validated by a microbiology laboratory.

Alternative procedures

Instead of immersing the packed olives in a water bath, packed olives are placed in a steam chamber that can achieve the required temperatures. Again, the contents must be brought up to the pasteurising temperatures, then maintained at temperatures from 70–80°C for the required time. In some processing plants the olives, packed in jars without their caps, are passed through a steam tunnel for a set time. The bottles are capped after passing through the steam tunnel.

Another common procedure for olives preserved by fermentation in brine follows. This procedure is carried out under aseptic (clean) conditions to reduce the risk of bacteria and moulds entering the containers.

(a) Clarify the fermentation brine (salt concentration 6–7% w/v and pH 4.0–4.5) by filtration.
(b) Heat the brine to 80°C and maintain this temperature for approximately five minutes.
(c) Add the hot brine to final containers prefilled with processed olives.
(d) Spray the tops of the containers with steam if possible.
(e) Screw on the lids.
(f) Send representative samples of the packaged table olives to a microbiology laboratory for safety clearance before sale.

Note: This procedure kills a number of potential pathogens and prevents any secondary fermentation or spoilage occurring in the final containers. If fermentation brines are used as packing solutions, they should be filtered and preferably pasteurised after checking and adjusting salt and acid levels. Further information on sterilising Spanish/Californian-style black olives is provided in Chapter 5.

5
Specific table olive processing methods

The types of table olives and table olive products available on the Australian market are reviewed in this chapter. The specific types of table olive processes and the potential problems that may occur in the production system are also introduced. The major processing procedures discussed are soaking in water, in brine, with lye (caustic soda/ sodium hydroxide), Californian/Spanish-style (caustic soda and oxidation), heat, and salt-dried olives. A detailed protocol for processing olives has been developed and is presented. An alternative method of debittering olives by microbial means rather than with the use of caustic soda is introduced. Safety and spoilage issues are examined. The chapter also explains how to overcome stuck fermentations. The topic of secondary table olive production – pitting and stuffing table olives, marinades, packaging processed olives in different solutions and spices, olive pastes and tapenades – is discussed in detail. Methods of preservation such as pasteurisation and sterilisation are further covered.

Processed table olives available in Australia

Knowing the types of table olive products available is a guide to important products that are consumed by Australians.

Table olives are sold from a number of diverse outlets and these are listed in Table 5.1. In season, raw unprocessed olives are also available for home processing from growers, continental delicatessens, wholesale fruit and vegetable markets and popular food markets in major Australian cities. Bulk quantities are available from food wholesalers and grower/processors (Fig. 5.1). In Australia, many households have planted at least one olive tree for home use. Others, especially Greek and Italian Australians pick olives off those planted as street trees, from feral olive trees and their own trees.

Table 5.1. Australian enterprises supplying processed and unprocessed table olives

Enterprise category	Processed olives			Unprocessed olives
	Loose	Packaged	Bulk commercial quantities	
Continental delicatessen	yes	yes	yes	yes
Local supermarket	no	yes	no	no
Major supermarket	yes	yes	no	no
Specialty/gourmet/cellar door	yes	yes	no	no
Food wholesaler	no	yes	yes	no
Grower/processor – boutique	no	yes	yes	yes
Grower/processor – small scale	no	yes	yes	yes
Grower/processor – medium scale	no	yes	yes	no
Wholesale fruit/vegetable market	no	no	no	yes
Popular food market	yes	yes	no	yes

Figure 5.1 Raw olives in cartons for the fresh fruit market.

Loose processed olives are mainly available from continental delicatessens, gourmet shops, popular food markets and more recently from state and national supermarket chains. Packaged olives are widely available in small and large supermarkets, continental delicatessens, gourmet shops, cellar doors and regional food shops. The latter provide a variety of locally produced products including olives and olive pastes that are often sought after by tourists.

Wholesale table olive trade in Australia

Table olive products are generally imported from Spain and Greece, with some products, such as *Kalamata* olives, produced or transformed into specialty styles in Australia. Container sizes vary from a few kilograms in cans and jars to plastic barrels containing up to 200 kg of olives (Table 5.2). Most olives available to Australian consumers are either lye treated, for example black and green Spanish-style, or naturally processed in brine, for example Kalamata-style, Greek-style black and green olives. They are sold in large, 150–200 kg barrels, and are often purchased by third parties for preparing specialty products to be sold loose in 100 g to 1 kg lots or packed into consumer size containers with or without further embellishment such as stuffings and marinades. Wholesalers also sell processed olives (Greek, Spanish and Italian) to retail establishments packaged in consumer size containers in carton lots. Some Australian processed olives are also available for wholesale purchase.

Table 5.2. Table olives available from Australian wholesale enterprises

Olive style	Container type	Quantity	Origin
Black Spanish			
Manzanilla, Hojiblanca	Cans, jars and plastic barrels	1.5 kg–200 kg	Spain
Sliced	Jars and cans	1–3 kg	Spain
Green Spanish-style			
Manzanilla, Sevillana, Chalchidikis	Cans, jars and plastic barrels	1 kg–200 kg	Spain, Greece
Cracked	Plastic barrel	12 kg	Greece
Destoned (pitted)	Jars and plastic barrels	2 kg–200 kg	Spain, Greece
Stuffed with pimento	Jars, cans and plastic barrels	1 kg–200 kg	Spain
Sliced	Jars and cans	1–3 kg	Spain
Kalamata-style (*Kalamon*)			
Plain	Plastic barrels	5–200 kg	Greece, Australia
Marinated	Plastic barrels	5 kg	Greece, Australia
Extra jumbo (large size)	Plastic barrels	12 kg	Greece, Australia
Continental mixture	Plastic barrels	5 kg	Greece, Australia
Home-style split	Plastic barrels	5 kg	Greece, Australia
Greek-style			
Naturally black – brine fermentation			
Conservolea (*Volos*)	Plastic barrels	200 kg	Greece
Manzanilla	Plastic barrels	200 kg	Australia
Salt-dried (date) black-ripe	Plastic barrels	50–100 kg	Greece, Australia
Green – brine fermentation			
Manzanilla	Plastic barrels	200 kg	Australia
Turning colour – brine fermentation			
Jumbo Kalamata	Plastic barrels	200 kg	Australia

Sale of loose table olives in Australia

Table olives are available loose for sale in small lots at larger national supermarkets, continental delicatessens, markets (Plate 23), and at specialty and gourmet food outlets A large number of different products are available (Table 5.3); however, most of these are based around the types and variety of olives that are available at the wholesale level.

Generally there is no indication as to the source or origin of these olives, such as whether they are Australian or imported. In supermarkets they are displayed in low temperature glass showcases with the name and style of the product, but mostly out of reach of customers. There is less control of the olives in continental delicatessens where customers can sample before they buy. Black Spanish-style olives are the least expensive, selling for as little as A$6/kg, whereas embellished olives, for example stuffed olives packed in olive oil with herbs, spices and sun-dried tomatoes, can retail at around A$40/kg. In this case, customers purchase the required quantity of olives, which are placed in an unlabelled plastic container and sealed. It is assumed that these olives will be consumed soon after purchase, as storage information is not provided.

Across a wide range of retail outlets innumerable table olive products are available, with the majority in marinades. National supermarkets often have the largest selection compared with continental delicatessens and smaller local supermarkets. Olives sold as mixtures of several varieties are marketed under trendy names such as 'Provencale', 'Mediterranean mix', 'Connoisseur' or 'Continental blend'. These contain olives that have been processed separately then mixed with olive oil, herbs and spices, before presentation for sale either as loose or packaged olives.

Table 5.3. Different types of loose table olive products available to consumers in Australia

Olive style/product	Olive style/product
Black, Kalamata	Marinated, Thai-style herbs
Marinated, chilli and oregano	Marinated, pitted
Marinated, chilli and garlic	Marinated, pitted, anchovy in oil
Marinated, lemon and garlic	Marinated, pitted, fetta cheese in oil
Marinated, with mixed herbs	Marinated, pitted, marinade, chilli and garlic
Destoned (pitted)	Marinated, pitted, pimento
Salt-dried black-ripe (date or shrivelled)	Marinated, pitted, pimento, marinade, herbs and spices
Black Greek-style	Marinated, pitted, sun-dried tomato in oil
Black Spanish-style	Marinated, sliced
Black Spanish-style, pitted	Cracked marinated, chilli
Black Spanish-style, sliced	Mixed selections (several olive varieties)
	Connoisseur marinated, herbs and spices
Greek-style donkey, herbs and spices	Continental marinated, herbs and spices
Green Spanish-style	Green and black Spanish-style, sliced
Marinated, chilli and garlic	Destoned (pitted) marinated, chilli and herbs
Marinated, marinade, lemon and vinegar	Destoned (pitted) marinated, herbs and spices
Marinated, marinade, lemon and garlic	Provencale, marinade, herbs and spices

Provencale olives. The Provencale mix referred to above should not be confused with the French Provencale olives. Traditional Provencale olives are prepared from whole or pitted green or black *Picholine* olives processed in brine and marinated in brine, olive oil, thyme, lavender and garlic. Some recipes include citric acid, giving the olives a slight citrus flavour. Cracked Provencale olives (Olives cassées de la Valée des Baux) are prepared from green *Salonenque* or *Verdale* olives harvested in the middle of the season. These olives are cracked (bruised), processed in brine, then flavoured with anise or fennel.

Packaged table olive products in Australia

There is no shortage of packaged table olive products available for sale from Australian supermarkets, continental delicatessens and specialty food shops (Table 5.4).

Packaged table olive products in small jars and tins include whole, pitted and sliced olives (with or without embellishments), antipasti with olives and pickled vegetables, and tapenades and olive pastes (Figs 5.2 and 5.3). Most products are imported, with many packed in Australia. Few products are of truly Australian origin indicating the lack of penetration of Australian table olives into the retail market. Australian products are available in regional areas where they are sold at specialty food shops, gourmet shops, specialty olive shops and from wineries.

Table 5.4. Packaged table olives available in Australia

Olive style/product	Olive style/product
Antipasti, Green Spanish-style olives and fetta cheese	Green Spanish-style, pitted, natural pepper
	Green Spanish-style, pitted
Antipasti, Mediterranean Gourmet	Green Spanish-style, pitted, pimento
Antipasti, Mediterranean Mix	Green Spanish-style, halves (bacchetta)
Antipasti, Tapas	Green Spanish-style, marinade, herbs and spices
Antipasti, Char grilled, Mediterranean	Green Spanish-style, marinade, Mediterranean
Black, Kalamata	Green Spanish-style, pimento, marinade, herbs and spices
Black, Kalamata, halves	
Black, Kalamata, marinade, chilli	Green Spanish-style, pitted
Black, Kalamata, marinade, garlic and vinegar	Green Spanish-style, pitted, pimento
Black, Kalamata, marinade, herbs and spices	Green Spanish-style, pitted, anchovy
Black, Kalamata, pitted	Green Spanish-style, pitted, blue vein cheese
Black, Kalamata, sliced	Green Spanish-style, pitted, fetta cheese
Black, Kalamata, marinade, balsamic vinegar	Green Spanish-style, pitted, Parmesan cheese
Black, Kalamata sliced, marinade, balsamic vinegar	Green Spanish-style, pitted, pimento, tuna
	Green Spanish-style, pitted, smoked salmon
Black heat-dried olives	Green Spanish-style, pitted, tuna
Black Greek-style	Green Spanish-style, Queen
Black Greek-style, low salt	Green Spanish-style, Queen, pitted
Black Greek-style, lemon and garlic	Green Spanish-style, Queen, pitted, pimento
Black Spanish-style	Green Spanish-style, sliced
Black Spanish-style, pitted	Green Spanish-style crushed, marinade, seasoned
Black Spanish-style, pitted, almonds	Green Spanish-style marinade, lemon and garlic
Black Spanish-style, pitted, anchovies	Mixed, cocktail, marinade, herbs and spices
Black Spanish-style, pitted, spices	Mixed, Mediterranean, marinade, herbs and spices
Black Spanish-style, sliced	Mixed, Provencale, marinade, herbs and spices
Green, natural fermentation in brine – Frizantina-style	Olive paste, black
	Olive paste, green
Green, natural fermentation in brine – Ligurian-style	Olive paste, tapenade
Green Spanish-style	

Figure 5.2 Antipasti with artichokes, peppers and green olives.

Figure 5.3 Australian grown and processed table olives. (Photo: Frankland River Olives, Western Australia.)

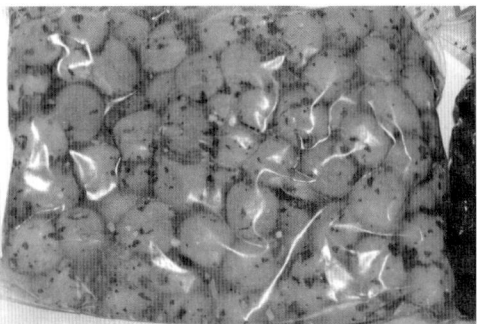

Figure 5.4 Vacuum packed green olives with herbs and spices.

Olives are packed mostly in glass jars and some, generally imports, in cans. A small number of products are vacuum packed in plastic bags (Fig. 5.4). Vacuum packed olives sold in 1–5 kg lots are popular in the food services industry as they are easy to store and handle. Most packaged table olive products use basic olive types with different embellishments such as garlic, chilli, Mediterranean herbs, lemon and vinegar.

Processing olives with water (water-cured)

Traditional processing methods involve subjecting raw olives to many water changes (or weak brines) over 10–14 days until they debitter. Green-ripe, turning colour or naturally black-ripe olives can be processed in this way. The bitter glycoside oleuropein is leached out of the olives and removed from the tank/container when soaking solutions are discarded. Generally no fermentation occurs using this method. Once debittered, after the last wash 10% w/v sodium chloride (salt) is added. Over time, when the brine equilibrates with the olives, final brine concentrations fall to approximately 6–7% w/v salt.

Figure 5.5 Turning colour *Azapa* olives processed by multiple soakings in water.

This method is popular with home processors because the olives are ready to eat within a few weeks from the start of processing (Fig. 5.5), but is unsuitable for serious commercial olive processing.

Disadvantages of this method are the large amounts of water required that need to be disposed of, and the increased risk of spoilage through microbial contamination. With water and low salt brines, proteolytic enzymes break down protein in the flesh to amino acids that further degrade to ammonia and hydrogen sulphide (H_2S). The resulting olives may have a urine (ammonia) and/or faecal or rotten egg odour (hydrogen sulphide) if not processed carefully. Over-soaking leads to soft olives with a 'washed out' taste.

Naturally black-ripe *Kalamata* olives that have been debittered by this method and embellished with olive oil, lemon and red wine vinegar gives the traditional Kalamata-style olive. However, in modern processing establishments a simple fermentation in brine (8–10% w/v salt) is used as the preferred method, rather than prolonged water soaking steps (see later for more detail on processing Kalamata-style olives).

With some traditional recipes, for example Ligurian (Benedictine-style), *Taggiasca* olives are soaked in water for weeks to months to debitter the olives. Salt, herbs and spices are then added to the debitterred olives. Today, most Ligurian-style olives are prepared by placing them in brine where they undergo a weak fermentation.

Packing specifications. The packing solution for olives produced by this method should contain at least 6% w/v salt, have a minimum acidity of 0.3% calculated as lactic acid and a pH of 4.3 or less to ensure their safety and preservation. The same conditions apply if a preservative is included or they are to be stored under refrigeration. Alternatively, if the olives are to be pasteurised or sterilised then the packing solution should have a pH of 4.3 or less. In this case, salt and acid concentrations are not specified, but should be at levels that do not compromise the safety and organoleptic qualities of the olives as determined by Good Manufacturing Practice. Outside of these parameters, the processor has to guarantee the safety of the olives by having them cleared by an accredited microbiology laboratory.

Note: With commercial table olive products, preservatives such as sorbic acid (or its salt potassium sorbate) and/or benzoic acid (or its sodium salt) are often added to packing brines to prevent the growth of oxidative yeasts, particularly after the containers are opened. Ascorbic acid, an antioxidant, is also added to processed green olives to prevent discolouration of the olives and the brine.

Step-by-step procedure for water-cured olives

(a) Use/accept quality raw olives (green-ripe, turning colour, naturally black-ripe).
(b) Store raw olives correctly before processing.
(c) Wash olives with potable water.
(d) Size grade and sort raw olives; remove damaged olives.
(e) Use whole, slit or cracked olives.
(f) Pack olives into containers with water; make sure olives are submerged and held in place with a grate.
(g) Remove and replace water daily for 10–14 days.
(h) After last water soaking step add brine 10% w/v sodium chloride (salt) to the brim.
(i) Monitor brine levels to achieve final levels of 6–7% w/v.
(j) Size grade and sort processed olives (optional).
(k) Pack the processed olives in a 6–7% w/v salt brine or alternatively in brine made with 3 parts 10% w/v salt + 1 part vinegar.
(l) Check pH and salt content to meet safety requirements.
(m) Pasteurise the packed processed olives if required.
(n) Send samples to a microbiology laboratory for testing.
(o) Label packed processed olives in accordance with food standard requirements.
(p) Implement a safety recall system for tracking faulty, contaminated or incorrectly labelled olives.

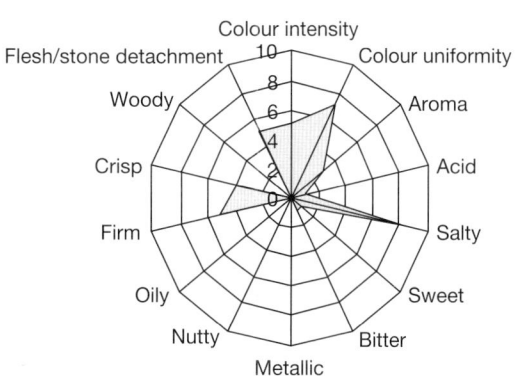

Figure 5.6 Organoleptic profile of water-cured green-ripe *Ascolana* olives packed in brine.

Nature of product. The nature of the product depends on the variety and maturation state. Olives may be firm to soft, and are usually salty with a little bitter taste. When embellishments such as olive oil, herbs, spices or other foods are added, aromas and flavours will change accordingly. The organoleptic profile of water-cured green-ripe *Ascolana* olives packed in brine is presented in a graphic form above (Fig. 5.6).

Processing olives with brine (brine-cured)

This is a traditional olive processing method that is gaining popularity as a commercial method even though some believe there are commercial limitations. There are many variations of this method depending on the country of origin, for example Greece, Italy, Turkey, Cyprus, Lebanon and Egypt. The principal differences are: sometimes several water soaking steps are used prior to brining, brine strengths vary, and there is variability in the bitterness of the final products.

Processing time is dependent on olive variety, maturation state, size and the Flesh:Stone ratio. All varieties at any maturation state can be processed by this method. Green olives can take up to one year to debitter whereas naturally black-ripe olives take as little as three months. Fermentation is slow because fermentable substrates (sugars) need to diffuse through the intact skin of the olive into the brine. Diffusion of substrates is faster when olives are subject to a short pretreatment with lye or blanched with near boiling water for one to two minutes. Debittering occurs by oleuropein diffusing into the brine (where it is diluted), and by the hydrolysis, or breakdown, of oleuropein into less bitter by-products, possibly by the action of some microorganisms. Small amounts of sugar are released during the hydrolysis of oleuropein, which also contribute to fermentation. Slitting, cracking or bruising the olives also speeds up the diffusion of fermentable substrates, reducing processing time.

Naturally fermented olives are either sold to consumers or the food services industry in bulk 5–20 kg buckets. Small consumer size packages – glass or plastic jars, tins or vacuum packs – are pasteurised before sale. Alternatively a preservative can be added. For example, Spanish researchers have found that fermented naturally black-ripe olives preserved with 0.05% potassium sorbate compared favourably with those pasteurised at 80°C for four minutes after nearly one year of storage.

Naturally black-ripe olives processed in brine

Black olives prepared by this method are often termed Greek-style black olives. Similar products are prepared in most Mediterranean countries, particularly Turkey. The following discussion, which relates to the processing of naturally black-ripe olives by spontaneous fermentation, in principle also applies to the natural processing of green-ripe and turning colour olives.

With this method, washed olives are placed directly into brine (8–10% w/v salt), after which anaerobic (no oxygen present) conditions are established and over time the olives take up salt and undergo spontaneous fermentation. In colder regions, brine with lower salt concentrations (6% w/v) is used to facilitate processing. An aerobic process is also available for preparing natural olives and is briefly described later in this chapter.

Features of anaerobic conditions

Anaerobic conditions:

- encourage Gram negative organisms to produce carbon dioxide;
- allow yeasts to produce alcohol and carbon dioxide;
- prevent oxidative organisms, particularly fungi, growing as a film on the surface of the brine; and
- provide the environment for fermentative bacteria to function and multiply.

Salt in the brine draws water-soluble components, sugars, organic acids and minerals out of the flesh into the brine. These substances provide substrates and nutrients required by fermentative bacteria and yeasts. During fermentation, sugars are converted to lactic and acetic acids, alcohol and other compounds that contribute to the taste of the olives. Overall, during processing by this method, the pH of the brine falls from near neutral (pH 7) to pH 4–5, depending on the amount and types of acid produced during fermentation (see Chapter 4). The levels of fermentative substrates also influence the final amounts of acid produced.

During long processing periods other water-soluble compounds diffuse into the brine including oleuropein. The net result is the debittering of the olives with oleuropein converted to by-products including hydroxytyrosol. Olives processed by the brine method are self-preserving because of the pH/salt balance achieved during processing. It takes six to eight weeks for the salt levels in brines and olives to equilibrate. In some cases, brine levels of oleuropein are high enough to inhibit the action of fermentative microorganisms such as lactic acid bacteria, *Lactobacillus pentosus* (formerly *Lactobacillus plantarum*). If this occurs, corrective action is needed to replace part or all of the brine with fresh brine.

The high salt levels used in this method inhibit proteolytic bacteria, which if unchecked can lead to spoilage. During processing, salt levels of 6–7% w/v are achieved with initial brines of 8–10% w/v salt. In centres where the olives are subject to high environmental temperatures (summer/autumn) during processing, the final brine strength needs to be increased to 8–10% w/v or more to prevent spoilage.

During processing, olives can lose weight, with olives from irrigated orchards affected more than those from unirrigated orchards. Also, large olives lose proportionally more weight than small olives.

A mixed flora of yeasts and bacteria are involved in the fermentation. Once the raw untreated olives are placed in the brine, and the container sealed, anaerobic conditions are established. This is achieved by excluding air from tanks/barrels, and through the activity of bacteria and yeasts. Any residual oxygen is consumed by oxidative organisms and carbon dioxide is released into the brine, resulting in an initial lowering of the pH. Carbon dioxide is produced by Gram negative organisms (species of the genus *Citrobacter*, *Achromobacter*, *Aeromonas* and *Escherichia*), by yeasts, and by the olives themselves. Note that raw olives are still alive and respiring (produce carbon dioxide) in the first few days after brining.

During fermentation, yeasts (including species of the genera *Saccharomyces*, *Hansenula*, *Torulopsis*, *Debariomyces* and *Candida*) and lactic acid bacteria (including species of the genera *Pediococcus*, *Leuconostoc* and *Lactobacillus*) are active. On the other hand, *Lactobacillus pentosus* (formerly *Lactobacillus plantarum*) is not the dominant organism because of the high salt levels and the inhibitory action of oleuropein and other phenolic compounds. The phenolic composition changes during processing because of the acid conditions created by fermentation. Phenolic glucosides in the flesh (oleuropein, verbascocide and luteolin 7-glucoside) are hydrolysed to new compounds. This results in lower oleuropein levels in the brine, which are always less than 1.5 mM, and a continuous increase in brine hydroxytyrosol levels. Higher temperatures favour the diffusion of polyphenols from the flesh into brines, facilitating the debittering process. So, maintaining fermentation temperatures at around 25°C is advantageous. After several months of storage the main residual phenols in the flesh are hydroxytyrosol, verbascocide, tyrosol, vanillic acid, caffeic acid, *p*-coumaric acid and oleuropein. Metabolic products formed during fermentation contribute to the organoleptic qualities of the olives.

Fermentation of naturally processed olives occurs in two stages (Table 5.5).

Table 5.5. Summary of events during the fermentation stage of naturally processed green-ripe olives

Parameter	Early			Late
Brine pH	Initial pH 7	pH 6	pH 5	Falls to pH 4.3–4.5
Brine salt levels	10% w/v, which falls over 6–8 weeks to 6%			
Gram negative organisms	Low	Maximum numbers by 3–4 days → disappear after 7–15 days Gas production: potential for gas pockets		
Lactic acid bacteria *Pediococcus* spp. *Leuconostoc* spp. *Lactobacillus mesenteroides* *Lactobacillus brevis*	Present when salt levels in the fermentation brine are low			
Yeasts	Present	Increase in number	Reach maximum numbers in 10–25 days	Yeasts dominate
Lactobacillus pentosus (*Lactobacillus plantarum*)	Present during the whole fermentation period as long as salt is 8% w/v or less. At lower salt levels they are the predominant organisms.			

Stage 1. When the olives are first placed in the brine a robust fermentation by a heterogeneous group of microflora occurs.
(a) Copious amounts of carbon dioxide are produced by Gram negative organisms (coliforms), yeast and fruit. The carbon dioxide formed creates enough pressure to cause brine to bubble and overflow.
(b) Water-soluble substances – minerals, sugars, food acids and oleuropein – begin to diffuse from the flesh into the brine. With black-ripe olives, the brine takes on a rich magenta colour due to anthocyanins, whereas with green-ripe olives the brine is yellow-pink due to oxidised phenolic compounds.
(c) Brine pH falls from 7 to 5.

Stage 2. A mild fermentation occurs, supported predominantly by yeasts and to a lesser extent by the following lactic acid bacteria (LAB):

(a) *Pediococcus* spp.
(b) *Leuconostoc* spp.
(c) Heterofermentative bacteria: *Lactobacilli* spp. (e.g. *Lactobacillus brevis*)
(d) Homofermentative bacteria: *Lactobacillus pentosus* (formerly *Lactobacillus plantarum*)

By manipulating the initial salt levels in the brine a predominantly yeast fermentation or a predominantly *Lactobacillus* fermentation can be effected.

Yeast fermentation. Yeast fermentation occurs when salt levels in brine are kept above 8% w/v; final acidity, measured as lactic acid, is 0.2–0.4% w/v with a pH range of 4.3–4.5. At high salt levels, yeasts produce acetic acid.

Lactic fermentation. Lactic fermentation occurs when salt levels in brine are kept at 3–6% w/v; final acidity, measured as lactic acid, is greater than 0.6% w/v with a pH range of 3.9–4.1.

Note: The final acid concentration in the brine depends on the type of fermentation and the available fermentable substrates, such as sugars.

Processing takes at least three months and up to 12 months depending on the variety, maturation level of the fruit, temperature, salt and pH levels of the brine. Green-ripe olives take longer to process than naturally black-ripe olives. As mentioned earlier, processing time can be reduced, particularly with green-ripe olives, by damaging the skin.

Advantages of the brine method are its simplicity and low water requirements compared with processing olives with lye (Spanish-style green olives). Water is only required for the initial rinsing step and for preparing fermentation brines. Washing water has only low levels of low-grade contaminants and so can be disposed of through grey water systems or standard sewerage systems or used for irrigation purposes. Water is further conserved if the fermentation brine is filtered and reused as packing brine. The major disadvantage with methods using spontaneous fermentation in brine is the relatively long processing times. However, once an annual cycle is established, a continuous supply of processed table olives is ensured.

Bulk product of naturally black-ripe olives in brine. After processing, indicative parameters of the brine are: pH 4.5–4.8; titratable (free) acidity as lactic acid, 0.1 to 0.6% w/v; and a final sodium chloride concentration of 10% w/v. If salt levels are lower, then more salt is added. If it is desired that salt levels are lower, then the brine needs to be acidified to pH 4.0–4.2. A small quantity of residual sugar is present.

After processing, the olives can be packed in the fermentation brine, in new brine or a combination of the two. Fermentation brines should be clarified by filtration and pasteurised before reuse. Naturally black-ripe olives processed by this method give the traditional Greek-style black olive. Adding red wine vinegar and olive oil to Greek-style black olives gives a product similar to Kalamata-style olives.

Packing specifications. The packing solution for olives produced by this method for sale to consumers should contain at least 6% w/v salt, have a minimum acidity of 0.3% calculated as lactic acid and a pH of 4.3 or less to ensure safety and preservation. The same conditions apply if a preservative is included or they are to be stored under refrigeration. Alternatively, if the olives are to be pasteurised or sterilised then the packing solution should have a pH of 4.3 or less. Salt and acid levels are not specified in this case, but should be determined by Good Manufacturing Practice, so as not to compromise the safety and organoleptic qualities of the olives. With products that have parameters outside of these levels, the processor has to guarantee the safety of the olives through the use of appropriate testing methods. Typically, packaged olives produced by the brine method have the following parameters: pH 3.6–4.5; titratable (free) acidity, 0.3–1.0% as lactic acid w/v, and sodium chloride 8–10% w/v.

Note: To ensure preservation in containers, the product can be pasteurised or a preservative added, for example sodium or potassium sorbate, to give a final equilibrium level equivalent to 0.05% sorbic acid. Sorbic acid, a preservative, prevents the formation of a layer of oxidative yeast at the air-brine interface a few days after olive containers are opened. Ascorbic acid can be added to cover solutions/brines for naturally processed green-ripe and turning olives to prevent discolouration on storage.

Gas pockets. When olives are processed by the traditional anaerobic fermentation method they can develop gas pockets ('fish eyes') because of the evolution of copious amounts of carbon dioxide by Gram negative bacteria, some of which probably colonise the flesh, particularly in naturally black-ripe olives, in the early stages of the process. Other microorganisms, including yeasts, have been implicated in gas pocket formation in olives. Gas collects in pockets under the skin and/or in the flesh, resulting in soft, wrinkled olives. Using contaminated water and processing under unhygienic conditions increases the likelihood of gas pocket defects. In addition to using potable quality water and undertaking processing under hygienic conditions, this problem can be prevented in two ways:

(a) Acidify the brine to pH 4.5 with acetic acid at the beginning of processing, hence bypassing the early gas producing phase.
(b) Undertake fermentation under aerobic conditions.

Aerobic fermentation of untreated olives. With aerobic fermentation of olives, the fermenter is fitted with a central column, through which air is passed through the fermentation brine. The brine (10–11% w/v salt) is aerated by bubbling air (0.1–0.3 volumes per fermenter per hour) into the fermentation tank for eight hours per day during active fermentation to remove carbon dioxide gas produced by the olives and microbial activity. The pH is maintained between 4.0–4.5 with acetic acid to prevent the growth of spoilage bacteria. As some oxygen from the air passing into the brine dissolves and residual carbon dioxide is expelled, aerobic conditions are established. Under aerobic conditions, facultative anaerobic organisms (those that can adapt to aerobic conditions) and oxidative organisms, rather than fermentative organisms, predominate. These include Gram negative bacteria (Enterobacteriacae) and yeasts that are present throughout the fermentation. (Facultative yeasts include: *Torulopsis candida, Debaryomices hansenii, Hansenula anomala* and *Candida diddensii*. Oxidative yeasts include: *Pichia membranaefaciens, Hansenula mraki* and *Candida bodinii*.) When the salt levels of the brine fall to 8% w/v salt or less, the lactic acid bacteria *Leuconostoc* and *Pediococcus* predominate, whereas three weeks later, *Lactobacillus* bacteria prevail. Implementing aerobic conditions speeds up processing. This is attributed to brine circulation and faster effective diffusion of fermentation substrates and oleuropein through the olive skin. Processing times can be halved by this method. A disadvantage of aerobic fermentation, however, is that enhanced growth of yeasts can occur, resulting in soft olives. (For more details refer to Garrido Fernández *et al.* 1997.)

Green-ripe olives processed in brine

Green-ripe olives processed by the anaerobic fermentation method give products similar to Greek-style green or Sicilian-style olives (Fig. 5.7). Adding herbs, spices and aromatics – for example lemon, fennel, garlic and oregano, or mixed herbs, mustard seeds and chilli – further enhances flavours.

Note: Another type of Sicilian olive, Castelvetrano-style, debittered with lye, is discussed later.

Turning colour olives processed in brine

Turning colour olives of the varieties *Taggiasca* or *Frantoio* processed by the anaerobic fermentation method give a Ligurian-type olive (Fig. 5.8). Turning colour *Jumbo Kalamata* olives processed by natural fermentation in brine, then packed in brine, olive oil and herbs and spices, is a popular Australian olive product. Large olives such as *Jumbo Kalamata* should be slit or bruised before processing to reduce processing time and the risk of gas pockets spoilage.

Figure 5.7 Green-ripe *Manzanilla* olives processed by spontaneous fermentation in brine.

Figure 5.8 Turning colour *Taggiasca* olives processed as Ligurian-style olives.

Step-by-step method for processing untreated olives in brine under anaerobic conditions

(a) Use/accept only quality raw olives (green-ripe, turning colour, black-ripe).
(b) Store raw olives correctly before processing.
(c) Spray wash olives with potable water.
(d) Size grade and sort raw olives.
(e) Use whole, slit or cracked olives.
(f) Pack olives into barrels/tanks with brine (10% w/v sodium chloride in potable water); make sure olives are submerged.
(g) Fill containers/tanks to brim and loosely seal.
(h) Monitor brine pH daily and salt weekly (pH of brine will fall from around 6–7 (initial) to pH 5 in the first few days).
(i) Replace lost brine so that there is as little air space as possible between surface and lid.
(j) Seal the barrel/tank so that anaerobic conditions are established. The brine pH will fall progressively to between pH 4 and 4.6 by about 40 days from the start of processing.
(k) Monitor brine pH levels weekly and salt levels monthly.
(l) If the brine pH does not fall, add sufficient food grade acid, for example lactic acid or acetic acid, to give pH 5.

(m) Monitor brine sodium chloride levels to achieve final levels of 6–7% w/v and pH 4.0–4.5. Olives are ready when the bitterness is acceptable. Approximate processing times: black, three months; turning colour, six months; and green, 12 months.
(n) Size grade and sort processed olives (optional).
(o) Pack the processed olives in a brine made up with 3 parts of 10% w/v sodium chloride + 1 part of vinegar (the final solution has a salt level of approximately 7% w/v salt).
(p) Check pH and salt content (see below for limits).
(q) Pasteurise the packed processed olives if required.
(r) Send samples to a microbiology laboratory for testing.
(s) Label packed processed olives in accordance with food standard requirements.
(t) Implement a safety recall system for tracking faulty, contaminated or incorrectly labelled olives.

Darkening processed traditional black olives in brines. If naturally black-ripe olives are pale in colour after processing by anaerobic fermentation, they can be darkened by one of two methods:

(a) Pass compressed air through the olive mass in the fermentation tank.
(b) Pack the processed black olives into small perforated crates (20–25 kg) or trays, leaving them exposed to air under hygienic and dust-free conditions for up to 48 hours.

Nature of the final product. This depends on the state of maturation and fermentation conditions. Green olives are grey-green in colour, firm, crisp, salty, with a distinct but acceptable bitter taste. Turning colour olives are buff coloured, less firm but crisp, salty and have a slightly bitter, nutty flavour. Black olives have variable colours from pale to dark purple, are salty, slightly sweet, mildly bitter, firm, but with a succulent texture. Brine-cured olives appear moist when removed from brine, and at the time of serving the addition of a small amount of olive oil enhances appearance and flavour. The organoleptic profiles of naturally fermented untreated olives are presented in a graphic form below (Figs 5.9–5.13).

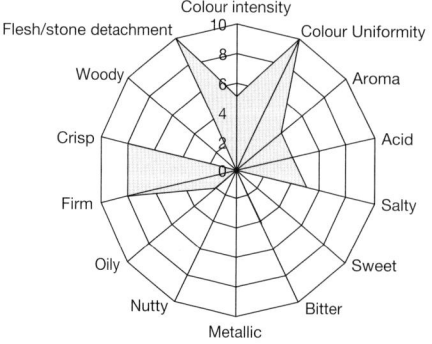

Figure 5.9 Organoleptic profile of naturally fermented (anaerobic) untreated green-ripe *Manzanilla* olives.

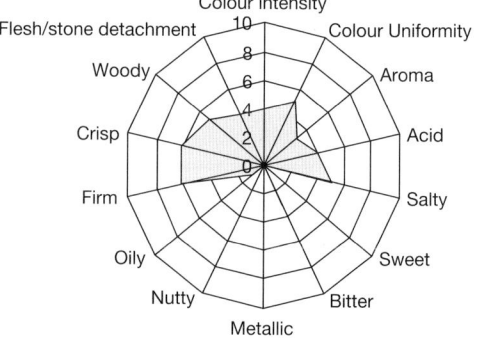

Figure 5.10 Organoleptic profile of naturally fermented (anaerobic) untreated turning colour *Jumbo Kalamata* olives.

Specific table olive processing methods

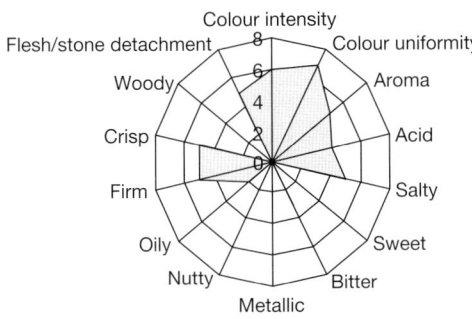

Figure 5.11 Organoleptic profile of naturally fermented (anaerobic) untreated black-ripe *Volos* (*Conservolea*) olives.

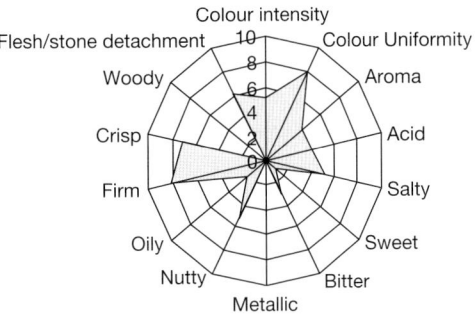

Figure 5.12 Organoleptic profile of naturally fermented (anaerobic) untreated turning colour *Kalamata* olives.

Bruised/cracked olives processed in brine

Method 1: Bruise olives before processing. Green-ripe or early turning colour olives (Plate 24) are given two slits (optional) then passed through a bruising machine. Machine harvested or marked olives can be used to prepare bruised olives if processing is undertaken immediately after harvesting. The olives are placed in 10% w/v salt brine where they undergo a weak fermentation. When the fermentation is over, the olives can be packed in filtered, pasteurised fermentation brine. To speed up the debittering process the bruised olives can be immersed in potable water for 10–12 days, changing the water every day or so. If the water is not changed, the risk of spoilage increases. Bruised green and turning colour olives discolour quickly because of the oxidation of polyphenols. They must be submerged in water or brine to prevent such discolouration. For small-scale processing, olives can be bruised by hitting them with a blunt instrument.

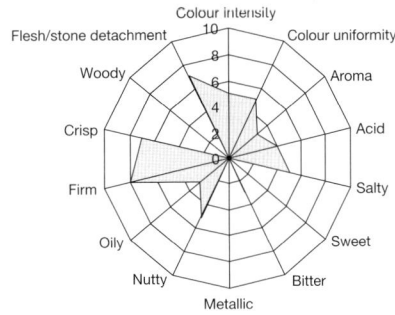

Figure 5.13 Organoleptic profile of naturally fermented (anaerobic) untreated turning colour *Frantoio* olives.

Method 2: Bruise olives during processing. If debittering whole green-ripe or turning colour olives is slow, they can be passed through a 'bruising machine' after 50 days of brining and then repacked in brine to continue processing. Bruising crushes the flesh while the stone is allowed to remain intact but still in contact with the flesh. *Verdale* olives are very suitable for this style of olive preparation. Care must be taken not to contaminate the olives or brine during transfer of olives and brine.

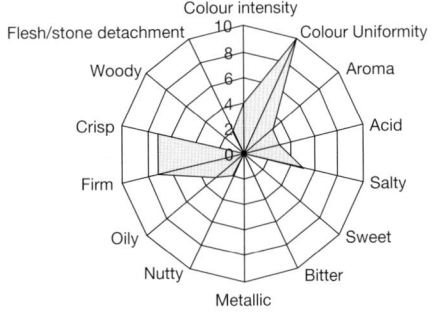

Figure 5.14 Organoleptic profile of naturally fermented (anaerobic) cracked untreated green *Verdale* olives.

Method 3: Bruise olives after processing. Any type of processed green or turning colour olive can be bruised by tapping with a blunt object or passing them through a 'bruising machine'. These bruised olives show no discolouration because the phenol oxidase enzymes that oxidise polyphenols in raw olive flesh are inactivated during processing. Care must be taken not to contaminate the olives during the bruising operation.

Nature of the final product. The final product is green to buff in colour, with salty, slightly bitter taste, crisp in texture with nutty flavours reminiscent of the olive stone and condiments if added. Of course the quality is very much influenced by the characteristics of the original processed olives. Aromatics, herbs and spices, such as oregano, fennel, coriander, chilli and lemon slices are added to extend the flavour of the olives. The organoleptic profile of naturally fermented cracked untreated *Verdale* olives is presented in a graphic form in Fig 5.14.

Processing Kalamata-style table olives

Kalamata-style olives are recognised all over the world as a high quality table olive. The processing method is a variant of that used for naturally black-ripe olives in brine that is commonly practised in Greece and other Mediterranean countries. The raw naturally black-ripe *Kalamata* olives are generally less bitter than other olive varieties.

Short method. 'Original' Kalamata-style olives are prepared from naturally black-ripe *Kalamata* olives, a variety originating from the Greek Peloponnese region, by the traditional short method. It is called the short method because the olives are ready to eat in a much shorter time than the long method. Olive varieties such as *Barnea*, *Leccino*, *Hojiblanca*, *Conservolea* and *Mission* (Californian) can also be processed as Kalamata-style olives when *Kalamata* olives are unavailable. The short method involves debittering the olives by multiple soakings in water or weak brine over a week or so. Once the olives are debittered they are packed in brine of 6–8% w/v sodium chloride with added wine vinegar. A layer of olive oil or other vegetable oil and slices of lemon are added to the container packed with olives. Processing times are reduced by slitting the olives before placing them in the fermentation brine. If the olives are not slit or split before processing, this can be done after processing so the wine vinegar and lemon flavours penetrate into the olive flesh.

Long method. A second method used to prepare Kalamata-style olives involves placing slit naturally black-ripe olives in 10% w/v salt in potable water until they debitter

Figure 5.15 *Kalamata* olives processed as Kalamata-style olives by spontaneous fermentation in brine in plastic barrels. Note there is no mould growth.

Figure 5.16 Slit *Kalamata* olives processed as Kalamata-style olives by spontaneous fermentation in brine.

(Figs 5.15 and 5.16). Depending on the processing facility, Kalamata-style olives can be processed in plastic barrels or large tanks. One system used in South Africa involves placing olives and brine in a squat fibreglass tank (one tonne) then covering the surface with thick plastic, held in place with a heavy fibre-cement lid to ensure anaerobic conditions are maintained. Regardless of the size of tank, processing can take around three months. During this time a weak fermentation takes place. After debittering, the olives are packed in brine, wine vinegar and olive oil as with the short method.

There are local variations in the process. The olives may be soaked in tanks of vinegar rather than adding vinegar. Indicative polyphenol levels remaining in the flesh of *Kalamata* olives after processing, which give the olives a slight bitter taste, range from around 500–1500 mg/kg of flesh with residual hydroxytyrosol levels at around 250–750 mg/kg of flesh.

Traditional short method for processing Kalamata-style olives

(a) Use/accept only quality raw, nearly black-ripe olives.
(b) Store raw olives correctly before processing.
(c) Spray wash olives with potable water.
(d) Size grade and sort raw olives.
(e) Slit olives lengthwise: two to three slits with a sharp knife or machine.
(f) Pack olives into barrels with water or weak brine (2% w/v salt in potable water); make sure olives are submerged.
(g) Remove and replace water or brine two to three times a day for five to eight days.
(h) Taste olives to determine their bitterness.
(i) Monitor brine sodium chloride levels to achieve final levels of 6–7% w/v.
(j) Size grade and sort processed olives (optional).
(k) Darken olives by exposing olives to air or by passing food grade air through the tank.
(l) Pack the processed olives in brine made up with 3 parts by volume of 10% w/v sodium chloride + 1 part by volume of red wine vinegar (alternatively with 4 parts by volume of 10% w/v sodium chloride + 1 part by volume of red wine vinegar).
(m) Check pH and salt content (see below for limits).
(n) Add extra virgin olive oil or quality seed oil (10 ml/100 ml of brine) and thin slices of lemon. (The number of slices depends on the size of the container. Half to one slice should suffice in consumer size containers.)
(o) Pasteurise the packed processed olives (if required).
(p) Send representative olive samples to a microbiology laboratory for testing.
(q) Label packed processed olives in accordance with food standard requirements.
(r) Implement a safety recall system for tracking faulty, contaminated or incorrectly labelled olives.

Long method for processing Kalamata-style olives

(a) Use/accept only quality raw, nearly black-ripe olives.
(b) Store raw olives correctly before processing.
(c) Spray wash olives with potable water.
(d) Size grade and sort raw olives.
(e) Slit olives lengthwise: two to three slits with a sharp knife or machine.

(f) Pack olives into barrels/tanks with brine (10% w/v salt in potable water); make sure olives are submerged.
(g) Fill barrels/tanks to brim and loosely seal to allow gases, formed during processing, to escape.
(h) Monitor brine pH daily and salt weekly. The pH of brine will fall initially from around pH 6–7 to pH 5 in the first few days.
(i) Replace lost brine so that there is as little air space as possible between the surface and the lid.
(j) Seal the barrel/tank so that anaerobic conditions are established and maintained.
(k) The brine pH will fall progressively to 4.0–4.5 by about 40 days from the start of processing.
(l) Monitor brine pH levels weekly and salt levels monthly.
(m) If brine pH does not fall, add food grade acid, for example lactic acid or acetic acid.
(n) Monitor brine salt levels to achieve final levels of 6–7% w/v and pH 4.0–4.6. Olives are ready when the level of bitterness is acceptable. This occurs within three months.
(o) Size grade and sort processed olives (optional).
(p) Pack the processed olives in brine made up with 3 parts by volume of 10% w/v sodium chloride + 1 part by volume of red wine vinegar (alternatively with 4 parts by volume of 10% w/v sodium chloride + 1 part by volume of red wine vinegar).
(q) Check brine pH and salt levels (see below).
(r) Add olive oil or seed oil to give a thin layer at the surface, for example 10 ml/100 ml of packing solution.
(s) Pasteurise the packed processed olives if required.
(t) Send samples to a microbiology laboratory for testing.
(u) Label packed processed olives in accordance with food standard requirements.
(v) Implement a safety recall system for tracking faulty, contaminated or incorrectly labelled olives.

Packing specifications. The packing solution for olives produced by this method should contain at least 6% w/v salt, have a minimum acidity of 0.3% calculated as lactic acid and a pH of 4.3 or less to ensure safety and preservation. The same conditions apply if a preservative is included or they are to be stored under refrigeration. Alternatively, if the olives are to be pasteurised or sterilised then the packing solution should have a pH of 4.3 or less. Salt and acid levels are not specified in this case, but should (determined by Good Manufacturing Practice) not compromise the safety and organoleptic qualities of the olives. Outside of these parameters, the processor has to guarantee the safety of the olives through the use of appropriate testing methods.

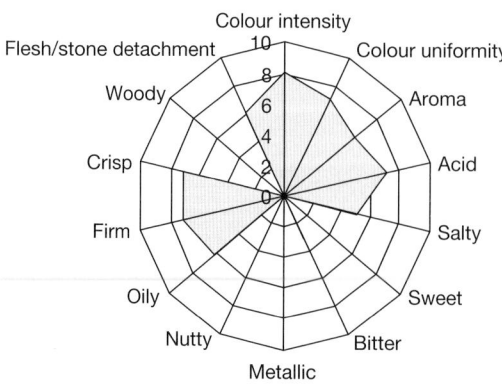

Figure 5.17 Organoleptic profile of Kalamata-style black-ripe *Kalamata* olives (olive oil and vinegar added).

Nature of final product. Kalamata olives have a homogeneous but variable chocolate to dark brown colour. The olives are firm and crisp with a distinct fruity taste that is also salty, slightly bitter and wine-like. When served on their own or in salads they have a distinct glistening appearance because of the added olive oil. Added condiments give extra flavour. The amount of wine vinegar can be reduced if a more subtle vinegar flavour is desired as long as the packing brine parameters are achieved. Unpleasant lemon flavour tones occur if too many lemon slices are added. The organoleptic profile of *Kalamata* variety olives processed as Kalamata-style olives is presented in Fig. 5.17. Note the high oil and acid levels compared with other organoleptic graphics, which are due to the added vinegar and olive oil.

Processing olives with lye (sodium hydroxide)

Spanish-style green olives. The commonest table olive produced by this method is the Sevillian-style, also known as Spanish-style green olives. Washed olives, generally green-ripe, are placed into tanks and soaked in a lye solution (1.3–2.6% w/v food grade sodium hydroxide in potable water) for five to seven hours or more to debitter. The concentration of lye used depends on the variety and the temperature of the operation. With higher processing temperatures the rate of lye penetration through the flesh is faster. When weak lye solutions are used, penetration through the flesh is slow. Under cold conditions, lye concentrations of up to 3.5% w/v need to be used. As some varieties are sensitive to high lye concentrations, for example *Ascolana Tenera* and *Sevillana*, then weaker solutions are used over a 9–10 hour period. Lye treatment is followed by a lactic fermentation to preserve the olives and produce the characteristic organoleptic features of this style.

Lye treatment. The lye is allowed to penetrate through three-quarters of the flesh, leaving a small volume of flesh around the stone unaffected. This part of the flesh provides the necessary sugars for subsequent fermentation and provides a slight bitter taste to the olives. Lye treatment debitters the olives by chemically converting the bitter oleuropein to non-bitter and less bitter chemical compounds including hydroxytyrosol and elenolic acid. Hydroxytyrosol diffuses rapidly from the flesh into the brine solution during fermentation where its concentration remains essentially unchanged during fermentation. Lye treatment also increases the permeability of the olive skin allowing a two-way interchange of soluble substances, especially fermentable substrates.

$$\text{Oleuropein} + \text{lye (sodium hydroxide)} \rightarrow \text{Hydroxytyrosol} + \text{elenolic acid}$$

When olives are pre-treated with lye, the level of lye penetration can be monitored by slicing the olives and observing colour changes of the olive flesh. The degree of lye penetration can be more easily visualised by placing several drops of phenolphthalein solution onto the cut flesh, changing its colour from green-brown to red (Fig. 5.18). Low concentrations of lye give the olives a less desirable green colour whereas high lye

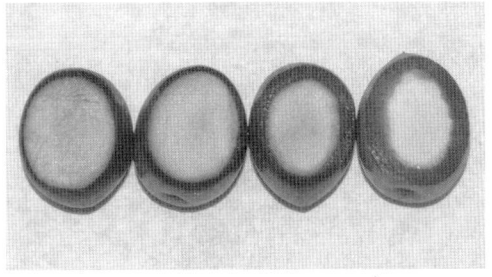

Figure 5.18 Penetration of lye into the flesh of green-ripe olives over five to six hours.

concentrations leach out water-soluble fermentable substrates and soften the olives because of textural changes in the flesh.

Lye treatment is followed by several rinse/wash cycles with potable water to remove excess lye:

Rinse. Drain lye solution from olives, add potable water to cover olives in barrels/tanks and drain immediately.

First wash. Add water to olives in barrels/tanks and drain after two hours.

Second wash. Add water to olives in barrels/tanks and drain after 10–20 hours.

In some centres, a first wash is eliminated for convenience and environmental considerations. The implications of this are high levels of lye in the wash water and olives. Partial neutralisation with food grade hydrochloric acid overcomes this problem.

Note: Over-washing results in loss of sugars required for subsequent fermentation, increasing the possibility of 'stuck fermentations' where microbiological acidification does not occur. If the wash periods are excessively long, the risk of bacterial contamination increases.

Brining the lye treated olives. Tanks with the lye treated olives are then drained of the last washing water and filled with brine (10–12% w/v salt in potable water), depending on the variety and maturation state of the fruit. As soluble substrates leave the flesh and accumulate in the brine, a culture medium suitable for fermentation develops. Furthermore, hydrolysis by-products of oleuropein, such as hydroxytyrosol and elenolic acid, pass into the brine. Indicative levels of phenols remaining in the flesh of *Conservolea* and *Chalchidikis* after processing by this method (based on research by others) range from around 150–550 mg/kg and 400–1200 mg/kg respectively. Residual hydroxytyrosol levels in flesh are around 150–500 mg/kg.

Table 5.6. Summary of events during the fermentation stage of Spanish-style green olives

Parameter	Early			Late
Brine pH	Initial pH 7	pH 6	pH 5	Falls to less than pH 4
Brine salt levels	10% w/v, which falls over six to eight weeks to 6%			
Gram negative organisms	Low	Maximum numbers by two days → disappear after 12–14 days Gas production		
Lactic acid bacteria *Pediococcus* *Leuconostoc*		Increase	Decrease with decreasing pH	
Yeasts		Increase	Yeasts persist	
Lactobacillus pentosus (*Lactobacillus plantarum*)			Maximum numbers by 7–10 days	Numbers diminish in 60–300 days

Although the initial brine has 10–12% w/v salt, this falls rapidly to around 6% w/v salt, depending on the variety being used and the Flesh:Stone ratio because of interchangeable water in the olives. When anaerobic conditions are established, lactic fermentation proceeds. The processed table olives can be kept in the fermentation brine until they are used or repacked for sale. Brine levels of bulk product should be maintained at 8% w/v salt or more to ensure effective preservation by topping up with concentrated brine.

Fermentation. A number of microbiological events occur in the brine during fermentation (Table 5.6). A starter culture of the fermentation organisms (Fig. 5.19) may need to be added as lye can destroy natural microflora on the olive skin. In long established processing facilities this may not be necessary, because fermentative microorganisms are widespread in the immediate environment. The salt levels used must not inhibit the growth of *Lactobacilli* but still inhibit the growth of spore-forming clostridial organisms when the pH is still high in the early stages of processing.

Early in the procedure Gram negative bacteria multiply. These organisms are found on the olives, in water and in the environment of the processing plant. These Gram negative organisms, species of the genera *Enterobacter, Citrobacter, Klebsiella, Flavobacteria, Aerochromobacter, Escherichia* and *Aeromonas*, require little in the way of nutrition. They reach maximum numbers two days after brining, producing copious amounts of carbon dioxide gas that lowers the brine pH from 7 to 6. Other microorganisms can also produce carbon dioxide.

Secondly, at pH 6 the lactic acid producing bacteria *Streptococcus* and *Leuconostoc* spp., also present in the brine, multiply and produce acid metabolites, further lowering the brine pH to around 5. Under these conditions Gram negative organisms disappear and *Lactobacillus pentosus* (formerly *Lactobacillus plantarum*) activity increases, resulting in lactic acid production from fermentable sugars such as glucose. Brine pH levels fall to

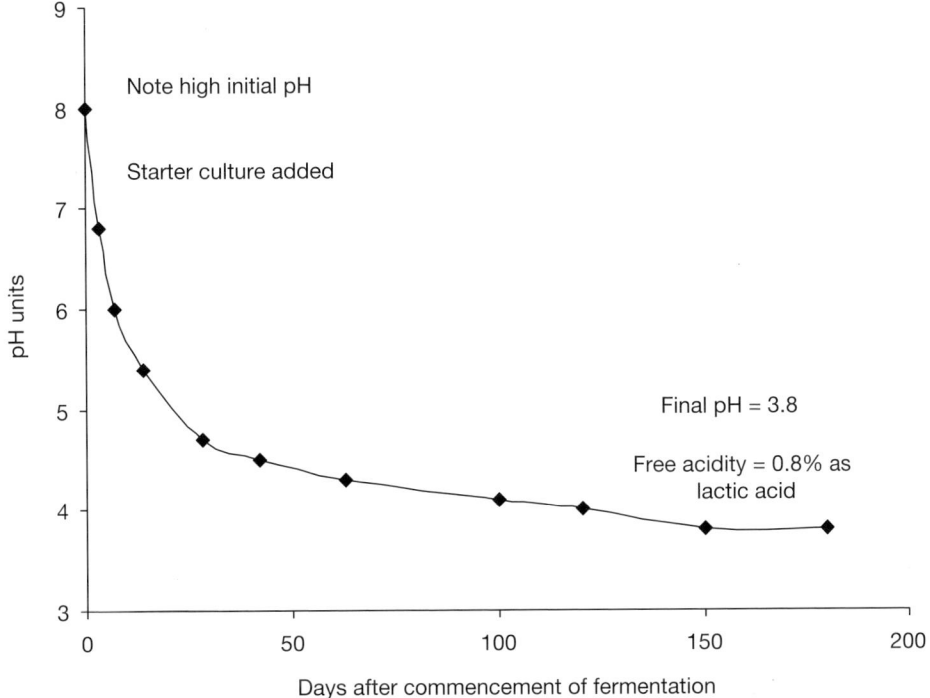

Figure 5.19 Typical changes in brine pH after the addition of a starter culture to lye treated green-ripe olives undergoing anaerobic fermentation.

less than 4. The maximum numbers of lactic acid bacteria are present around 7–10 days after brining, and then they gradually decline in number over 60–300 days.

In the third phase, the fermentable substrates from the olives are depleted and acid formation ceases. At this stage yeasts appear with the lactic acid bacteria. Fermentative yeasts can improve the organoleptic qualities of the olives but do not cause spoilage. These yeasts can include species of the genera *Hansenula*, *Candida* and *Saccharomyces*. However, oxidant yeasts are undesirable because they can consume lactic acid, thereby raising the brine pH and compromising the process.

Note: Lye treatment kills the natural microflora on the olive fruit and it may be necessary to introduce starter cultures of fermentative organisms to the olives in brine to establish fermentation. Sources of organisms for starting new fermentations come from the immediate environment in long established processing plants, brine from active fermentation tanks or starter cultures. Adding cultures of lactic acid bacteria to the lye treated olives and brine 24 hours after brining markedly reduces the risk of spoilage by swamping natural populations of spoilage organisms (coliforms) and rapidly reducing the brine pH. If coliform growth is not controlled in the early stages of brining, the risk of gas pocket formation increases and faecal odours can develop in the olives. The latter problem renders the olives inedible. As mentioned earlier, lowering the brine pH with food grade acids can also reduce the problem of gas pockets.

The level of coliforms in brine can be determined visually using a microscope. Using a Gram stain, Gram positive *Lactobacilli* stain blue and Gram negative coliforms are red. By using this simple test, early olive spoilage during processing can be prevented.

Green olives prepared by the Spanish-style method are ready to eat within four to five weeks, although a longer period of standing (up to three months) is recommended so that the olives equilibrate in the fermentation brine to enhance the organoleptic characteristics of the olives.

The main disadvantage of this method is the large amount of water required and the amount of waste-water produced. Also, significant technical skills, particularly in chemistry, are required to process olives with lye. The processed olives can be packed in the fermentation brine, new brine or a combination of both. When packing brines with low levels of salt are used, the packaged products need to be pasteurised.

Characteristics of lactic acid bacteria.   As lactic acid bacteria are the predominant fermentative organisms when preparing Spanish-style green olives, it is of value to understand the relevant features of this group of organisms. They grow under anaerobic conditions; multiply in water and brine up to 8% w/v sodium chloride (inhibited above 8% w/v sodium chloride); are active in slightly acid conditions and until the pH drops to 3.5–3.8 and a (free) acidity of 1.2% w/v lactic acid or more is reached; and are active between 15°C and 27°C but are inactive below 15°C and above 30°C (therefore they are classified as a mesophilic-type organism).

Salt is used to inhibit undesirable microorganisms. The desirable lactic acid bacteria are reasonably salt tolerant (8% or less) whereas coliforms and other spoilage organisms are not.

Processing tips for treating green-ripe olives with lye

(a) Prepare lye solutions well in advance and allow to cool to around 15–20°C before adding to olives. When the sodium hydroxide dissolves in water, heat is produced and excessively hot lye solutions can cause skin blistering and sloughing of olives.
(b) Keep olives submerged during lye treatment to prevent skin discolouration due to oxidation of polyphenols. After processing in lye the olives sink in the liquid because of an increase in their relative density.
(c) Maintain brine temperature between 15°C and 25°C (<30°C) by air conditioning or by pumping the brine through a suitable heat exchanger. The main organisms active in brine during fermentation are lactic acid bacteria.
(d) Check total titratable (free) acidity and pH regularly; the acidity should increase and the pH should fall.
(e) Ensure enough salt and acidity is present in the brine to preserve olives of this type. Check salt. Without adjustment this falls to around 5–7% w/v depending on the initial salt concentration in the brine. Salt levels in the bulk product are adjusted to 8% w/v or more by adding a saturated solution of salt or dry salt.
(f) Replace part of the brine with a fresh salt solution and add lactic acid if the acidity needs to be adjusted.
(g) Final total (free) acidity should be 0.8–1.0% w/v calculated as lactic acid in the bulk product.
(h) Allow fermented olives to remain in the fermentation brine for one to two months to allow equilibration, stabilisation and development of the organoleptic characteristics of the olives.

It is possible to use other alkaline compounds, such as potassium hydroxide or lime, to prepare the lye solution; however, the resulting products may have a residual chemical bitterness. Although some traditional olive processing methods in the past have used wood ash to prepare lye solutions, such practice is of little importance in commercial table olive production as processing would be difficult and expensive to control. Home processors using wood ash to prepare lye solutions must ensure that the wood is not contaminated with chemicals, paints or toxins.

Step-by-step method for processing green-ripe olives with lye

(a) Use/accept only quality raw olives (green-ripe). Store raw olives correctly before processing.
(b) Spray wash olives with potable water (optional).
(c) Size grade and sort raw olives. Use whole olives.
(d) Pack olives into barrels/tanks with lye (1.3–2.6% w/v sodium hydroxide in potable water depending on variety, lye concentration and environmental temperature). Make sure olives are submerged.
(e) Allow lye treatment to proceed for five to seven hours or more.
(f) Monitor penetration of lye into olive flesh using phenolphthalein indicator.
(g) Drain solution from the olives. Rinse olives immediately with potable water, then with two static washes: one for two to three hours and a second for 10–20 hours.
(h) Add brine (10–12% w/v salt) to the brim of the barrel/tank and loosely seal.

(i) Check brine pH.
(j) Acidify the brine to pH 6.2–6.5 by adding food acids (lactic acid, acetic acid or hydrochloric acid) or pass carbon dioxide gas through the tank/container.
(k) Check sugar levels in the brine. Add dextrose if required to give brine concentrations of 1–2% w/v.
(l) Add starter culture from actively fermenting lots or commercial lactic acid bacteria culture (Vege-Start™) (Fig. 5.19).
(m) Monitor brine pH levels daily and salt levels weekly. The pH of the brine will fall from around 6–7 to pH 5 in the first few days. (Note the initial pH of the brine may be 8 or more depending on the effectiveness of the washing steps after lye treatment.)
(n) Check free acidity in the brine.
(o) Replace lost brine, ensuring minimal air space between the surface of the liquid and lid of the container.
(p) Seal the barrel/tank so that anaerobic conditions are established. The brine pH will fall progressively to between pH 4.0–4.5 in about 40 days from the start of processing.
(q) Monitor brine pH weekly and salt monthly. If pH does not fall this could indicate a 'stuck' fermentation. Check dextrose and/or starter culture and food acid (see below).
(r) Monitor brine sodium chloride and pH levels to achieve final levels of around 8% w/v and pH 4.0–4.2 or less. Add salt if required. Olives are ready when the bitterness is acceptable; the approximate time for this to occur is six weeks.
(s) Remove discoloured olives by hand or a photosensitive machine.
(t) Size grade and sort processed olives (optional) with a divergent cable machine.
(u) Pack the processed olives in brine.
(v) Pasteurise the packed processed olives if required.
(w) Send representative samples to a microbiology laboratory for testing.
(x) Label packed processed olives in accordance with food standard requirements.
(y) Implement a safety recall system for tracking faulty, contaminated or incorrectly labelled olives.

Potential problems: stuck fermentations and corrective actions
- Olive brine pH is too high, incomplete washing of lye treated olives: wash olives and add food grade acid.
- Salt levels in the brine are too high, inhibiting lactic acid bacteria: reduce salt levels.
- Insufficient sugar: add dextrose to the brine to give a concentration of 1–2% w/v (10–20 g/l).
- Brine pH is too high: add food grade acid to lower pH to around 5.
- Lack of fermentation microorganisms: add a starter culture of lactic acid bacteria.

Stuck fermentations and their management
If the fermentation becomes 'stuck' (nothing is happening), the cause should be determined and corrected.

Evidence for stuck fermentation includes no increase in total acid or a fall in pH. Stuck fermentations can be prevented. A simple approach is to exchange brine with that from a normally fermenting container of the same size, so that each container finishes up with 50% of the brine from the other container. Fermentation should continue or proceed in both containers.

Alternatively, the following parameters should be assessed and corrected:

- Check pH. If it is too high, fermentative organisms are inactive. Add food acid or pass carbon dioxide gas through the container/tank.
- Check the temperature of the brine and adjust temperature if necessary.
- Measure soluble sugars in the brine and add dextrose in two portions, a week apart, to give a titratable (free) acidity of at least 0.8% w/v of lactic acid by fermentation.
- Check levels of lactic acid bacteria. If necessary, add pure starter culture or brine from an active fermentation tank (Fig. 5.20).
- If yeasts predominate (check microscopically), discard brine and add fresh brine with a starter culture.
- If total (free) acid is less than 1% w/v of lactic acid, add lactic acid to raise levels to 1%; pH levels should be between pH 3.7 and 4.0 (Fig. 5.20).
- Check salt levels in brine and raise these to 8% w/v if necessary. If environmental temperatures are high (>30°C) then the salt levels are raised to at least 8.5% w/v to prevent *Zapateria* spoilage by microorganisms such as *Propionibacteria* (Fig. 5.21).

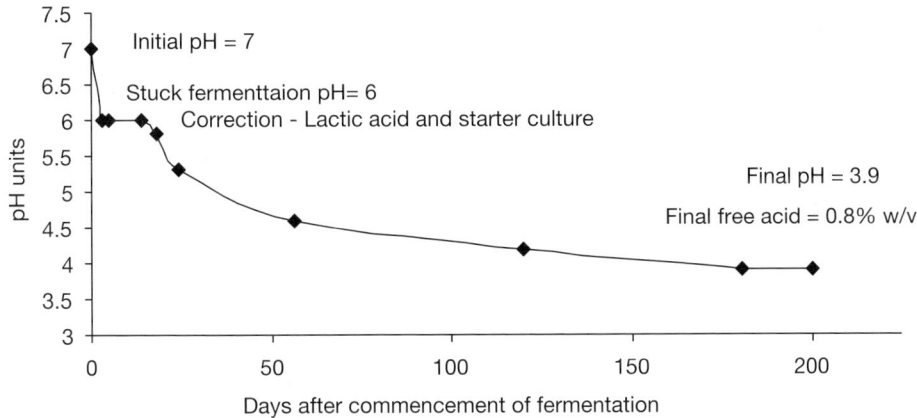

Figure 5.20 Typical changes in brine pH of a 'stuck' fermentation of lye treated raw green-ripe olives before and after correction with lactic acid and starter culture.

Bulk fermented product. The final bulk fermented product has typical brine parameters: pH 3.8–4.2; free acidity 0.8–1.2% w/v as lactic acid; combined acidity 0.09–0.11M; sodium chloride 7–8% w/v. If the pH is kept below 3.5, then a lower salt level of at least 5% w/v can be used.

Packing specifications. The packing solution for olives produced by this method should contain at least 5% w/v salt, have a minimum acidity of 0.5% calculated as lactic acid and a pH of 4.0 or less to ensure safety and preservation. When a preservative is used or the olives are to be refrigerated, the packing solution should contain at least 4% w/v salt, have a minimum acidity of 0.4% calculated as lactic acid and a pH of 4.0 or less to ensure safety and preservation. Alternatively, if the olives are to be pasteurised or sterilised then the packing solution should have a pH of 4.3 or less. Salt and acid levels are not

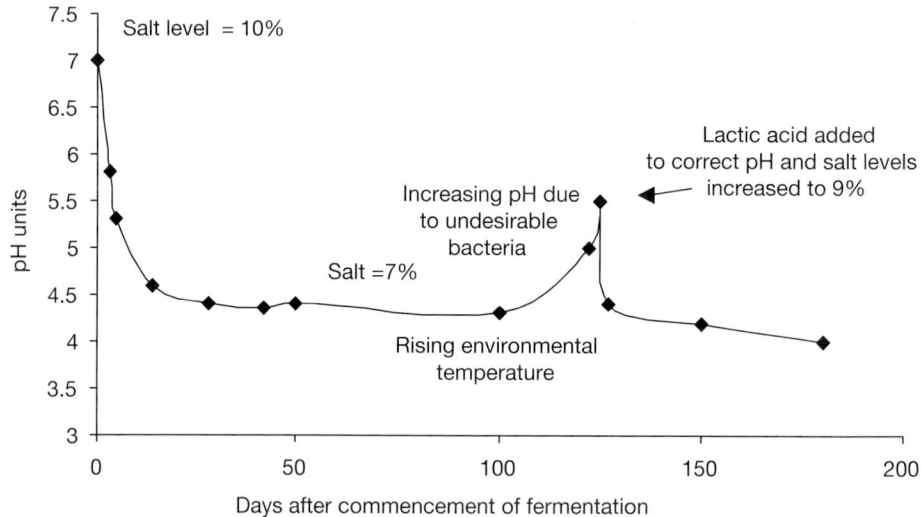

Figure 5.21 Possible pattern for undesirable bacterial activity during brine fermentation of table olives causing an increase in brine pH.

specified in this case, but levels should not compromise the safety and organoleptic qualities of the olives determined by Good Manufacturing Practice. With products that have characteristics outside of these parameters, the processor has to guarantee the safety of the olives through the use of appropriate testing procedures.

Packed products produced by this method have typical brine parameters: pH 3.2–4.1; titratable (free) acidity 0.4–0.6% w/v as lactic acid; combined acidity 0.02–0.07M; and sodium chloride 5–7% w/v. The salt levels can be reduced to 2% w/v sodium chloride if the packed products are pasteurised. Ascorbic acid can be added to cover solutions/brines for lye processed green-ripe olives to prevent their discolouration on storage.

New alternative directions to lye treatments. Research undertaken in Italy and elsewhere is being directed towards eliminating the use of lye as a debittering agent. One procedure is to use a specific microorganism such as the yeast *Candida veronae*, which has oleuropein hydrolysing properties. This treatment facilitates subsequent lactic fermentation by *Lactobacillus pentosus* (formerly *Lactobacillus plantarum*). An alternative procedure is to use *Lactobacillus pentosus* strains isolated from brine of naturally ripe olives. These have beta-glucosidase activity that can split oleuropein into an intermediate aglycone and then through an esterase to hydroxytyrosol and elenolic acid. Using such procedures, debittering occurs after three weeks.

Debittering olives by microbial breakdown of oleuropein:

β-glycosidase esterase + oleuropein → oleuropein aglycone hydroxytyrosol + elenolic acid

glucose → food acids

Another problem that faces processors, particularly at sites with low environmental temperatures, is poor fermentation rates and prolonged processing times. To overcome this, researchers are investigating the use of *Lactobacillus pentosus* (formerly *Lactobacillus plantarum*) strains isolated from cold fermentation brines as starter cultures to process green table olives at 9–15°C.

Nature of the final product. Spanish-style olives are an even olive green colour, firm to touch, crisp with a mild salty taste, with little bitterness and a slightly acidic flavour. The organoleptic profile of green *Sevillana* olives processed by the Spanish-style method is presented in Fig. 5.22.

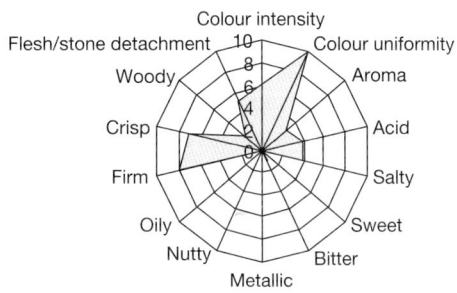

Figure 5.22 Organoleptic profile of green *Sevillana* olives processed by the Spanish-style method.

Lye treated green-ripe olives without fermentation

Different procedures for processing green-ripe olives with lye, but without a fermentation step, are used at the local or regional levels in traditional olive growing countries. Some of the more popular ones are presented below.

Spanish-style green 'look alikes'. Initially the process is the same as producing Spanish-style green olives. After the excess lye is removed, the washed olives are placed in 5–6% salt brine for two days. The olives are then drained and placed in salt brine (10.5–11.5% w/v salt) with food acid (hydrochloric acid, lactic acid or citric acid) to lower the pH to around 4. The olives are packed in brine with a final salt content of 7% w/v and a pH of 4.5. With salt concentrations of less than 2%, the pH of the brine should be 4 or less. The final product is pasteurised.

Step-by-step procedure for lye treated green-ripe olives without fermentation
(a) Use/accept only quality raw green-ripe olives.
(b) Store raw olives correctly before processing (optional).
(c) Spray wash olives with potable water.
(d) Size grade and sort raw olives.
(e) Use whole olives.
(f) Pack olives into barrels/tanks with lye (1.3–2.0% w/v sodium hydroxide in potable water); make sure olives are submerged.
(g) Allow lye treatment to proceed for 5–10 hours.
(h) Monitor penetration of lye into olive flesh using phenolphthalein indicator.
(i) Drain lye solution from the olives.
(j) Wash olives immediately with potable water then two static washes: one for two to three hours and a second for 10–20 hours.
(k) Add brine (10.5 to 11.5% w/v salt) to the olives in the barrels/tanks and food acid to lower the pH to 4.
(l) Size grade and sort processed olives (optional).
(m) Pack the processed olives in a brine (7% w/v salt) and pH 4 or less.
(n) Pasteurise the packed processed olives.
(o) Send representative samples to a microbiology laboratory for testing.
(p) Label packed processed olives in accordance with food standard requirements.
(q) Implement a safety recall system for tracking faulty, contaminated or incorrectly labelled olives.

Nature of the final product. Lye treated olives in brine without fermentation are an even green colour, firm to touch, crisp with a salty taste, a little bitterness and an acidic (lactic/citric) flavour. Some consumers comment on the sweetness of this type of table olive, which may be due to the presence of sugars that would have been utilised if a fermentation step was involved.

Picholine-style table olives (Olive de Nimes). Green-ripe *Picholine* olives from southern France (Languedoc and Luques) are processed in lye then packed in brine. Similar olives are prepared in Algeria and Morocco from other varieties and specifically after a lye treatment. Similar to Spanish-style green olives, the olives are placed in a brine of 5–6% w/v salt for two days, during which the flesh absorbs salt. The brine is then drained and the olives packed in a brine of 7% w/v salt brine and enough citric acid added to give pH 4.5. A concentrated aqueous solution of citric acid can be used and the final product can have maximum citric acid concentration of 1.5% w/w. The olives, which have an intense green colour, are ready to eat after 8–10 days. For long-term storage of bulk product, depending on the environmental temperature, brine with 8–10% w/v salt is required. The bulk product can be stored in a 3% brine solution under refrigeration at 3–7°C. When the final product is packed in consumer size containers it is pasteurised.

Step-by-step method for *Olive de Nimes*
(a) Use/accept only quality raw green-ripe *Picholine* olives.
(b) Store raw olives correctly before processing.
(c) Spray wash olives with potable water.
(d) Size grade and sort raw olives.
(e) Use whole olives.
(f) Pack olives into barrels/tanks with lye (1–2% w/v sodium hydroxide in potable water); make sure olives are submerged.
(g) Allow lye treatment to proceed for 8–10 hours.
(h) Monitor penetration of lye into olive flesh using phenolphthalein indicator.
(i) Drain lye solution from the olives.
(j) Wash olives immediately with potable water, then two static washes: one for two to three hours and a second for 10–12 hours.
(k) Place the olives in brine of 5–6% w/v salt for two days.
(l) Drain the brine.
(m) Pack the olives in brine of 7% w/v salt brine and citric acid corrected to pH 4.5.
(n) The olives, which are an intense green colour, are ready to eat after 8–10 days.
(o) For long-term storage, place olives in a brine of 8–10% w/v salt and pasteurise.
(p) Send representative samples to a microbiology laboratory for testing.
(q) Label packed processed olives in accordance with food standard requirements.
(r) Implement a safety recall system for tracking faulty, contaminated or incorrectly labelled olives.

Nature of the final product. Picholine olives in brine have an even green colour, are firm to touch, crisp with salty taste, and have a slight bitterness and an acidic (citric) flavour.

Castelvetrano-style olives. Castelvetrano-style olives are popular in Sicily. They are different to the Sicilian-style green olives that undergo a natural fermentation without a pretreatment with lye. With *Castelvetrano* table olives, a local variety, *Nocellara de Belice*, is used. After washing the raw green olives, they are placed in a 2–3% lye solution in a suitable plastic barrel or drum. After one hour, coarse salt (3–4 kg/100 L barrel) is added and agitated to mix and dissolve the salt. The barrel is sealed and further agitated to ensure mixing and to form the processing brine. The olives are ready to eat after two weeks. When the barrel is opened, the lye/salt brine is drained and the olives are washed to remove excess lye. Under ambient storage conditions the *Castelvetrano* olives have a shelf-life of only a few months, especially under hot conditions. Deteriorated olives lose colour and develop malodours. As this olive style is a seasonal product, long-term storage is not required. Future research, however, may refine the process and include steps for long-term storage under refrigeration or in packing solutions. The final product is salty and lacks bitterness.

Californian/Spanish-style black olives

This method/style of olive, inappropriately called 'black-ripe olives', was originally developed in California. It has also been adopted in Spain and some north African table olive producing countries. Fresh green olives or turning colour olives (often *Manzanilla*) are soaked in several caustic soda (lye) solutions of different strengths until the lye penetrates the flesh through to the stone. By using this process the olives debitter quickly. After each lye treatment, the lye is replaced with potable water and air is passed through tanks. The olives turn a brown/black colour through the oxidation/polymerisation of polyphenols (hydroxytyrosol and caffeic acid).

Although consumers often malign this type of olive as: 'the dreaded canned black-ripe olive'; 'they are basically rusted then petrified'; 'horrid little black things', if eaten slowly, not unpleasant subtle spicy flavours can be perceived.

After the last lye treatment, olives are drained and washed several times with water to remove residual lye. If the wash waters are acidified with hydrochloric acid or with carbon dioxide, fewer washes are required. They are then immersed in an iron salt, 0.1% w/v of ferrous gluconate, which stabilises the colour. The olives are washed to remove excess iron, then packed in a 2–3% w/v food grade sodium chloride solution and sterilised. The IOOC Table Olive Standards (2004) stipulates that residual iron in the olives should not exceed 150 mg/kg. No fermentation takes place during this method and the processed olives are preserved by sterilisation.

This method of olive processing is expensive because of the equipment required, the cost of inputs (especially water), and the high level of technical expertise needed. In large establishments, raw olives are processed in large horizontal tanks, for example 10 tonne stainless steel, or polyester and fibreglass tanks, that can be fed through different inlets for water, lye solution, brine and compressed air and can be drained easily during processing. Specifically designed rotating tanks that facilitate even processing are used in some centres. Processing is undertaken at around 20°C. The strengths of lye used depend on the olive variety, maturation state, size and whether they have been previously stored in brine before lye treatment.

In some centres, pre-brined olives are purchased from other sources, often from different regions or even different countries. Pre-brined olives if stored long enough in brine undergo a weak fermentation. Olives that have been previously stored in brine require lower strengths of lye for processing as they are already debittered.

At different olive processing centres, local modifications are made based on environmental considerations, the available equipment, olive varieties, source of olives and levels of technical expertise. Olives processed by this method have much lower residual phenolic levels than those processed as Greek-style naturally black olives, Kalamata-style olives or Spanish-style green olives. A marked decrease in olive flesh sugar (nearly thirty-fold) occurs during processing due to the extensive lye treatments and washing steps.

Step-by-step method for preparing Californian/Spanish-style black olives
(a) Use/accept only quality raw olives (green-ripe to early turning colour).
(b) Store raw olives correctly before processing.
(c) Spray wash olives with potable water (optional).
(d) Size grade and sort raw olives.
(e) Use whole olives.

At this point, olives can be treated immediately with lye or placed in 10% w/v sodium chloride holding brine and used later. Olives are given three lye treatments (0.5 to 1.5% w/v sodium hydroxide) for short periods.

Lye treatment 1
(a) Pack olives into a container/tank with lye solution 1.
(b) Allow olives to interact with 1.5% w/v lye solution just long enough to penetrate the skin. This takes one to two hours.
(c) Drain the lye solution for later use as Lye treatment 2.
(d) Add water to the tank and pass compressed air through the mass for 24 hours to darken the olives. (Oxidation and polymerisation of colourless phenolic compounds occurs under alkaline conditions.)
(e) Drain water and allow olives to dry and stand for 3–12 hours until the skin turns a uniform black colour. In some centres the olives are kept in water.

Lye treatment 2
(f) Add the used lye from Lye Treatment 1 and allow it to penetrate 1 mm into the flesh where oleuropein is hydrolysed. Debittering of the flesh begins.
(g) Pass compressed air through the tanks to mix the mass.
(h) Drain the lye solution for later use as Lye treatment 3.
(i) Add water to the tank/olives and pass compressed air through the mass for around 20 hours.

Lye treatment 3
(j) Add the used lye from Lye treatment 2 enriched with fresh lye and allow time to penetrate through all the flesh down to the stone. This takes four to six hours.
(k) Check penetration of lye through the flesh.
(l) Drain the lye solution.

Washing step
(a) Wash the olives with potable water and aerate for three to four days until the residual lye is removed.
(b) Drain and replace water two to three times a day.
(c) Check pH, which should fall from pH 11 to 9. Add hydrochloric acid, lactic acid or carbon dioxide to facilitate removal and neutralisation of lye.
(d) Use warm water (80°C) to prevent softening and gas pocket (fish eye) spoilage.
(e) The final pH should be around 7.0 (6.0–7.8).

Colour enhancing step
(a) Add ferrous gluconate or ferrous lactate to the last wash step to give final concentrations of 0.1% w/v or 0.05% w/v respectively.
(b) Leave the olives in the iron solution for up to 24 hours. The iron reacts with tannins in the olive flesh to form iron tannate.
(c) Drain the iron solution and remove the excess iron by further washing with water.

Brining step
(a) Add a fresh brine of 3% w/v salt to the washed olives and allow this to equilibrate for three days.
(b) Bulk pasteurise either by injecting steam into the brine until the temperature reaches 90–95°C or heat at 60°C for 45 minutes to prevent spoilage by aerobic bacteria.
(c) Allow the olives to cool naturally.

Further steps
(a) Size grade the olives if required.
(b) Pack the processed olives into a brine (2–3% w/v sodium chloride).
(c) Sterilise the packed processed olives in cans or glass jars (see Table 5.7).
(d) Send representative samples to a microbiology laboratory for testing.
(e) Label packed processed olives in accordance with food standard requirements.
(f) Implement a safety recall system for tracking faulty, contaminated or incorrectly labelled olives.

Final products of Californian-style black olives. As the olives produced by this method are sterilised, the packing solution is based on principles of good manufacturing practice. Typically, the pH is 5.8–8.0 and sodium chloride is 1–5% w/v (depending on the commercial product). The maximum iron level can be up to 0.15 g/kg of fruit.

Processing notes. The lye concentration used depends on the olive variety, maturation state and processing temperature. Olives are processed chemically and no fermentation occurs with this method. Advantages are that the olives retain the firmness of green-ripe/turning colour olives and, as processing only requires a few days, they are ready for the market within one to two weeks of harvest. Disadvantages are the large volumes of water required for lye treatments and washing as well as the disposal of the resulting wastewater. Australia currently imports large amounts of these olives. They are commonly available on supermarket shelves and the destoned form is frequently used in cooking and on pizzas. This type of olive is not currently produced in Australia.

Processing olives by this method should only be undertaken where complete processing and canning lines are available and where food scientists are employed to direct and control production.

Sterilisation. Californian/Spanish-style black olives require sterilisation to ensure safety. Sterilisation is undertaken in an autoclave under the indicative conditions shown in Table 5.7.

Table 5.7. Indicative sterilisation conditions for table olives
(After Garrido Fernandez, A., Fernandez Diez, M. J., and Adams, M. R. (1997).)

Can capacity	Thermal treatment	
	Temperature °C	Time (mins)
1 kg or less	115–116	60
3 kg or less	115–116	70
1 kg or less	121.1	45
3 kg or less	121.1	50

In principle, sterilisation involves exhausting the air from cans containing olives and brine using 93–95°C heat for five minutes, creating an anaerobic environment. This procedure results in an internal temperature of 71–77°C. The cans are then sealed and sterilised according to the above conditions. Such conditions assume an initial temperature of between 21°C and 71°C. For glass jars, a type of autoclave called a 'retort' is used with compressed air, otherwise the pressures that develop in jars during sterilisation would blow off the lids.

All facilities using sterilisation techniques should seek expert advice from engineers and food scientists when establishing conditions for specific products. If the olives are not properly sterilised, the anaerobic conditions inside the containers provide the environment for *Clostridium* spp., including *Clostridium botulinum*, to grow leading to spoilage, food poisoning and potentially the more lethal condition, botulism, in consumers.

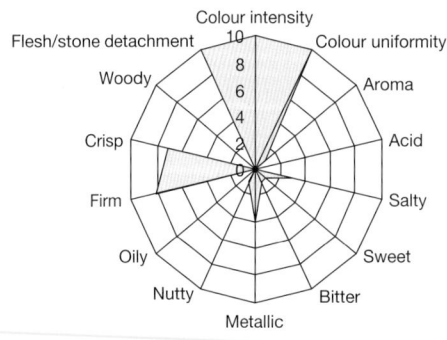

Figure 5.23 Organoleptic profile of lye treated *Manzanilla* olives followed by oxidation (Spanish/Californian-style black olives).

Use of Californian/Spanish-style black olives. These olives are available whole or pitted to eat as snacks and aperitifs, sliced for pizzas, and made into pastes to be used in hot and cold collations.

Nature of the final product. Californian/Spanish-style black olives are a uniform dark colour, firm with compact textured flesh. They have no bitterness, have slight aromatic qualities, are low in salt and have a slightly sweet taste with subtle liquorice overtones. The organoleptic profile of this type of table olive is depicted in Fig. 5.23.

Processing dried table olives (shrivelled olives)

Heat-dried or salt-dried olives are popular with consumers (see also Chapter 4). Dried olives have a low water activity compared with the original raw olives that renders them self-preserving.

Both fresh and processed olives in brine can be dried by heating. However, when processed olives are heat-dried the salt concentrates in the olive flesh. If the salt concentration is too high the olives are inedible. To avoid this problem the olives can be pre-soaked in potable water for one to two days to remove salt prior to drying.

Note: Dehydration is an effective means of inhibiting the growth of microorganisms. Most moulds can grow on foods with as little as 16% moisture. Few organisms will grow at moisture contents of 5%. Bacteria and yeasts generally require high moisture levels, usually greater than 30%. Fruits dried to 16–25% moisture are susceptible to mould if exposed to high humidity and air. Some pathogenic toxin-producing bacteria can withstand the less than favourable conditions of dried foods.

Traditional Greek products include Greek-style salted and Throuba-style olives.

Greek-style salted olives. *Megaritiki* olives are allowed to partially dry naturally on the tree, and then dried with coarse salt. Other varieties such as *Kalamata, Manzanilla, UC13A6* or *Leccino* can also be used to produce salt-dried olives.

Throuba-style olives. *Thrubolea* olives, growing mainly in Greece (Crete, the Aegian Islands and the Attica region), are thought to undergo debittering and transformation by a fungal enzyme. The olives, which change colour from black to copper green, are then sun-dried and salted. There is little evidence actually supporting the role of a fungus in debittering this type of olive.

Salt-dried olives

Salt-dried olives are prepared by packing naturally black-ripe olives in alternating layers with dry coarse salt (equivalent to 10–20% w/w of the weight of olives) in slatted containers that allow drainage of the vegetable water drawn out by the salt. The resulting olives, or 'date olives', are shrivelled in appearance and have a salty bitter-sweet taste. Salt is also taken up by the olive, which acts as a preservative. Processing time is around four to six weeks and the olives are best eaten within three months of processing. Addition of olive oil enhances the flavour of the olive; however, oxidation of the oil can give the olives a slight rancid taste.

Step-by-step method for preparing salt-dried black olives

(a) Use/accept quality raw olives. (Use naturally fully black-ripe olives, for example *Kalamata, Manzanilla, UC13A6* or *Leccino*.)
(b) Store raw olives correctly before processing.
(c) Spray wash olives with potable water.
(d) Size grade and sort raw olives (optional).
(e) Use whole olives.
(f) Slit or prick olives (optional). This speeds up the process.

(g) Allow olives to stand for two days in a cool place; they will lose some water through transpiration.
(h) Pack olives in alternating layers of coarse salt (10–20% of olive weight).
(i) If required, after a few days, mix the olives periodically to ensure good contact with salt.
(j) Allow the released black vegetable water to drain. Do not allow olives to sit in black water as this delays debittering.
(k) Add more coarse salt if required (coarse salt should always be visible).
(l) Olives are ready to eat in one to two months.
(m) Expose the olives to air to darken.
(n) Pack the olives in plastic containers or cans with potassium sorbate (an antifungal agent) under vacuum or in 10% brine.
(o) Send representative olive samples to a microbiology laboratory for testing.
(p) Label packed processed olives in accordance with food standard requirements.
(q) Implement a safety recall system for tracking faulty, contaminated or incorrectly labelled olives.

Salt-dried olives, which have more than 10% salt, may be unpalatable to some consumers. Even under high salt conditions these products are never sterile and can support mould growth if exposed to the environment, and have a short shelf-life. In some centres they are stored in a small quantity of olive oil.

The levels of phenols in salt-dried olives will depend on the variety, maturation state and the quantity lost after processing. Researchers in Greece have found that residual phenol levels in *Thassos* are between 600 mg/kg and 800 mg/kg but hydroxytyrosol levels are much lower than other olive styles, presumably because there is no fermentation step or acid conditions during processing. The sugar content in salt-dried olives decreases to about half that of fresh fruit. Some sugars and other water-soluble compounds are lost with the liquid drawn out of the olives during salting. A similar product produced by table olive processors in South Africa is called '*Kalahari*'. After salting, these olives are dipped in 5% acetic acid or the vinegar equivalent.

Nature of the final product.   Black salt-dried olives have a brown to black wrinkled appearance, a pleasant chewy texture (like prunes) and a slightly bitter-sweet and very salty taste. Their intense flavour makes them very popular with consumers who like

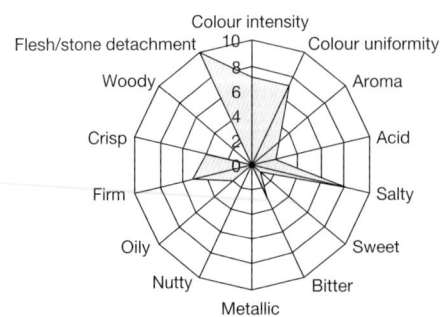

Figure 5.24 Organoleptic profile of salt-dried black-ripe *Manzanilla* olives.

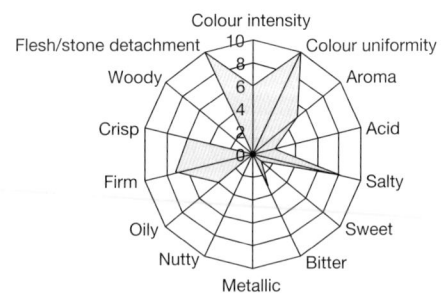

Figure 5.25 Organoleptic profile of salt-dried green-ripe *Manzanilla* olives.

full-flavoured olives. As they might be too salty for some consumers, soaking them in water for 24 hours before eating reduces the salt content. When exposed to air, particularly if olive oils have been added, salt-dried olives can develop a slight rancid taste. Green-ripe or turning colour olives can also be salt-dried using the above method for black salt-dried table olives. When packed in salt, green-ripe olives take on a green/brown colour and gradually shrivel. Because the green-ripe or turning colour olives are firmer than naturally black-ripe olives, the final products are crisp in texture and firm to touch. The organoleptic profile of salt-dried black-ripe and green-ripe *Manzanilla* olives is presented in Figs 5.24 and 5.25.

Heat-dried table olives

Heat-dried olives are dehydrated by placing naturally black-ripe olives in the sun or in an oven set to give gentle heat (40–50°C). Higher temperatures will cook the olives. Using this method the olives lose moisture and bitterness by chemical degradation and vaporisation resulting in a slightly bitter-sweet product. A small quantity of salt is added to enhance their flavour and keeping qualities.

Step-by-step method for preparing sun/heat-dried olives

(a) Use/accept quality raw olives. (Use naturally black-ripe olives, for example *Kalamata*, *Manzanilla*, *UC13A6* or *Leccino*.)
(b) Store raw olives correctly before processing.
(c) Spray wash olives with potable water.
(d) Size grade and sort raw olives (optional).
(e) Use whole olives.
(f) Slit or prick olives (optional). This speeds up the process.
(g) Weigh a sample of olives (original weight).
(h) Dry olives in sun or by heat (40–50°C). This can take several days.
(i) Monitor weight of a sample of olives.
(j) Re-weigh olive sample periodically. Olives are ready when their original weight falls by 15–20%.
(k) Check water activity if laboratory facilities are available. Values should be less than 0.8.
(l) Mix in solid salt to taste.
(m) Pack the dried olives with salt in plastic sachets or other containers, such as glass or under vacuum. Alternatively, pack dried olives without salt into olive oil.
(n) Send representative samples of olives to a microbiology laboratory for testing.
(o) Label packed processed olives in accordance with food standard requirements.
(p) Implement a safety recall system for tracking faulty, contaminated or incorrectly labelled olives.

Nature of the final product. Heat-dried olives have a brown-black wrinkled appearance, and a pleasant chewy texture with a slightly bitter but distinctly sweet prune-like taste (if not salted), particularly with *UC13A6*. The organoleptic profile of heat-dried black-ripe *UC13A6* (Californian Queen) olives is presented in Fig. 5.26. Heat-dried olives embellished with olive oil and herbs are shown in Plate 25.

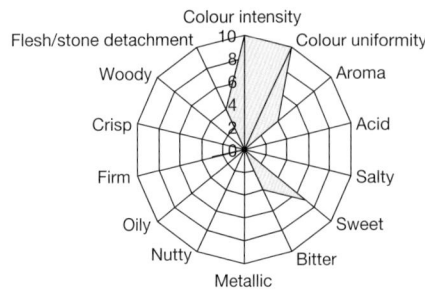

Figure 5.26 Organoleptic profile of heat-dried black-ripe *UC13A6* (Californian Queen) olives.

Chocolate-coated dried olives. A novel presentation of this style of olive is to dip small pitted dried olives, for example *Koroneiki*, into melted chocolate and allow the chocolate to solidify. This product complements chocolate-coated raisins, nuts and coffee beans. The final product has a distinct chocolate flavour with a subtle bitter-sweet taste of the olive. Salted olives are used if a salty taste is desired.

Ferrandina-style table olives

A variation of heat-dried olives is *Ferrandina* olives, a traditional style from Italy. Initially naturally black-ripe olives are blanched in water at 95°C for one to two minutes, which damages the skin (making it more permeable) and kills surface microorganisms. After draining the water, the olives are salted by soaking them in brine (10% w/v salt) for a few days (or packed in dry salt for two to three days). Olives drained of brine (or dry salted) are dried using gentle heat (40–50°C) in an oven over a few days or on racks exposed to the sun (which takes longer) until the bitterness of the olives falls to an acceptable level. The residual moisture falls to around 15–20% w/v and the water activity of less than 0.8 aids in their preservation.

Step-by-step method for processing *Ferrandina* olives

(a) Use/accept only quality raw olives. (Naturally black-ripe olives, for example *Kalamata*, *Manzanilla*, *UC13A6* or *Leccino*.)
(b) Store raw olives correctly before processing.
(c) Spray wash olives with potable water.
(d) Size grade and sort raw olives (optional).
(e) Use whole olives (slit or prick to speed up the process).
(f) Add hot water (95°C).
(g) Drain hot water after one to two minutes.
(h) Soak olives in brine (10% w/v sodium chloride) or pack in coarse salt for two to three days.
(i) Drain brine or gently wash off coarse salt.
(j) Weigh a sample of olives.
(k) Dry olives in an oven at 50°C over two to three days until the residual moisture falls to around 15%.
(l) Monitor the weight of a sample of olives.
(m) Re-weigh the olive sample periodically. (The olives are ready when they are debittered and the residual moisture is around 15–20% and water activity is less than 0.8.)
(n) Pasteurise processed olives for long-term storage.
(o) Send representative samples of olives to a microbiology laboratory for testing.
(p) Label packed processed olives in accordance with food standard requirements.
(q) Implement a safety recall system for tracking faulty, contaminated or incorrectly labelled olives.

Nature of the final product. *Ferrandina* olives have a black wrinkled appearance, a chewy texture and a slightly bitter salty taste.

Toasted olives. A similar product produced by olive growers in Italy is 'Toasted Olives' (*Olive Tostata*). The olives are dried to a firmer, more chewy consistency than *Ferrandina* olives.

Dried olives in olive oil. Dried olives, salted or plain, are often packed in olive oil as a specialty item. Olives are ready to eat after three to four months.

Post-processing table olive operations (secondary processing)

Post-processing operations include: preparing packing solutions, secondary table olive processing, preservation, packaging and labelling.

Packing solutions for table olives

Packing solutions are prepared by dissolving food grade salt in potable water, with or without the addition of vinegar (or other suitable food acid such as citric acid), olive oil, herbs, spices and aromatics (see below). The brine should be clean and free of abnormal odours and tastes. Fermentation brines can be used to store olives in bulk containers.

Once processed, most table olives are packed in brines or marinades. Fermentation brines can be used as packing solutions as long as they are filtered and pasteurised and they meet the salt, acid and pH levels required for a particular table olive product. Brine in glass containers should be clean and transparent (except where they have been embellished). The minimum sodium chloride concentration, minimum acidity and maximum pH levels accepted in packing solutions for different types of olive products, following the Codex Alimentarius (1987)/IOOC (2004) Table Olive Standards, are given in Table 5.8.

Table 5.8. Final packing/covering solutions/conditions for common table olive products
(After Codex Alimentarius (1987)/IOOC (2004) Table Olive Standards)
GMP, Good Manufacturing Practice

Product	Minimum sodium chloride content % w/v			Maximum pH limit			Minimum acidity % w/v calculated as lactic acid		
	Specific chemical attributes	Preservative Refrigerate	Pasteurise Sterilise	Specific chemical attributes	Preservative Refrigerate	Pasteurise Sterilise	Specific chemical attributes	Preservative Refrigerate	Pasteurise Sterilise
Treated olives e.g. Spanish-style green	5	4	GMP	4	4	4.3	0.5	0.4	GMP
Natural olives Green Turning colour Black	6	6	GMP	4.3	4.3	4.3	0.3	0.3	GMP
Dehydrated and/or shrivelled olives	10	10	GMP	GMP	GMP	GMP	GMP	GMP	GMP
Olives darkened by oxidation Californian-style Spanish-style	GMP	GMP	GMP	GMP	GMP	GMP	GMP	GMP	GMP

If fermentation brines are used to prepare packing/covering solutions, and the final product is not pasteurised, they should be heated at 80°C for 10–15 minutes to denature fermentative organisms that may initiate secondary fermentations when the olives are packed in their final containers.

Secondary table olive processing

Secondary processing of table olives involves all processes after the olives have been processed and debittered (primary processing) including: adding vinegar and olive oil; adding herbs, spices and marinades; pitting and stuffing table olives; making olive pastes and tapenades; and adding other pickled vegetables to processed olives.

Adding vinegar and olive oil

A simple packaging procedure is to add vinegar and extra virgin olive oil to the olives in brine. Additions contribute to the organoleptic properties of the olives. The addition of vinegar to brine increases its acidity as well as lowering the pH of table olive products, improving their keeping qualities.

Recipe for packing solutions with brine, vinegar and olive oil

(a) Make up a 10% brine solution (7–8% brine can also be used if the 10% brine is considered too salty).
(b) Use 3 parts of this brine and add 1 part of vinegar of choice (one litre of the brine/vinegar solution will require 750 ml of 10% brine and 250 ml of vinegar).
(c) Pack the olives in the containers and add the brine/vinegar solution nearly to the shoulder of the container.
(d) Check the salt content, pH and free acidity of the packing liquid.
(e) Add a layer of oil (enough to cover the surface of the solution, for example 10 ml in a 300 ml container).

Vinegar. As vinegar is a common addition to table olive products, the following information will be of interest to both commercial and home processors. Like the olive, vinegar production can be traced to ancient Egyptian, Greek and Roman civilisations. Some 'vinegars' are prepared synthetically, then diluted with water to have acetic acid levels similar in strength to fermented vinegars. Some countries will not allow these chemical derived 'vinegars' to be labelled vinegar.

Vinegar is a dilute solution of acetic acid and often produced by a two-stage microbiological process.

Stage 1. Conversion of fermentable sugars to ethanol by yeasts (usually *Saccharomyces* spp.).

Stage 2. Oxidation of ethanol to acetic acid by bacteria (usually *Acetobacter* spp.).

Vinegars from different sources, depending on the sugar source, have numerous flavour components, including a very small amount of alcohol (ethanol), that give the table olives desirable flavours and aromas when combined with the acetic acid.

With respect to table olive processing, vinegar has a number of functions: preservation, flavouring, a solvent for herbs and spices, and provides additional polyphenols.

Vinegar has a high antimicrobial action because it is a weak acid. Weak acids in solution are present in two forms, unionised (acetic acid) and ionised (acetate + hydrogen ions). The proportion of the two forms depends on the pKa of acetic acid, which is 4.76, and the pH of the solution it is put into. At pH values around the pKa of acetic acid most of the acetic acid is in the unionised form. As little as 0.1% of the unionised form will inhibit the growth of most food-poisoning and spore-forming bacteria. At a concentration of 0.3%, mycotoxigenic moulds are prevented from growing. Together with salt, acetic acid also lowers the water activity (W_A) of brines. The antibiotic activity of acetic acid is not only due to its acid properties. The unionised form can pass into microorganisms where it comes into contact with the essentially neutral (pH 7) cytoplasm and rapidly ionises releasing acid that lowers the pH inside the cells, inactivating essential enzymes.

Different types of vinegar can be used and their composition will obviously influence the flavour of the olives. Most vinegars contain 5–6% (0.8–1.0 M) acetic acid with a pH of less than 3. Apart from white vinegar, other vinegars listed below have other components that can contribute to the organoleptic characteristics of the olives. White vinegar, available in bulk or at supermarkets, is essentially a solution of acetic acid in water. It has a sharp acid/sour taste and pungent aroma. Other vinegars can have pigments, flavour compounds of the source material, for example malt, apple, grape, and varying amounts of residual fermentable substrates, such as sugar. Adding vinegar with fermentable substrates to table olives after packaging can initiate secondary fermentations if the final products are not pasteurised.

Types of vinegars. Types of vinegars include white and brown (often made synthetically), malt, honey, cider, wine (red and white), and balsamic vinegar.

Lighter coloured vinegars are generally used with green olives whereas coloured vinegars are used with black olives.

Extra virgin olive oil is the preferred oil to be added to table olives. However, if table olives with olive oil are stored under refrigeration, the oil solidifies and some consumers may believe the table olives are unsafe to eat. For this reason many processors/packagers of table olive products prefer to use seed oils, such as canola, or high oleic sunflower oil, that do not exhibit this effect. The effect of refrigeration on olive oil should be explained on the label.

Note: Some commercially available vinegar that contains 15% acetic acid is useful for adjusting brine pH during table olive processing. Glacial acetic acid can also be used, but this product is highly corrosive and needs careful handling. These should not be used for preparing packing solutions.

Adding herbs, spices and marinades to table olives

Herbs and spices can be added to final containers with table olive products and brine or as specific marinades. The objective is to embellish primary processed olives with additional flavours and aromas as well as enhance their nutritional qualities. Fresh herbs and spices such as garlic, chilli, basil and oregano should be pre-dried and free of harmful and spoilage microorganisms unless they are added just prior to consumption. Like all inputs, all herbs and spices should be controlled and backed up with specific input profiles as well as specifications for their use included in the final product profile.

Adding herbs and spices. It is best to use dried herbs rather than fresh herbs to prevent spoilage and the development of 'off' flavours, particularly if the processed olives are to be stored for more than a month or so. Herbs and spices provide additional aromas and flavours to the olives, have a mild preservative action and provide antioxidants that have health benefits. Olive oil, herbs and spices can be added to all processed olives including salt-dried, sun-dried and heat-dried olives. The flavours of the herbs and spices should be subtle and balanced. Also, the full effect of the additions can take up to a few weeks to stabilise. Although tastes and aromas depend on individual preference, as a guide 3–5 grams (about a heaped teaspoonful) of an individual herb or spice can be added to a 300 gram container of olives in brine. Olive oil tends to strengthen the flavours and aromas of herbs and spices, as many of their aromatic properties are oil soluble. Concentrated extracts of herbs and spices can also be used. The best flavour is obtained if dried whole spices are ground in a spice mill or coffee grinder when needed. Table 5.9 shows some of the herbs, spices and aromatics that can be used to embellish processed table olives.

Table 5.9. Typical herbs, spices and aromatic substances added to table olives

Herbs, spices and aromatics		
Basil	Cinnamon quills	Lemon oil
Bay leaf	Coriander seeds	Preserved lemon peel
Black pepper	Fennel seed	Dried orange peel (zest)
Capers	Garlic	Marjoram
Cardamom seed	Ginger	Mustard seed
Chilli	Lemon grass	Oregano

Preserved lemons. Often preserved lemon is used as part of the embellishment of table olives, particularly in north Africa. Preserved lemons are expensive to buy so it is worth processing them for use in table olive products. The following simple method can be used:

(a) Divide six lemons (900 g–1000 g) into quarters.
(b) Sprinkle the flesh with coarse raw salt.
(c) Place 30 g of salt into a sterilised glass jar.
(d) Press the lemons so they release their juice into a clean glass jug.
(e) Place quartered lemons in layers in jar and press down.
(f) Add coarse raw salt between each layer.
(g) Pour lemon juice into the jar.
(h) Add coarse salt to the top of the lemons.
(i) Shake the jar daily to ensure mixing and salting.
(j) Lemons are ready after 20–30 days.
(k) Preserved lemons keep for up to one year.

Before adding to olive recipes, the lemon quarters are removed and rinsed with boiled, cooled tap water. The pulp is scooped out and discarded. The preserved lemon peel can be used whole or chopped coarsely or finely.

Simple combinations of herbs and spices that can be added to processed olives are:

- hot mixtures: chilli and mixed herbs;
- Mediterranean flavours: garlic, oregano and lemon slices (fresh or preserved);
- Greek flavours: rosemary, marjoram, wine vinegars;
- Italian herbs: basil, garlic, rosemary, marjoram, parsley, sage, thyme, chilli;
- Asian flavours: chilli, ginger, cardamom seed and lemon grass, with or without added sugar;
- tapas and antipasti: olives, peppers, small whole onions, cucumber, capers; and
- olives with added whole pickled peppers, salted capers, pickled onions, sun-dried tomatoes.

Note: When the olives are packaged as tapas or antipasti, the olives and vegetables are processed separately then mixed. For commercial production of aromatised olives, herbs, spices and olive oil should never be added during the primary processing stage.

Embellishing olives for immediate consumption. The starting materials for producing aromatised table olives are all types of primary processed olives: green, turning colour, black and dried. Olives in brine are rinsed with tap water and drained well. In principle, the amounts of olives to be prepared or served are placed in a container (jar or bowl) and the additions made. When olive oil is used, this is added to the olives first, followed by the herbs and spices. Slitting, cracking or destoning the olives allows for better penetration of herbs and spices. Olives and additives are then gently mixed. Olives can be eaten immediately; however, they are more flavoursome several days later. The prepared olives will keep in a refrigerator for at least one month. Combinations of aromatics can be used. A touch of wine, balsamic vinegar, orange zest or lemon juice can also be added. When served in a bowl, a sprig of fresh rosemary or oregano gives the finishing touch. Different ways of presenting olives are shown in Plates 26–28.

The following recipes are not prescriptive, so ingredient quantities may be increased, added to, or deleted. The authors acknowledge that everyone has his or her own and often 'better' recipe for preparing table olives! The ideas presented in the recipes can be translated to brine marinades, but fresh herbs and spices should be replaced with dried ones. For olives in marinades, place olives and other ingredients into a suitable container with lid and shake gently. Remove the lid and add salt brine vinegar. Salt brine vinegar can be prepared as follows:

$$10\% \text{ w/v salt brine (3 parts by volume)} + \text{vinegar (1 part volume)}$$
$$= \text{approximately } 1.25\% \text{ acetic acid}$$

$$10\% \text{ w/v salt brine (4 parts by volume)} + \text{vinegar (1 part volume)}$$
$$= \text{approximately } 1.0\% \text{ acetic acid (The second mixture is used if olives with a lighter vinegar/wine-like taste are desired.)}$$

The addition of vinegar increases the free acidity of the brine. The final pH of the covering solution in packaged olives should be less than 4.3 and for commercial purposes the olives should be pasteurised. Pasteurised, aromatised table olives have a shelf-life of at least two years. Aromatised table olives that are not pasteurised have a three month shelf-life if a combination of the preservatives sorbic acid (500 ppm) and benzoic acid (1000 ppm) are used. The reasons for the short shelf-life of this type of product are the effects of vegetative microorganisms and enzymatic action of the added herbs and spices.

Recipes

Calabrese green cracked olives

Processed green cracked olives	approximately 1 kg
Chopped garlic	20 g
Chopped oregano	5 g
Crushed dry red chilli	5–10 g
Chopped fennel	5 g
Whole roasted fennel seeds	5 g
Extra virgin olive oil	80 ml

Chilli garlic marinated mixed olives

Processed green and black olives*	approximately 1 kg
Chopped garlic	20 g
Dried Italian mixed herbs	5 g
Crushed dry red chilli	5–10 g
White wine vinegar	40 ml
Extra virgin olive oil	80 ml

* mix naturally processed green *Manzanilla* and black *Kalamata* olives. Initial flavours are subtle with strong garlic overtones developing gradually.

Connoisseur/Continental marinated olives

Processed mixed olives*	approximately 1 kg
Chopped oregano	5 g
Crushed garlic	20 g
Chopped dry red chillies	5 g
Small whole red chillies	2
Lemon slices	4 thin half slices
White wine vinegar	40 ml
Extra virgin olive oil	80 ml

* mix equal proportions of olives of different sizes, colour and variety. Use naturally processed green olives (*Manzanilla, Picholine*), turning colour olives (*Verdale, Jumbo Kalamata*) and naturally black-ripe olives (*Kalamata, Leccino*). Initially the olives have subtle aromatic flavours, followed by the hot chilli taste.

Cypriot-style marinated green cracked olives

Processed green cracked olives	approximately 1 kg
Crushed dry coriander seeds	10 g
Chopped dry oregano	10 g
Chopped garlic	20 g
Lemon juice	to taste
Extra virgin olive oil	80 ml

Greek-style marinated green cracked olives

Processed green and black olives	approximately 1 kg
Crushed dry rosemary	10 g
Whole dried hot chillies	4
Chopped garlic	20 g
Lemon slices	4 quarters
Extra virgin olive oil	100 ml

Italian-style marinated black shrivelled olives

Toasted Italian olives (or dried olives)	approximately 1 kg
Chopped garlic	20 g
Cracked dried fennel seeds	5 g
Extra virgin olive oil*	80 ml

* some recipes use sunflower oil

Moroccan green olives

Processed unpitted green olives	approximately 1 kg
Preserved lemon peel strips	20 g
Crushed paprika	5 g
Chopped garlic	20 g
Roasted whole cumin seeds	5 g
Cracked red pepper	5 g
Roasted whole coriander seeds	5 g
Roasted whole fennel seeds	5 g
Extra virgin olive oil	100 ml

Moroccan black olives

Processed destoned black olives	approximately 1 kg
Preserved lemon slices	2–3
Bay leaves	4
Vinegar	40 ml
Whole dried cloves	5 g
Chopped cinnamon quills	10 g
Extra virgin olive oil	80 ml

Neapolitan black olives

Processed black olives	approximately 1 kg
Lemon juice	40 ml
Chopped marjoram	5 g
Chopped oregano	5 g
Extra virgin olive oil	80 ml

Niçoise olives (fennel and orange scented olives)

Processed small black olives*	approximately 1 kg
Chopped orange rind	10 g
Dried fennel flower or seed	10 g
Chopped garlic	20 g
Extra virgin olive oil	80 ml

* Picholine, Frantoio, Leccino or Koroneiki.

Oriental-style olives

Processed green olives*	approximately 1 kg
Quarter slices of orange	4 pieces
Quarter slices of lime	4 pieces
Chopped lemon grass	5 g
Chopped ginger	5 g
Cracked coriander seeds	5 g
Chopped chilli	5 g
Chopped garlic	20 g
Extra virgin olive oil	80 ml

* olives processed by spontaneous fermentation in brine. For a sweeter taste, sugar or honey can be added.

Tunisian-style marinated black olives

Processed black olives	approximately 1 kg
Harissa*	10–15 g
Extra virgin olive oil	100 ml

* harissa is a spice mixture containing red chillies, dry roasted cumin and coriander seeds, chopped garlic, salt and olive oil. Ingredients are placed in a food processor and blended into a paste. It is also available commercially as a powder. Similar flavours can be achieved by using the individual herbs and spices.

Antipasto

Antipasto, popular with Italians, are foodstuffs eaten as openers prior to having an entrée or the main meal. They consist of vegetables, olives, meats and fish. A vegetarian antipasto combination includes olives, cheese and the pickled, dried or salted vegetables listed below. The proportions of each will depend on consumer preferences.

Mix equal proportions from the following selection:

- naturally processed *Kalamata* olives (drained);
- sun-dried tomatoes;
- fetta cheese;
- pickled artichoke;
- pickled caperberries; and
- crisp pickled vegetables (carrots, celery and cauliflower).

Add extra virgin olive oil to taste.

Oil cured olives

To process dried olives, such as *Leccino* olives, soak in extra virgin olive oil (or sunflower oil) for several months.

For home use, pack black-ripe olives, without salt, in extra virgin olive oil then place them in a freezer for several months. The number of olives required is then removed, allowed to thaw and consumed immediately. The thawed olives are wrinkled in appearance. Salt, herbs and spices can be sprinkled over the olives to give additional flavours.

Pitting and stuffing table olives

Pitting and stuffing is a traditional way of presenting olives and is very popular with consumers. Green olives processed by the Spanish method (usually *Manzanilla* and *Sevillana*) and packed in glass jars are often sold in the destoned form or stuffed with pimento, sun-dried tomato, anchovy, nuts or cheeses. Pitted and stuffed olives are also placed in marinades of herbs and spices and sold loose as gourmet items.

Note: When destoned table olives are stuffed by hand, this should be undertaken under strict sanitary conditions (see Chapter 4). All equipment and surfaces should be cleaned and sanitised. Surfaces can be tested for microbial contaminants by taking regular swabs. Some quick tests have been developed for some microbes that can be done in-house; otherwise swabs have to be sent to an appropriate laboratory for testing.

Large 'Queens' rather than smaller olives are selected for destoning. Popular varieties for this purpose are green-ripe *Sevillana*, *Manzanilla* and *Chalchidikis*. Destoning and stuffing can be undertaken manually or with an automatic destoning/stuffing machine. Because of the high capital cost of equipment, boutique and small-scale table olive processors still use manual methods. Some table olive processors purchase imported destoned table olives from wholesalers, then stuff these by hand.

Pitted olives, for example black *Manzanilla* and *Kalamata*, are used in preparing cold and cooked foods including pizzas. Destoned and stuffed olives are popular at functions and parties, as there are no stones to dispose of. Naturally processed black olives, such as *Kalamata*, are too soft to maintain their shape during pitting and are less suited to eat as appetisers.

Common stuffing materials include pimento, peppers, jalapas, chilli, anchovy, cheese (various), sun-dried tomatoes and almonds. New fillings, such as dried apricot and pickled vegetables, are being developed. The dimensions of stuffing material have to match the cavity created by the pitting machine. If too narrow the stuffing falls out, whereas if it is too wide, the olive can bulge and crack. When pitted and stuffed olives are mass produced, fillings are prepared so that they can be fed easily into the olive by machine. In the case of pimento, the peppers are ground into a paste, mixed with guar gum and sodium alginate and allowed to solidify into a form that can be used in an automatic stuffing machine. Anchovy fillings are prepared in a similar way.

Hand stuffed olives are considered a superior product, as they have fresher tastes and aromas than those that are mass-produced. All operations should be undertaken under hygienic conditions and disposable gloves worn by operators to prevent physical and microbiological contamination of the olives.

Where fresh fillings are used, for example fetta cheese, anchovies, vegetables, herbs and spices, the final products should be refrigerated or pasteurised, otherwise their shelf-life is short and the risk of contamination or spoilage increases.

Stuffed marinated olives. Use large green olives, such as *Sevillana*, *Manzanilla* or *Hardy's Mammoth*, and destone with a hand or commercial pitter. Insert stuffing into the cavity, for example fetta cheese, pieces of pickled paprika, blanched almonds, tuna or anchovy. To serve, add olive oil and chopped oregano (or mixed herbs).

Step-by-step method for stuffing table olives

The olives are pitted then filled with pimento, anchovies, almonds, cheese, onions, capers or other desired filling.

(a) Process olives.
(b) Destone olives or buy destoned processed olives.
(c) Insert stuffing.
(d) Pack olives in jars with packing brine/marinade (see Table 5.8).
(e) Pasteurise to prevent secondary fermentation.
(f) Send representative samples of olives to a microbiology laboratory for testing.
(g) Label packed processed olives in accordance with food standard requirements.
(h) Implement a safety recall system for tracking faulty, contaminated or incorrectly labelled olives.

Note: Stuffed olives containing seafood, cheeses or nuts may cause allergic or other food related reactions in susceptible consumers, so containers should be labelled with appropriate warnings.

Olive pastes

Olive pastes originated in ancient times. They were probably made from whole olives or the remaining mash after olive oil production. Olive pastes are very popular with consumers, particularly in Italy and France, and are used to spread on dry biscuits or bread as appetisers or as condiments for pasta, fish or meat dishes. They can be made from most olive varieties, for example *Manzanilla*, *Kalamata* or *Leccino*. Pastes are prepared from processed green-ripe (green paste) to fully black-ripe olives (black paste). Tapenade is a French olive paste that generally contains salted anchovies and capers.

The organoleptic characteristics of olive pastes and tapenades depend on the quality of the processed olives used. Olives processed by spontaneous fermentation, such as *Kalamata*, Greek-style black and Sicilian-style green olives, will have a more distinct flavour than lye treated olives, for example Spanish-style green and Californian/Spanish-style black olives. Poorly processed olives with 'off' flavours should not be used, otherwise the paste will be tainted.

Destoned processed olives should be washed under running potable water then dried thoroughly to keep the amount of water in the paste as low as possible to aid preservation.

Olives are crushed into a paste in a food processor or similar device. For medium to large-scale manufacture of olive pastes, specialised equipment is used. The olives are fed via a hopper into a perforated stainless steel cylinder with small holes (2–3 mm). The olives are carried along the length of the cylinder by a screw that forces them against the walls. The olive flesh passes through the holes in the cylinder, while skin and stones are expelled at the end of the cylinder. The paste is allowed to stand so that vegetable water can drain, after which extra virgin olive oil is added (5–10% w/v).

Aromatics such as herbs, spices and essential oils (thyme, rosemary and lemon) can be added to the olive paste to enhance flavour.

Because of their high salt content, olive pastes are fairly stable, especially when kept in a cool place and the surface is covered with a layer of olive oil. For commercial purposes, olive paste preparations should be pasteurised with or without a preservative, such as potassium sorbate.

Olive vegetable paste. Olive vegetable pastes can be prepared from olives and various vegetables. For immediate consumption, fresh herbs, spices and other vegetables can be used; otherwise they should be dried or processed. Where fresh vegetables are used, more vinegar or lemon juice may be added to enhance the flavour of the paste. Vegetables should be pickled in salt/vinegar before incorporating into the paste. For commercial purposes the paste should be pasteurised.

The recipe for green olive vegetable paste that follows is an example of the idea.

Green olive vegetable paste
Destoned processed green olives	700 g
Celery	100 g
Cauliflower	100 g
Garlic (equivalent to 15 fresh cloves)	6 g
Carrots	100 g
Oregano	5–10 g
Flat leaf parsley	10 g
Red wine vinegar	10 ml
Balsamic vinegar	10 ml
Extra virgin olive oil	GMP*
Cracked pepper	to taste

* sufficient olive oil is added to give the desired consistency. This recipe makes approximately 1 kg of paste.

Two recipes for preparing aromatised black olive paste follow.

Black olive paste with oregano

Destoned processed black olives	800 g
Garlic	10 g
Chopped oregano	20 g
Lemon juice	60 ml
Extra virgin olive oil	GMP*

* sufficient olive oil is added to give the desired consistency. This recipe makes approximately 1 kg of paste.

Black olive paste with fennel and mint

Destoned processed green olives	800 g
Chopped fennel	20 g
Roasted cracked fennel seed	20 g
Chopped mint	10 g
Whole hot chilli	2
Orange or lemon juice	30 ml
Red wine vinegar	40 ml
Extra virgin olive oil	GMP*

* sufficient olive oil is added to give the desired consistency. This recipe makes approximately 1 kg of paste.

Typically, olive products have been mainly consumed by those living around the Mediterranean and in Middle Eastern countries. Olives are now also popular in Asian and south Pacific countries, which has led to innovation and the development of new products to meet the demands of different palates. An indicative recipe for one such product, Sweet Kalamata Olive Paste, is given below. This product needs to be pasteurised.

Sweet Kalamata olive paste

Destoned processed *Kalamata* olives (drained)	610 g
Sugar	390 g
Wine vinegar or lemon juice to taste	Approx. 6 ml

Tapenade

Tapenade is an olive paste popular around the Mediterranean region, especially in France, and now internationally. It is used as spreads and dips. The basis of tapenade is the ground flesh of processed green, turning colour or black olives, to which capers, anchovies and other foods and spices are added. The addition of capers differentiates tapenades from other olive-based pastes. Various tapenades are presented in Plate 29.

Traditional French tapenade from Provence contains the flesh of Olives of Nice, 2% capers, thyme, savory and a small quantity of extra virgin olive oil. Olives of Nice are processed in brine for 6–12 months. The original recipe can be embellished with additional herbs and spices as well as other foodstuffs.

In addition to destoned, processed, black, turning colour or green olives, preserved capers and olive oil (preferably extra virgin olive oil) are added. The following ingredients are often included: garlic, salted anchovies (or tuna), lemon juice, cracked pepper, aromatics (herbs and spices) and foodstuffs (pine nuts, chilli, sun-dried tomatoes).

Almost any processed olive can be used to make tapenade, but it is very important that the tapenade is not too salty. Pitted processed olives should be drained and dried over

24–48 hours before crushing into a paste. Alternatively, commercially available olive paste can be used instead of pitted olives.

Salted anchovies in oil should be used, or alternatively the anchovies can be replaced wholly or partly with tuna. For commercial production, dried herbs and spices should be used. For home preparation or for immediate use, fresh herbs, spices and garlic can be used as long as the tapenade is stored under refrigeration. Cracked pepper is added to taste. Depending on the oil content of the processed olives, the amount of extra virgin olive oil required is gauged so that the desired final consistency is achieved (not too runny and not too dry).

Boutique level producers can make tapenade using a standard food processor whereas small- and large-scale producers will need to use commercial grade equipment.

Note: As there is substantial handling of ingredients during the preparation of tapenades, all aspects of hygiene and health and safety requirements must be observed to prevent contamination with potential food poisoning organisms.

Tapenade

Destoned processed black olives (drained)*	900 g
Capers (drained)	180 g
Garlic (equivalent to 15 fresh cloves)	6 grams
Extra virgin olive oil	GMP**
Cracked pepper	to taste

* destoned green or turning colour olives can be used; **, sufficient olive oil is added to give the desired consistency. Also, anchovy fillets can be added. This recipe makes approximately 1 kg of tapenade.

Note: Tapenades containing seafood or nuts may cause allergic reactions in susceptible consumers, so containers should be labelled with appropriate warnings.

Step-by-step method for making tapenade

(a) Destone processed olives or use commercial destoned olives.
(b) Drain destoned olives if required.
(c) Rinse olives with potable water.
(d) Check that there are no stones.
(e) Place the olives, anchovies (if included), capers and garlic into the food processor.
(f) Apply short sharp impulses to the mixture to give a moderately coarse paste.
(g) Add sufficient olive oil and mix in to give a slightly granular firm paste (not runny).
(h) Pack into containers.
(i) Pasteurise.
(j) Send representative samples of olives to a microbiology laboratory for testing.
(k) Label packed processed olives in accordance with Food Standards.
(l) Implement a safety recall system for faulty packed tapenade.

Note: Additions can be made to the basic tapenade according to taste or commercial requirements.

The following tapenade is a variant of the basic recipe above. For immediate consumption, fresh herbs, spices and other additives can be used, otherwise they should be dried or processed.

Tapenade-style salsa

Destoned processed black *Kalamata* olives	450 g
Destoned green or turning colour olives	450 g
Capers	180 g
Garlic (equivalent to 15 fresh cloves)	6 g
Sweet red peppers	50 g
Sun-dried tomatoes	100 g
Flat leaf parsley	10 g
Red wine vinegar	10 ml
Balsamic vinegar	10 ml
Cracked pepper	to taste
Extra virgin olive oil	GMP*

* sufficient olive oil is added to give the desired consistency. This product should have a pourable consistency. This recipe makes approximately 1 kg of salsa.

Packaging olive pastes and tapenades. Olive pastes and tapenades can be packed in 200 g to 1 kg glass jars.

(a) Undertake all steps hygienically.
(b) Wash and dry jars.
(c) Pack the paste into jars.
(d) Tap the jars to settle the paste and to eliminate air pockets.
(e) Leave a space of up to one centimetre from the top of the paste to the rim of the jar.
(f) Firmly close the lid.

Preservation of olive pastes and tapenades. When freshly prepared under hygienic conditions, olive pastes and tapenades can be stored under refrigeration for at least two weeks and up to three months in unopened containers. It is preferable, however, to reduce the risk of spoilage or food poisoning by pasteurising olive pastes or tapenades. Pasteurisation involves heating the pastes to a temperature that will kill spoilage and food poisoning microorganisms without altering their organoleptic properties.

The equipment required is a stainless steel tank fitted with a heater and thermostat. The tank is filled with water and jars placed individually by hand or by batches in stainless steel wire or perforated baskets.

The following is an indicative pasteurising procedure and should only be used as a guide. Conditions will vary according to the size of the containers.

(a) Raise the temperature of the water in the tank to 75°C.
(b) Immerse the packed jars of paste in the water and allow the contents to equilibrate (this can take up to 20 minutes).
(c) Over the next 25 minutes the water temperature is increased to 80°C (higher temperatures denature the paste).
(d) The heater is then turned off and the jars allowed to cool to 60°C before removal.
(e) Sample jars should be sent to a microbiology laboratory for safety clearance.

Note: Using this procedure, the shelf-life of the olive pastes and tapenades is at least 18 months. Other procedures involve bulk pasteurisation of the olive paste or tapenade, then packing aseptically while hot into washed and dried containers. Before this method is used, a food consultant should be engaged and the appropriate research and development undertaken.

Final table olive products

Table 5.10 gives a summary of the methods available for processing table olives.

Table 5.10. Summary of common table olive processing methods

Procedures shown in parentheses are optional. ¹ *Frantoio, Taggiasca* – Ligurian-style.
² Sicilian-style olives can also be prepared with dry salt.

Sevillian-style	Spontaneous fermentation Greek-style¹	Kalamata-style	Sicilian-style²	Californian/Spanish-style black	Salt-dried	Heat-dried
Treated	Untreated			Treated	Untreated (Treated)	
Green-ripe	Green-ripe Turning colour Nearly black-ripe	Nearly black-ripe	Green-ripe	Green-ripe to turning colour	Green-ripe Turning colour Nearly black-ripe	Nearly black-ripe
Hand harvest			Hand harvest (facilitated manual harvest)			
Spray wash – optional for lye treated olives						
Sort and size grade						
Whole	Whole or slit	Slit	Bruised or stoned	Whole Can be fresh or prestored in salt brine	Whole, slit or pricked	Whole, slit or pricked
Lye solution 2%	Brine 10% salt (Presoak in water for 10–14 days with water curing method)			Lye solution Several strengths	(Treat in weak lye)	Blanch (Treat in weak lye)
Washed				Oxidise at alkaline pH	(Wash)	(Wash)
Place in salt brine 10% salt				Wash	Add 10–20% dry coarse salt	Add dry coarse salt or soak in salt brine
Anaerobic fermentation (Water cured: weak fermentation)				Add ferrous gluconate		Heat dry
				Bulk pasteurise		
(Size grade)						
Pack in fermentation brine or fresh packing brine	Pack in fermentation brine or fresh packing brine	Pack in fermentation brine or fresh packing brine	Pack in fermentation brine or fresh packing brine with vinegar and olive oil	Pack in weak salt brine	Pack in brine or without	
(Pasteurise)				Sterilise	(Pasteurise)	

Processed table olives should retain some level of bitterness and fruitiness. The final salt and pH levels of packed olives depend on the style, variety and ripeness of the fruit as long as safety requirements are met. In all cases, processed table olives should be determined as microbiologically safe by an accredited laboratory before being offered for sale. Procedures should have checkpoints for calculations and quantities used, and monitoring procedures for pH and salt levels. Once packed, representative samples of the olives must be taken and the following parameters assessed: brine pH and free acidity, brine salt levels, brine and fruit microbiology, and organoleptic characteristics.

The final product may be preserved by one or more of the following procedures: brining, pH control, pasteurisation, sterilisation, or drying to prevent microbiological spoilage. Refrigeration is an option for short-term storage and preservatives can be used.

6
Quality and safety

This chapter deals with the quality and safety of the finished processed table olive products and the chemical and microbiological testing that is required to guarantee the quality of the final product. The physical properties of table olive products that need to be monitored, including firmness, Flesh:Stone ratio and colour, are discussed. The chemical analyses that are required to develop the nutritional panel (mandatory for commercially packed olives) are presented. (The chemical testing carried out by the authors (Kailis and Harris 2004) showed that some of the mineral content of olives is leached out during processing.) A general nutritional label is given along with two worked examples and a typical label for tapenade. The concentrations of brine solutions are listed along with the pH required for the storage of processed olives to prevent microbiological spoilage. The importance of microbiological testing is emphasised as problems in this area can have rapid and devastating effects on the consumer as well as the manufacturer. The types of microbiological tests available are provided and we discuss the need for all table olive producers to have a product recall system in place. The successful production of table olive products rests on their organoleptic and safety qualities. A number of organoleptic tests and a scoring regime that is consistent with the IOOC Table Olive Standards (2004) are provided. Spoilage and deterioration of table olive products that can occur during production, or appear later in the final product, are also discussed.

Introduction

Processed table olives must be edible, tasty, nutritious and safe to eat. These objectives can only be achieved if processors follow practices that ensure such qualities are retained. As indicated in Chapter 3, quality starts at the planning stage so that appropriate varieties relevant to processing method, style and consumer preference are used. Olives selected for processing should be produced by Good Agricultural Practices (GAP); processed by principles of Good Manufacturing Practice (GMP) (namely with respect to product specifications, controlled methods and the use of potable water and food grade ingredients); and produced at premises using equipment and personnel that can deliver Good Hygienic Practices (GHP).

As indicated in Chapter 5, there are different methods for producing table olives. The end product of each method has specific characteristics that can be assessed by laboratory tests (physical, chemical and microbiological) and organoleptic evaluation. Furthermore, as there is also the risk of introducing physical, chemical or microbiological contaminants during processing and packaging, careful control during these processes, backed up by specific testing procedures, can minimise risk. Control of microbiological quality is of major importance in the prevention of food-borne illness and to reduce the risk of spoilage.

In Australia, Food Standards Australia New Zealand (FSANZ) is responsible for national food product standards. Although there are no specific Australian standards for table olives, this authority has limit requirements for certain ingredients in foodstuffs, regulates food additive use and sets tolerance limits for contaminants such as toxins, heavy metals and pesticide residues. Table olives must meet these requirements. However, Australia has encompassed many aspects of the Codex Alimentarius (1987)/IOOC (2004) Table Olive Standards, which have sections relevant to olive oil and table olives.

The International Olive Oil Council Codex Alimentarius (1987)/IOOC (2004) Table Olive Standards have prescribed standards specific to table olives. Ongoing developments by these bodies include refining the standards, particularly those for evaluating organoleptic attributes. As Australia has encompassed IOOC standards for olive oil it is logical for Australia to encompass international standards for table olives. Table olive producing and importing countries such as Spain, Greece and the USA also have prescribed standards administered through specific agencies responsible for foodstuffs. If Australian processors intend exporting olives, international or country specific standards will need to be followed.

General qualities of packed table olives

The following general qualities apply to table olives packed for wholesale and retail trade. Where relevant, olives (and brine) should be:

- sound;
- clean;
- at the correct maturation state for the style/method;
- free from malodours and abnormal tastes;
- free from defects that affect their edibility or preservation;
- free from foreign matter other than added herbs, spices or aromatics;
- free from signs of deterioration or abnormal fermentation;
- size graded;
- of a single variety (unless packed as specialty line); and
- uniform in colour or colour consistent with style/method.

In addition, brines need to be formulated with food quality and approved ingredients, and be clear (that is, not cloudy or hazy) and without malodours.

Quality and safety evaluation of processed olives

According to the Codex Alimentarius (1987)/IOOC (2004) Table Olive Standards, table olives are the sound fruit of specific varieties of the cultivated olive *Olea europaea* L., harvested at the correct stage of ripeness and processed to produce safe, edible, nutritious and marketable products. Therefore, table olives are a unique type of food. The general physical characteristics expected of processed table olives are detailed in Table 6.1.

Table olive growers, processors and traders must, therefore, be familiar with the accepted quality characteristics of processed table olives. Important general characteristics include the Flesh:Stone ratio (olives with a higher Flesh:Stone ratio are preferred) and the overall appearance of the olives. Olive size is important; uniform size is highly desirable unless they are packed as specialty items where several varieties are packed together, for example in a continental mix. Processed olives should be firm and retain the textural characteristics of the flesh and positive organoleptic characteristics. Olives that are soft (due to spoilage, or are over-ripe) and have fibrous or granular flesh (unripe) are less desirable. Olives should have colours consistent with maturation and style. Consumers expect black olives to be dark brown to black. Buff to pale brown table olives that were intended to be black are an indication that the olives were picked and processed too early or have lost colour during processing.

Table 6.1. General quality characteristics of processed table olives

Characteristic	Description
Flesh:Stone ratio	Must be no less than 3 (75% flesh) in black olives and 4 (80% flesh) in green olives.
Size (uniformity)	Olives should be of uniform size. See Table 6.2 for Codex Alimentarius/IOOC size grades and tolerances.
Flesh	Fine texture, not fibrous or granular, good aroma and flavour.
Firmness	At processing should be firm enough to prevent damage from handling.
Colour	Should be of appropriate colour: green olives, straw green; olives turning, pale pink; black olives, black or magenta.
Stone	Small, preferably smooth and easily detached from the flesh (freestone).
Appearance	Sound, no injury and/or defects.

The following sections focus on four levels of testing: (*1*) physical testing and evaluation of processed table olives, (*2*) chemical testing of table olives and brines, (*3*) microbiological testing of table olives and brines, (*4*) organoleptic evaluation of processed table olives and brines.

Physical testing/evaluation of processed table olives

Variety plays an important role in the quality of table olives. Olive varieties such as *Sevillana, Manzanilla, Kalamata, Leccino, Barouni* and *Mission* (Californian) are ideally suited for table olive production because their fruit has a good shape and size, favourable Flesh:Stone ratio, firm delicate flesh (not coarse or woody) and fine skin.

The flesh should separate easily from the stone, but this is not such a problem when the olives are deliberately bruised or cracked, pitted or stuffed.

See Chapter 2 for other table olive varieties.

Olive size. When packed for sale, processed olives should be of a uniform size. Size sorting is done by rough grading by hand or mechanical sorter of the raw olives prior to processing and again following processing. The olives can be categorised according to the number of olive fruit in one kilogram. A representative sample of olives (1 kg) is weighed and the number of olives counted.

Numerous categories have been detailed in the Codex Alimentarius (1987)/IOOC (2004) Table Olive Standards for international trade, ranging from 60–70 pieces/kg (large olives) to 381–410 pieces/kg (small olives) (see Table 6.2). Some table olive producing countries also use their own local categories.

Table 6.2. Summary of the size scale for grading table olives as set by the Codex Alimentarius (1987)/IOOC (2004) Table Olive Standards

Olive size range	Interval
60/70–111/120	10
121/140–181/200	20
201/230–381/410	30
Above 411	50

The size ranges apply to whole, bruised, slit, pitted, halved and stuffed olives and, in the case of imported olives, are often reported on the label. Of course, when processed olives of several different varieties are mixed as part of a specialty continental mixture, these ranges cannot apply. In the case of pitted olives, the size shown is the one corresponding to the original processed whole olive. For trade purposes, the number of stoned olives in a one kilogram sample is multiplied by a correcting factor set by the producing country to obtain the size range.

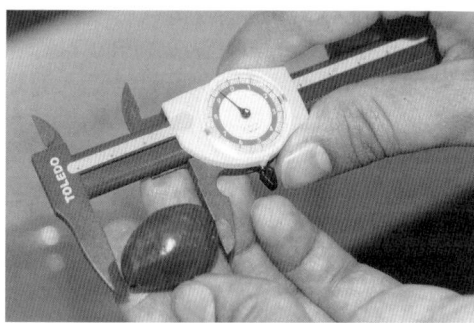

Figure 6.1 Determining olive dimensions using a hand-held micrometer.

The variation within a size range is also specified to ensure uniformity of olive size. Variability is determined by taking a representative sample of 100 olives from the batch. From this sample two olives, one with the smallest and one with the largest horizontal diameter, are removed. The horizontal diameters of the remaining 98 olives are measured. The difference between the horizontal diameters of the remaining olives may not exceed 4 mm if the international standard is to be met. The length and diameter of the olives can be measured with a hand-held micrometer (Fig. 6.1).

Other grading systems exist in which the olives of different sizes are given descriptive as well as numerical classifications (pieces/kg):

Bullets	351–380	Jumbo	181–200
Fine	321–350	Extra jumbo	161–180
Brilliant	291–320	Giants	141–160
Superior	261–290	Colossal	121–140
Large	231–260	Super colossal	111–120
Extra large	201–230	Mammoth	101–110

Flesh:Stone ratio. This parameter is easily determined by taking a representative sample of 50–100 olives, weighing the whole fruit, removing and weighing the stones and applying the following formula. The Flesh:Stone ratio is calculated for both raw and processed olives.

$$\text{Flesh:Stone ratio} = (\text{weight of olives} - \text{weight of stones}) \div \text{weight of stones}$$

The resulting value is expressed as a ratio where a value of five or more is considered desirable. The Flesh:Stone ratio is influenced by the variety and the amount of moisture in the flesh (which can be influenced by the level of natural precipitation or irrigation). Generally, the higher the moisture content in the olive flesh, the greater is the Flesh:Stone ratio. During processing, olives lose some of their mass (5–10%), hence the Flesh:Stone ratio is reduced. When olives are salt or heat-dried the Flesh:Stone ratio decreases because of moisture loss.

Note: As raw olives lose moisture during storage, they should be processed as quickly as possible as the Flesh:Stone ratio will also fall.

Olive firmness. The hardness of olives (see also Chapter 2) can be felt by pressing the olive between the fingers. In the laboratory, firmness can be assessed with instruments. Values obtained with a Shor 00 hardness tester for raw olives are higher than for processed olives (Table 6.3). Processed green olives are generally firmer than processed naturally black-ripe olives. If olives are overtreated with lye or are affected by some types of microbial spoilage, they tend to be softer than when processed correctly.

Note: Overtreatment with lye includes any or all of the following situations: lye solution concentration too high, prolonged soaking in lye solution or processing temperature too high.

Table 6.3. Indicative arbitrary pressure units as per Shor 00 apparatus for raw and naturally processed olives (measured by the authors)

Maturation state	Raw olives Shor 00 hardness units	Naturally processed olives Shore 00 hardies units
Deep green	100+	Too green to process
Green-ripe (straw colour)	80–90	70–80
Turning colour	70–80	60–70
Naturally black-ripe	50–70	50–60

Californian/Spanish-style black olives have a similar level of firmness as processed green-ripe olives because raw green-ripe olives are the starting material. Pressure measurements are not undertaken with salt or heat-dried olives as structural changes occur during drying, hence pressure measurements become meaningless.

Olive consistency. Olive flesh consistency or texture can be assessed subjectively or objectively. Subjectively it is one of the parameters that can be evaluated by a tasting panel or expert tasters. This is discussed below. With respect to instrumental evaluation of olive texture, sheer compression and compression measurements are made and a texture index is determined using the following formula:

$$\text{Texture index} = \text{sheer compression} \div \text{compression}$$

A major problem with texture measurements is the comparability of results between centres. Factors identified as affecting texture include: olive variety, brine characteristics, olive size and thermic processes (pasteurisation). There is reasonable correlation between subjective evaluation of texture and values obtained using laboratory instruments. For more information the reader is directed to Fernández *et al.* (1997).

Olive colour. The colour of olives is mostly assessed subjectively. There are some basic rules. Green-ripe olives processed by the Spanish-style have a distinct olive green hue, whereas lye treated green-ripe olives without fermentation take on a deep green colour (Plate 30). Californian/Spanish-style black-ripe olives take on a uniform black colour because of the oxidation of polyphenols in the skin and flesh. The colour of naturally black-ripe olives processed by soaking in water or brine or by spontaneous fermentation are brown to purple-black depending on the variety, and the coloration can vary between specific batches. Naturally processed green-ripe olives vary in colour depending on the processing procedure and variety. *Verdale* olives tend to be straw coloured whereas *Manzanilla*, *Sevillana* and *Hardy's Mammoth* tend to be olive green. If processed green-ripe olives are exposed to air, they take on a grey-green colour.

With the objective of standardising the colour description of olives, methods using reflectance and colour measurements have been applied. Although these techniques are commonly used in food and non-food industries for quality evaluation, their application to processed olives needs further investigation. For more information the reader is directed to Fernández *et al.* (1997).

Stone characteristics. Consumers prefer olives that have freestones that are small, round and smooth, rather than those with stones that are large, have rough surfaces or sharply pointed ends. Stone characteristics are often linked to the freestone/clingstone nature of the fruit and the ease of removing the flesh when in the mouth (see Chapter 2).

Physical quality criteria for table olives

According to the Codex Alimentarius (1987)/IOOC (2004) Table Olive Standards, based on the level of defects, three trade categories are identified: 'Extra', 'First' and 'Second'. The permitted levels of defects (Table 6.4) and definitions of defects for olive trade preparations follow.

Extra (or fancy) olives. This category includes high quality olives endowed to the maximum extent with characteristics specific to variety and trade preparation. As long as the overall favourable aspects and organoleptic characteristics are not compromised, a low level of defects is permitted.

Olives in this category include whole, split or stoned olives of the best varieties, providing their size exceeds 351/380. 'Best varieties' is not defined but could include most of the larger varieties detailed in Chapter 2.

First, 1st, choice or select olives. These are good quality olives with a greater level of defects allowed compared with 'Extra' olives. Olives in this category include all types of preparations and styles except for chopped or broken olives and olive pastes.

Second, 2nd or standard olives. Olives in this category are of a lesser quality than 'Extra' or 'First' categories, and permitted to have a greater level of defects as long as they meet the Codex Alimentarius (1987)/IOOC (2004) Table Olive Standards composition standards.

Table 6.4. Tolerance of defects
(After Codex Alimentarius (1987)/IOOC (2004) Table Olive Standards.)

Category	Extra/Fancy			First/1st/Choice/Select			Second/2nd/Standard		
	Green	Olives darkened by oxidation	Turning colour and black olives	Green	Olives darkened by oxidation	Turning colour and black olives	Green	Olives darkened by oxidation	Turning colour and black olives
Whole, stoned or stuffed olives Maximum tolerances as % of fruit									
Blemished fruit	4	4	6	6	6	8	10	6	12
Mutilated fruit	2	2	3	4	4	6	8	8	10
Shrivelled fruit	2	2	4	3	3	6	6	6	10
Abnormal texture	4	4	6	6	6	8	10	10	12
Abnormal colour	4	4	6	6	6	8	10	10	12
Stems	3	3	3	5	5	5	6	6	6
Cumulative maximum of tolerance of these defects	12	12	12	17	17	17	22	22	22
Maximum tolerances as units/kg or fraction									
Harmless extraneous material	1	1	1	1	1	1	1	1	1
Stoned (pitted) or stuffed olives Maximum tolerances as % of fruit									
Stones (pits) and/or stone (pit) fragments	1	1	2	1	1	2	1	1	2
Broken fruit	3	3	3	5	5	5	7	7	7
Defective stuffing Place-packed	1	1	1	2	2	2	–	–	–
Random-packed	3	3	3	5	5	5	7	7	7

1. Defects are assessed on a minimum representative sample of 200 olives.
2. For olives presented in the following styles – halved, quartered, divided, sliced, chopped or minced, broken, salad olives (except when prepared with whole olives) and olive paste – the presence of a stone (pit) or stone (pit) fragment shall be tolerated for every 300 g of net drained content of olive flesh.

Description of physical defects for processed table olives

The following explanations refer to Table 6.4.

Blemished fruit. Olives with marks on the skin that are more than 9 mm² (approximately 6 mm diameter) in surface area and that may or may not penetrate through the flesh. This includes insect or microbial damage that occurs after processing.

Mutilated fruit. Olives damaged by tearing the epicarp (skin) to such an extent that a portion of the mesocarp (flesh) becomes visible. (This does not apply to bruised olives.)

Broken fruit. Olives damaged to such an extent as to affect their normal structure. (This does not apply to bruised olives.)

Shrivelled fruit. This applies to whole, destoned, stuffed, halved and quartered olives where the appearance is materially affected, for example olives in strong brines, or olives with gas pockets. This defect does not include dried olives, which wrinkle during processing.

Abnormal texture. This includes olives that are soft/mushy or fibrous/woody.

Abnormal colour. This includes olives whose colour is different from the trade type and/or the average colour of the batch.

Stems (except for style with stems). Stems still attached to fruit and more than 3 mm in length. Table olives with stems that are a specialty product are excluded.

Harmless extraneous material. This defect includes vegetable matter not harmful to health or aesthetically undesirable, for example leaves and stems, but not herbs and spices.

Defective stuffing. This defect includes the following: blemished tissue, defective colouring of stuffing material, olives with no stuffing, olives incompletely stuffed and olives stuffed other than along the main axis.

Stone (pit) or stone (pit) fragments (except for whole olives). This includes whole stones or stone pieces weighing at least 5 mg.

Figure 6.2 Damaged olives that need to be removed before processing.

Figure 6.2 depicts a number of damaged olives that should be removed before processing and certainly after processing.

Chemical quality criteria for table olives

Chemical quality criteria for table olive products involves the evaluation of brine and olive flesh.

Brine analysis

Reducing sugar levels. Reducing sugars (glucose and fructose) present in olive flesh are the principal carbon source for lactic acid bacteria and other fermentative microorganisms. Reducing sugars are those sugars containing an atomic group, such as an aldehyde or ketone, which is easily oxidised. All the monosaccharides are reducing sugars, as are some disaccharides, but not sucrose. However, when sucrose is metabolised to glucose and fructose, these products are then available for fermentation. Sugar levels can be determined in several ways. There are some relatively fast semiquantitative methods using proprietary pharmaceutical products normally used to assess reducing sugar levels in the urine of diabetic subjects. Accurate laboratory methods use either a glucose oxidase enzyme system to measure glucose or high-pressure liquid chromatography to measure individual sugars.

Tests should be undertaken routinely during fermentation and storage of olives (and on packing solutions) as a control tool. A specific situation where brine sugar levels are measured is the case of suspected blocked fermentations. Samples should be removed aseptically using a clean sterile ladle to avoid contamination of the olives.

Note: When olives undergo complete fermentation, the concentration of reduced sugars is generally low to zero. With weak or incomplete fermentation, reduced sugars are detectable in the fermentation brine. Sugars sometimes added into packing solutions or released from additives, such as herbs and spices, are also detected by chemical testing.

Clinistix™ and Diastix™ test strips

Clinistix™ and Diastix™ test strips (Bayer, NSW, Australia) are designed for the semiquantitative measurement of glucose levels in the urine of diabetic patients. The strips do not measure other reducing sugars. Strips are coated with the enzyme glucose oxidase and a colour indicator system. The latter changes colour when glucose reacts with the enzyme and, depending on the glucose concentration, a range of colour shades develop.

The test strips can also be used to semiquantitatively determine the amount of glucose in fermentation brines. A test strip is dipped into a representative well-mixed sample of brine. The strips are removed and then read against a colour chart. To avoid contamination, the strips are never directly dipped into brine. The brine sample is discarded after testing and is never returned to the processing barrel/tank.

Note: A laboratory version of the above strip tests is also available. It is also based on the glucose oxidase system, using a liquid system and UV spectroscopy.

Clinitest™ tablets

Clinitest™ tablets are designed for the semiquantitative measurement of reducing sugar levels in the urine of diabetic patients. The tablets do not measure individual sugars. The test is based on the classic Benedict's copper reduction. Each tablet contains ingredients that react to reducing sugars, as well as sodium hydroxide (caustic soda) that generates heat (helping the reaction) when it comes in contact with aqueous solutions.

This test is based on a coloured copper complex forming when reducing sugars react with copper salts. Soluble copper sulfate (cupric sulfate) is converted to a copper complex (cuprous oxide) when it reacts with reducing sugars. The greater the level of copper complex, orange in colour, the higher the level of reducing sugars.

For assessing brines a representative sample of brine is removed from the processing tank/barrel and a 0.5 ml sub-sample, measured with a graduated pipette, is placed in a pyrex glass test tube (13 mm internal diameter × 145 mm height) and a Clinitest® tablet added. Heat is generated when the sodium hydroxide in the tablet interacts with water and a vigorous frothy reaction occurs. When the reaction is completed the test tube is shaken gently and the colour of its contents compared with the colours on the supplied Clinitest™ colour chart. The chart colour that gives the closest match to the test sample is recorded.

Note: As Clinitest™ tablets contain sodium hydroxide (caustic soda), users should avoid contact with the skin, eyes, mucous membranes (lips, mouth) and clothing.

High pressure liquid chromatography

The levels of individual sugars can be determined by taking a representative sample of brine from which a 10 ml aliquot is taken and made up to 100 ml with distilled water in a 100 ml volumetric flask. A sub-sample of the diluted brine is filtered through a 0.45 micron filter into a sample vial from which a portion is automatically removed for analysis. The sugars in the sample are then separated on an amino-bonded column and detected with a refractive index detector. The levels of individual sugars are calculated using pure sugar standards. This procedure requires a high level of technical expertise and expensive equipment, so would normally be undertaken in a registered chemical laboratory.

Sodium chloride. Salt levels are measured as part of the control procedures during fermentation, storage and of packing solutions for table olives. The amount and correct balance of salt and acidity (pH) are essential for the growth of fermentative organisms and the inhibition of food spoilage and/or food poisoning organisms. Sodium chloride levels in fermentation brines can be determined by a number of methods. Approximate values can be obtained with a conductivity meter, salinometer or a salt refractometer. Exact values can be measured by a titration method, flame photometry or an ion selective electrode. All three methods give comparable results for olives that have not been treated with lye. With the last two methods corrective factors need to be applied. Equipment should be kept in good working order and calibrated regularly against standard solutions.

Salt refractometer

Portable salt refractometers (Fig. 6.3) come in at least two measuring ranges, 0–10% w/v salt and 0–28% w/v salt. It is advisable to have both because the one that reads to 10% cannot detect higher salt levels, so if errors are made in the amount of salt used to make brines they cannot be detected. Only instruments with ATC ratings should be purchased as these have built-in automatic temperature compensation systems covering the range 10–30°C.

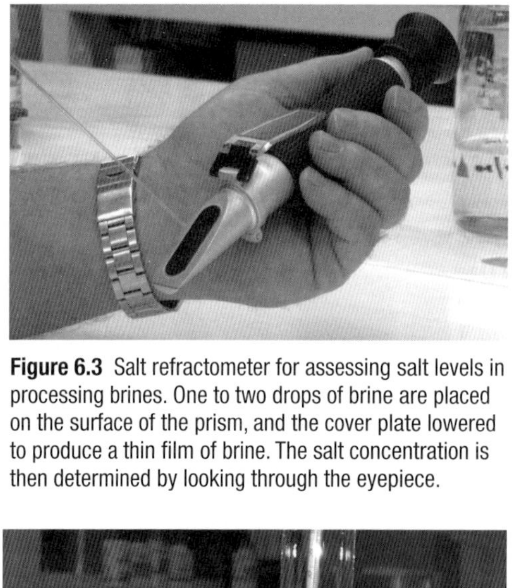

In principle, once the refractometer is calibrated with standard solutions, one to two drops of brine or packing solution are placed on the prism surface and a daylight cover plate is closed, spreading the liquid evenly over the prism surface. After 30 seconds, the refractometer is pointed towards a bright light source, and values read as a percentage on the scale. The prism and daylight cover plate are cleaned after each reading and before the instrument is stored for future readings.

Figure 6.3 Salt refractometer for assessing salt levels in processing brines. One to two drops of brine are placed on the surface of the prism, and the cover plate lowered to produce a thin film of brine. The salt concentration is then determined by looking through the eyepiece.

Salometer

Salometers, or salinometers, are a type of hydrometer made of glass and consist of a weighted lower end and a scale at the upper end which reads from 0–100% (note this is not salt concentration) (see Fig. 6.4). The salometer needs to be calibrated before use with water (0% salt) and 5% and 10% salt solutions. Salometers are inexpensive but very fragile. Because they are 30 cm long, a relatively large volume of solution is needed to make a measurement. A salometer should never be placed directly in solution, fermentation or storage tanks.

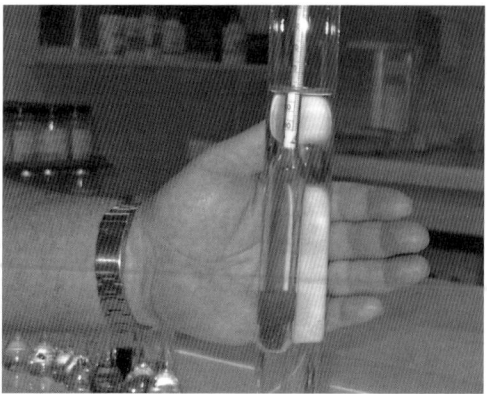

Figure 6.4 Salometer for assessing salt levels in processing brines.

Note: Although salometers are commonly used, the authors recommend that, if possible, the use of glass instruments be avoided in table olive processing facilities.

Volhard titration method for sodium chloride estimation

Using the titration method, duplicate samples of well-mixed brine are titrated with a standard silver nitrate solution using potassium chromate as the indicator, which changes from yellow to pink.

$$Na^+Cl^- + Ag^+NO_3^- \rightarrow AgCl + Na^+NO_3^-$$

Sodium chloride + Silver nitrate → Silver chloride (precipitate) + Sodium nitrate

The detailed procedure is as follows. A 1 ml sample of brine is pipetted into a 250 ml glass conical flask (Erlenmeyer flask). Approximately 50 ml of distilled water is then added to the brine sample in the flask and the contents mixed well. Approximately 1 ml of a 5% aqueous potassium chromate solution is then added to the flask. The mixture at this stage is clear and bright yellow in colour. The contents of the flask are then titrated using a burette with a standard 0.1 Molar silver nitrate solution until the first permanent red tinge is observed. During the titration a white precipitate of silver chloride is also formed. The quantity of silver nitrate solution needed to reach the end point is determined. To calculate the salt concentration in the brine, the following equation is used:

Percent sodium chloride (g/100 ml) = silver nitrate (ml used) × 0.5845 (for 1.0 ml brine)

A typical calculation is as follows:

If 9 ml of 0.1M silver nitrate solution is required,

% sodium chloride in the brine = 9 × 0.5845 = 5.26% w/v

Note: For this assay it is important that all glassware and equipment is washed with distilled water as tap water can contain significant amounts of sodium chloride.

pH measurements of olive brine. As controlling brine acidity during olive fermentation is critical, regular measurements of brine pH are essential. pH levels can be determined in several ways, including commercial pH strips and 'stick' pH meters (Fig. 6.5). Ensuring that the brine is well mixed, a sample of brine, for example 50 ml, is removed aseptically from the containers/tanks for testing. Several brands of pH strips are available. The Merck Company has developed strips with several ranges and can measure pH to one decimal point.

Figure 6.5 Portable stick-type pH meter.

Accurate pH measurements using a pH meter

The acidity of brines is measured by inserting a glass pH electrode into a brine sample. Readings are made on the associated meter (potentiometer). Portable pH meters can be used in the table olive processing facility, and more accurate desktop pH meters are used in analytical laboratories. The pH electrode must be well maintained and washed and cleaned with distilled water after use. pH meters should be calibrated against standard pH solutions before use, and if used regularly, at least once per day using standardised buffers. Common pH standards are 4 (acid), 7 (neutral) and 10 (alkaline). When not in use, the tip of the electrode is stored in 3 Molar potassium chloride solution. If the pH meter has been turned off, a 15 minute warm-up time is required.

Total titratable (free) acidity. Total titratable (free) acidity is a measure of the total acid in the brine. The measurement includes food acids produced during fermentation, for example lactic and acetic acids, and any externally added acid.

This test, based on a titration, is undertaken on brines during fermentation and on storage and packing solutions.

A representative brine sample is titrated with 0.2 Molar sodium hydroxide until reaching an end point detected by a colour indicator. Methods using a pH electrode as the detector are also available. These are better suited to the coloured brines associated with processed natural black olives in brine (Kalamata-style olives and Californian/Spanish black olives).

Titration method with colour indicator (phenolphthalein)

An accurately measured 10 ml aliquot of a representative brine sample is pipetted into a 250 ml conical flask (Erlenmeyer flask). Approximately 1 ml of 1% w/v ethanolic solution of phenolphthalein indicator solution is then added to the flask. The solution at this stage is clear and colourless. The contents of the flask are then titrated by adding small volumes, via a burette, of standard 0.2 Molar sodium hydroxide until the first permanent magenta/red tinge is observed. The flask and contents are shaken throughout the procedure to ensure efficient mixing of brine and sodium hydroxide solution.

The quantity of sodium hydroxide solution required to reach the end point is determined. Duplicate and even triplicate samples of brine should be tested. The following formula is used to determine titratable acidity as lactic acid.

Titratable acidity calculated as lactic acid for a 10 ml brine sample:

Titratable acidity g/100 ml (% w/v) = $0.9 \times$ volume (ml) $\times$ 0.2 Molar (sodium hydroxide).

A typical calculation follows:
For example, with a titration volume of 5 ml, the titratable acidity as

$$\% \text{ w/v} = 0.9 \times 5.0 \times 0.2 = 0.9\% \text{ w/v as lactic acid.}$$

Titration method with pH meter

An accurately measured 25 ml aliquot of a representative brine sample is pipetted into a 150 ml glass beaker containing a magnetic stirrer. The beaker and contents are then placed on the magnetic stirrer base and the pH electrode inserted towards the wall of the beaker so that it is well away from the butterfly magnetic stirring bar. The stirrer is then turned on and standard 0.1 Molar sodium hydroxide solution is introduced dropwise using a burette. The volume of 0.1 Molar sodium hydroxide required to achieve pH 8.1 is determined. Duplicate and even triplicate samples of brine should be tested.

Titratable acidity as lactic acid for a 25 ml brine sample:

Titratable acidity g/100 ml (% w/v) = 0.36 × volume (ml) × 0.1 Molar sodium hydroxide.

A typical calculation follows:
For example, with a titration volume of 15 ml, titratable acidity as

$$\% \text{ w/v} = 0.36 \times 15.00 \times 0.10 = 0.54\% \text{ w/v as lactic acid.}$$

Evaluation of table olive brines

Brine salt concentrations, pH and acid levels are important parameters used to monitor olives during and after processing. Lactic fermentation produces more free acid and a lower brine pH than mixed fermentations where lactic and acetic acids are produced

Table 6.5. Chemical evaluation of table olive brines (Kailis and Harris 2004)

Source	Statistical parameters	pH	Salt % w/v	Free acidity as % w/v lactic acid
Olives processed by authors				
Natural green	Mean	4.64	6.73	0.34
	Std deviation	0.40	0.69	0.24
	Median	4.50	6.71	0.26
	No. of samples	49	55	56
Natural turning colour	Mean	4.60	6.85	0.40
	Std deviation	0.19	0.47	0.06
	Median	4.55	6.87	0.43
	No. of samples	6	7	7
Natural black	Mean	4.71	6.63	0.40
	Std deviation	0.76	0.76	0.21
	Median	4.70	6.82	0.37
	No. of samples	15	19	19
Australian processed olives Bulk (all methods)	Mean	4.04	5.95	0.68
	Std deviation	0.46	2.90	0.41
	Median	3.95	4.99	0.62
	No. of samples	28	28	28
Australian processed olives Packaged olives (all methods)	Mean	3.71	5.60	0.81
	Std deviation	0.35	1.20	0.48
	Median	3.70	5.66	0.70
	No. of samples	50	51	51
Bulk imported olives (all methods)	Mean	4.42	7.49	0.50
	Std deviation	1.05	0.91	0.24
	Median	4.40	7.28	0.39
	No. of samples	5	5	5
Loose olives imported (all methods)	Mean	4.19	7.75	1.07
	Std deviation	0.18	0.41	0.39
	Median	4.18	7.73	1.19
	No. of samples	4	4	4

(see Chapters 4 and 5). The balance of salt, pH and acid are also important in the preservation and safety of the olives. In some centres the salt levels are maintained at 10% w/v sodium chloride after processing, particularly if prolonged storage times are envisaged. It is then up to the packer to develop packing brines that meet the stipulated requirements.

All trade preparations of table olives not meeting the standards may only be marketed if they are made according to traditional methods and the food safety is guaranteed by an official body, which authorises their distribution and sale. The presence of propionic acids or their salts may be observed in table olives that have undergone natural fermentation in accordance with Good Manufacturing Practice. Propionic acid has a characteristic sharp, sweet odour and has some antifungal activity.

Data presented in Table 6.5, derived from testing by the authors (Kailis and Harris 2004), gives processors an idea of the pH, salt and free acid levels for commonly available processed olives (local and imported) in Australia. The authors also processed olives by spontaneous fermentation in brine (10% w/v salt) for a number of varieties and at different maturation states and the summary results are presented.

Table olive flesh analysis

Proximate analysis and fatty acid profiles are undertaken to determine the chemical and nutritional characteristics of table olive flesh. Proximate analysis measures moisture, oil, protein and ash content of the olives. The total carbohydrate content is determined by difference (see below). Data from the proximate analysis and fatty acid analysis are used to prepare the nutritional panel required for the label on retail containers as required by FSANZ (Food Standards Australia New Zealand). The nutritional panel also includes the sodium level in the olives. If the olives are packed in olive oil, or have stuffings, the nutritional values will be different to just the processed olives. A specific example is with *Kalamata* olives where added olive oil gives a higher fat value per olive compared with olives in brine.

Proximate analysis of table olive flesh. Proximate analysis involves the steps indicated in Table 6.6. Data must be converted to an 'as received value'. Initially, a sample of flesh is dried to obtain the moisture content. The dried olive flesh is then defatted to determine the oil content. A sample of defatted olive flesh is then analysed for nitrogen (to calculate the protein content) and a further sample is ashed to determine the overall mineral content.

Carbohydrate levels are determined by difference.

i.e. carbohydrate % w/w = 100% − % moisture − % oil − % protein − % ash

A sub-sample of dried flesh is also used to determine macro- and micro-nutrient levels. Proximate analysis data and other compositional parameters measured by the authors (Kailis and Harris 2004) are given in Tables 6.7 and 6.8 for a number of olive varieties and processing methods.

Moisture level in processed olive flesh. The level of moisture in processed olive flesh (around 60–80% w/w) depends on several factors: variety, maturation level, growing environment and processing method (Table 6.7). Fully irrigated olives generally have higher water content than those grown without irrigation. The oil content of processed olive flesh shows an inverse relationship with moisture content; thus, more water equates with less oil.

Table 6.6. Summary of proximate analysis methodology for processed olive flesh
Ash, fat and protein levels must be calculated to an 'as received' basis

Activity	Result
Remove and weigh olive flesh from a representative batch of olives	Total weight of olive flesh
Dry olive flesh to constant weight	Calculate % w/w moisture content
Remove the oil from the dried olive flesh by hot solvent extraction (soxhlet)	Calculate % w/w oil content Determine fatty acid profile of oil
Measure nitrogen content in a sub-sample of dried olive flesh	Calculate % w/w protein content
Ash a sub-sample of dried olive flesh	Calculate % mineral content
Carbohydrate levels	Calculated by difference

Total fat levels in processed olive flesh. The 'fat' in processed olive flesh is essentially olive oil. Between 75% and 85% of the caloric value of table olives is predominantly oleic acid (monounsaturated fatty acid). A number of studies (for example Wahlqvist and Kouris-Blazos 2001) have shown that people who live in certain Mediterranean regions (including Crete) and consume a Mediterranean diet show a decreased incidence of coronary heart disease and several cancers. Monounsaturated fatty acids help to reduce levels of low-density lipoproteins (LDL), or 'bad' cholesterols, preventing harmful arterial plaque forming on artery walls. While table olives have beneficial monounsaturated fatty acids, the salt content of most processed table olives may be detrimental, particularly in those who should be on a low-salt diet.

Table 6.7. Composition of olive flesh from olives processed by spontaneous fermentation in brine (data from Kailis and Harris 2001)

Component	Kalamata Black-ripe	Manzanilla Green-ripe	Manzanilla Black-ripe	Verdale Green-ripe	Overall range
Moisture	61.0	69.5	69.5	78.0	61.0–78.0
Oil (total fat)	25.5	18.4	19.0	11.0	11.0–25.5
Saturated fatty acids %	12.5	17.5	18.5	18.5	12.5–18.5
Polyunsaturated fatty acids %	14.0	5.5	10.1	16.5	5.5–16.5
Monounsaturated fatty acids %	70.4	76.5	71.1	65.1	65.1–76.5
Carbohydrate					
Total %	8.0	7.0	6.0	6.0	6–7
Soluble sugars %	0.10	0.25	0.30	0.10	0.10–0.30
Protein %	2.1	1.2	1.1	1.5	1.1–2.1
Minerals					
Phosphorus %	0.02	0.02	0.01	0.02	0.01–0.02
Potassium %	0.3	0.2	0.15	0.25	0.15–0.3
Sodium %	1.9	1.8	2.2	1.7	1.7–2.2
Calcium %	0.05	0.04	0.03	0.05	0.03–0.05
Magnesium %	0.01	0.01	0.01	0.01	0.01
Sulfur %	0.03	0.02	0.02	0.02	0.02–0.03
Boron mg/kg	10.9	4.5	3.6	3.4	3.4–10.9
Copper mg/kg	3.9	1.5	2.4	1.5	1.5–3.9
Iron mg/kg	5.1	3.2	3.4	3.7	3.2–5.1
Manganese mg/kg	1.1	0.7	0.8	0.7	0.7–1.1
Zinc mg/kg	2.4	2.2	1.9	2.0	1.9–2.4
Ash (minerals) %	3.9	3.5	4.2	4.1	3.5–4.2

Quite wide variations in oil content occur between different processed table olive samples due to varietal differences, with a trend to lower levels when olives are treated with lye. It is also possible that lye reacts with a portion of the oil in the flesh, producing soaps and reducing the oil content as well as impacting on the organoleptic properties of the olives (giving them a soapy taste). Olives with high oil levels, for example *Frantoio*, *Leccino* and *Kalamata*, when processed in brine by spontaneous fermentation (natural) at the green-ripe or turning colour state, have a rounded nutty flavour that is absent in varieties such as *Jumbo Kalamata* or *UC13A6*. The oil in olives is an excellent alternate source of fat where animal fat is low or excluded from the diet.

Fatty acid levels. The oil in olive flesh consists predominantly of triacylglycerols – compounds containing glycerol and three fatty acids – the proportions of which can be measured. After the oil is extracted by hot solvent extraction, an aliquot is transmethylated in concentrated methanolic potassium hydroxide solution where volatile fatty acid methyl esters (FAMEs) are formed. FAMEs are separated by gas chromatography using a BPX–70 microbore column, their relative proportions determined, and then categorised as saturated, monounsaturated and polyunsaturated fatty acids.

$$\text{Triacylglycerols} + \text{Methanol} \rightarrow \text{Fatty Acid Methyl Esters (FAMEs)} + \text{Glycerol}$$

These three classes of fatty acids show some variation between varieties for both non-lye and lye treated olives. However, the levels of monounsaturated fatty acids (the major group) show the greatest consistency between different table olive samples. The levels of different fatty acids in olive flesh give an indication of the nutritional quality of olives. From a positive nutritional point of view, around 90% of the monounsaturated fatty acid content in table olives is the health benefiting oleic acid. Significant levels of beneficial polyunsaturated fats are also present in table olives, ranging from 5.5%–16.5% w/w in olives processed by spontaneous fermentation in brine and 3.7%–13.5% w/w in lye treated olives.

Table olives prepared by drying with salt or heat show similar results to olives processed in brine, except the components are more concentrated. Salt levels in these types of table olives depend on the pre-treatment levels of brine or solid salt.

Protein levels in processed olive flesh. The protein content of flesh from olives processed by spontaneous fermentation in brine is around 1–2% w/w but lower if the olives are treated with lye. A possible explanation is that solubilised proteins are lost from the flesh during lye treatment and through the subsequent multiple washes. Although protein levels are relatively low in table olives compared with animal meat and dairy products, they are in useful quantities where animal protein is low or excluded from the diet.

Carbohydrate and sugar levels in processed olive flesh

Total carbohydrate levels are of the same order of magnitude for both brine and lye treated olives.

Soluble (free) sugar levels in the flesh of processed table olives range from 0–0.3% w/w. Naturally processed table olives always retain a small amount of soluble sugar, whereas lye treated olives have none (Table 6.8). The variation in soluble sugar levels could reflect the amount of original free sugar in the olive flesh before processing, or as a result of the efficiency of the processing procedure. The free sugar level is a required parameter on the nutritional label. If sugar is added to processed table olives as a secondary processing procedure, the levels in the flesh will increase depending on the amount of added sugar. Furthermore, heat-dried olives become sweeter as they retain their original sugars during processing. Some soluble sugars are lost with salt-dried olives when the liquid drawn out with the salt is discarded. Varieties with high levels of sugars enhance fermentation.

Table 6.8. Composition of processed olive flesh from olives processed with lye (data from Kailis and Harris 2004) [1]Olives treated with ferrous gluconate. %, weight/weight.

Component	Chalchidikis Sevillean-style	Manzanilla Spanish-style black[1]	Manzanilla Sevillean-style	Sevillana Sevillean-style	Overall range
Moisture %	67.0	76.0	74.0	79.0	67.0–79.0
Oil %	15.0	14.0	17.0	11.0	11.0–17.0
Saturated fatty acids %	17.2	18.1	19.1	18.4	17.2–18.1
Polyunsaturated f. a. %	10.7	3.7	7.6	13.0	3.7–13.0
Monounsaturated f. a. %	72.1	78.3	73.3	68.6	68.6–78.3
Carbohydrate					
Total %	7.8	5.9	5.7	6.6	5.7–7.8
Soluble sugars %	0	0	0	0	0
Protein %	0.4	1.1	1.2	1.4	0.4–1.4
Minerals					
Phosphorus %	0.01	0.02	0.01	0.01	0.01–0.02
Potassium %	0.06	0.02	0.06	0.06	0.02–0.06
Sodium %	2.2	1.0	1.8	1.8	1.0–2.2
Calcium %	0.10	0.05	0.04	0.10	0.04–0.10
Magnesium %	0.005	0.010	0.010	0.010	0.005–0.010
Sulfur %	0.03	0.02	0.01	0.02	0.01–0.03
Boron mg/kg	1.6	2.4	6.3	1.8	1.6–6.3
Copper mg/kg	2.4	1.8	0.7	0.9	0.7–2.4
Iron mg/kg	3.8	150.6[1]	2.8	4.1	2.8–4.1
Manganese mg/kg	0.60	1.90	1.00	0.06	0.06–1.90
Zinc mg/kg	1.3	3.0	1.6	1.2	1.2–3.0
Ash (minerals) %	5.7	2.4	2.4	2.3	2.3–5.7

Mineral content of processed olive flesh

Ash levels are of the same order of magnitude for both brine and lye treated olives.

The levels of 11 minerals, both macro- and micro-, in processed olive flesh are presented in Tables 6.7 and 6.8. The levels for each olive variety are of a similar order except for boron, and in the case of Black Spanish-style olives the iron level is markedly higher because these olives are treated with ferrous gluconate or ferrous lactate to stabilise their black colour. This level of iron in the table olive will not cause any harm when consumed and will contribute to daily iron requirements and so could be of particular benefit to women, vegetarians and vegans. It should also be noted that the potassium level is lower in processed olives than in the raw olives.

In general, it would appear that the processing technique has a greater effect on the macro-minerals than the micro-minerals. With phosphorus, potassium and magnesium, there appears to be a leaching of the minerals from the olives. Calcium and sulfur are maintained at similar levels. This could indicate that the calcium is strongly bound in the flesh and the sulfur is part of the sulfur amino acid methionine.

Salt levels in processed olives. Sodium levels will vary according to the strength of sodium chloride in the processing and packaging brines and, for salt-dried olives, the amount of salt used. Sun-dried and heat-dried olives, if prepared from raw olives, contain negligible amounts of salt. The maximum recommended daily intake (RDI) for sodium is around 2.3 g/25 g serve of olives. For a 25 g sample of commercially available table olives in Australia the sodium content is of the following order of magnitude: Spanish-style green olives (whole, destoned and stuffed) give 19–20% of the RDI; Spanish-style black olives (whole and destoned) give 13–14% of the RDI, and Kalamata-style olives (whole, destoned and stuffed) give 14–15% of the RDI.

Some olives can, therefore, provide more than 20% of the RDI for sodium. Consumers that are watching their salt intake should avoid eating large quantities of table olive products where there are more than 1000 mg of sodium per 100 grams of edible portion stipulated on the label.

Microscopic examination of fermentation brines

Microscopic examination of fermentation brines, particularly in the early stages of fermentation, can be used to detect the presence Gram negative bacteria that can cause spoilage and be harmful to health. On the basis of staining techniques discovered by Dr H. C. Gram in the late 1800s, bacteria can be divided into two broad categories: Gram positive (stain purple/blue) and Gram negative (stain red).

The source of Gram negative organisms can be the fruit and poor quality water. The presence of Gram negative bacteria, such as coliforms, indicates the potential for olive spoilage, such as gas pocket fermentation and food poisoning, so needs to be controlled. Coliform bacteria are indicator organisms found in soil, decaying vegetation, animal faeces and untreated surface water. They are not normally present in deep groundwater or in surface water that has been treated. These indicator organisms may include pathogens but generally do not cause disease in healthy people. People at potential risk such as the very young, pregnant women and those with compromised immune systems may be affected.

Adjustment of pH and/or the addition of a starter culture of lactobacilli at the beginning of processing can overcome this problem. Microorganisms such as bacteria and yeasts in the fermentation brine can be observed microscopically. Fermentative organisms (Gram positive lactobacilli) stain blue and are easily differentiated from Gram negative coliforms, which are red and have a different shape.

Microbiological evaluation of table olives

During routine monitoring for hygiene, food hazard potential, processing and spoilage, the levels of indicator microorganisms should be assessed. Microorganisms to be assessed are ones that are likely to survive under the processing conditions and those that are easy to monitor. Prescribed tests and limits for olive flesh and brine are given in Table 6.9.

Generally, fermentation brines are tested during processing, whereas storage brines and olive flesh are tested with packaged olives.

Olives and brine must be free of microorganisms and parasites or below amounts that would be harmful to health. Furthermore, they must not contain substances such as bacterial or fungal toxins in amounts harmful to health.

Recent investigations undertaken by Greek researchers have been directed to establish the sites of microbial activities associated with fermented table olives. They found that during fermentation of black olives, Enterobacteriacae and *Pseudomonas* spp. in the brine decreased, while lactic acid bacteria and yeast populations increased. Scanning electron microscopy revealed that a yeast-rich biofilm developed on the epicuticular wax of the olive skin during fermentation. Yeasts also predominated in the stomatal openings, but bacteria were more numerous in number in the intercellular substance and the sub-stomatal flesh. The implications are that the microorganisms associated with olive skin and flesh may experience a different growing environment than that prevailing in the fermentation brine (Nychas *et al.* 2002).

Levels of various microorganisms in brine and olive flesh are determined by the standard procedures shown below.

Lactic acid bacteria
   Culture medium: MRS agar (De Man, Rogosa, Sharpe) Oxoid CM361 (Oxoid, UK)
   Incubation conditions: 48 hours at 37°C anaerobically; spread plates
   Gram stains, biochemical identification in API50CHL kits (Bio-Merieux, France)

Clostridium species
   Culture medium: TSC agar (Tryptose Sulphite Cycloserine agar) Oxoid CM587 with supplement SR88
   Incubation conditions: 24 hours at 37°C anaerobically; spread plates
   Confirmation of growth: Gram stain, followed by purification on non-selective agar and biochemical identification in API ID32A kits (Bio-Merieux)

Enterobacteriaceae
   Culture medium: Violet Red Bile Agar with glucose at 1% Oxoid CM107
   Incubation conditions: 24 hours at 30°C aerobically; pour plates
   Confirmation of growth: gas production in LT broth with 1% glucose added (Oxoid CM451) for 48 hours

Yeasts and moulds
   Culture medium: DG18 agar (Oxoid CM729)
   Incubation conditions: five days at 25°C aerobically; spread plates
   Confirmation of growth: microscopic examination

All dilutions are plated in duplicate. Sampling steps are undertaken using aseptic procedures. Results are expressed in colony forming units per millilitre (cfu ml^{-1}) or per gram (cfu g^{-1}) for brine and olive flesh respectively.

Table 6.9. Limit levels of microorganisms in brine and olive flesh

Organism	Brine limits cfu ml^{-1}	Olive flesh limits cfu g^{-1}	Test method
Lactic acid bacteria	<10	<100	American Public Health Association 15.413
Yeasts	<10	<100	Australian Standard 1766.2.2
Moulds	<10	<100	Australian Standard 1766.2.2
Clostridium spp.		<100	
Clostridium perfringens		<100	Australian Standard 1766.2.8
Clostridium botulinum		None	American Public Health Association 36.6
Coagulase positive *Staphylococcus aureus*		<100	Australian Standard 1766.2.4
Escherichia coli		<3	Australian Standard 1766.2.3
Enterobacteriaceae	<1	<10	Australian Standard 1766.2.3
Listeria monocytogenes		None	Australian Standard 1766.2.15
Sample size	100 ml	100 g	

Microbiological analysis of processed olive flesh and brines

Enterobacteriaceae (a marker organism) or *Escherichia coli* are used to monitor the potential for food poisoning. Lactic acid producing bacteria and yeast counts are indicators of fermentation, and mould counts are used to detect fungal contamination. For routine monitoring, levels of indicator organisms are assessed for hygiene, food hazard potential, processing and spoilage. Olives preserved by sterilisation, such as Spanish/Californian-style black olives, must receive a processing treatment sufficient in both time and temperature to destroy *Clostridium botulinum* spores. During processing, brine pH and salt levels should also be monitored regularly.

Sampling procedure for olive brine. A 100 ml sample is removed from the tank, bulk container, tin or jar. Prior to taking the sample ensure that the contents of tanks, barrels or containers are well mixed. Samples are removed aseptically into a sterile or previously unused container. The samples are kept cool for transport to the laboratory.

Sampling procedure for olive flesh. A 100 g sample is removed from the tank, bulk container, tin or jar. Prior to taking the sample the contents are well mixed. Olives are removed aseptically into sterile, or clean, previously unused jars. The samples are kept cool for transport to the laboratory.

The tests listed in Table 6.10 should be undertaken at the end of processing; however, if problems occur during processing, such as stagnant initial pH levels or stuck fermentations, these tests should be undertaken as part of the management of the process. Samples of olives packed in containers without pasteurisation should be tested to ensure the absence of harmful organisms, yeasts and moulds. The addition of 300 ppm sorbic acid (as potassium sorbate 400 ppm) reduces the chance of yeast or mould growth occurring during storage and/or after opening.

Where packed olives are pasteurised, testing for *Escherichia coli* is undertaken for health and safety purposes. *Escherichia coli* should be destroyed by pasteurisation as its presence in a supposedly 'pasteurised' product indicates either inadequate heat treatment or post-pasteurisation contamination.

Table 6.10. Microbiological tests for table olives

Test	Background reason for testing	Olives in processing tanks Brine	Processed olives in brine Packed olive flesh	Processed pasteurised olives Packed olive flesh	Processed dried olives Loose olive flesh
Enterobacteriaceae	Indicator of: Hygiene Food hazard	Do test	Test optional		
Escherichia coli	Indicator of: Contamination Food hazard		Do test	Do test for verification of health safety	
Coagulase positive Staphylococci	Food hazard/toxin		Do test	Other tests indicate microbe activity	Do test
Clostridium perfringens (Spore forming – more resistant)	Anaerobic spoilage bacteria Food hazard/toxin		Do test	Do test	
Clostridium botulinum	Can cause serious illness but not tested routinely	No test	No test	No test	No test
Listeria monocytogenes	Potential food hazard		Do test		Do test
Lactic acid bacteria	Processing indicator	Do test	Test optional		
Yeasts	Processing indicator	Do test	Do test	Do test (indicator of pasteurisation effectiveness)	Do test
Moulds	Indicator of: Contamination	Do test	Do test	Do test	Do test
Water activity estimation	Tests moisture level in heat-dried olives				Do test if there is a problem; at the end of process

The Codex Alimentarius (1987)/IOOC (2004) Table Olive Standards indicates that proprionic bacteria are the reference organisms for pasteurisation (see Chapter 4). *Clostridium perfringens* is an indicator of microbial spoilage as well as a food hazard, yeast counts are an indicator of pasteurisation effectiveness and mould counts are an indicator of microbial contamination. Several tests are recommended for dried olives and an additional physical test for water activity indicates the level of moisture remaining in the flesh after drying.

Note: Because food safety is a major issue, semiquantitative simple to use tests are available that can be undertaken in the processing facility as a guide, for example tests for Enterococci, *E. coli*, *Salmonella* and listeria. New tests are being developed. Where positive, suspected samples should be sent to a registered food laboratory for confirmation.

Organoleptic evaluation of table olive products

The appeal and allure of table olives, eaten on their own or in prepared food, is attributed to their aroma, flavour and texture. Flavour for table olives, as experienced by individuals, varies from bitter to sour, to piquant (sharp, spicy), to sweet. The flavour varies with olive variety, the processing method and by the contribution of embellishments (vinegar, olive oil, herbs and spices).

Organoleptic analysis of processed table olives serves a number of functions, such as providing a description of the table olive product; means of monitoring processing, for example debittering; a starting point for the evaluation of product abnormalities, defects and spoilage; and a focus for table olive competitions that seek excellence.

Changes in organoleptic characteristics are more easily discerned by tasters and consumers rather than by applying objective, physical, chemical or microbiological assessment. Official table olive organoleptic criteria are currently being developed by the IOOC using statistically robust methodology based on defined attributes and intensity measurements on a 10-point scale (see below). When these are implemented at the international level, consideration must be given to include these as part of the official assessment of table olives in Australia.

It is the view of the authors that credible organoleptic information on table olive products can be undertaken by a small panel of experts: an 'organoleptic panel' of 10–12 persons, or a group of 30 to 40 consumers.

A simple method for evaluating table olives used by the authors and based on the work of Marsilio (2002) is presented below.

In this evaluation, five olives from each sample were assessed by:

- visual observation: colour, uniformity blemishes;
- touch and pressure: olive firmness and softness;
- smell: for positive and negative odours; and
- taste, feel and kinaesthetic characteristics in the mouth.

The following specific attributes were assessed and the olive products graded on a scale of 1 to 10.

For positive attributes.   Appearance, colour, aroma, acid/vinegary/winey, salty, bitter, sweet, firmness, crispness, flesh/stone attachment, stone size.

For negative defects.   Poor appearance, skin hardness, off odour, rancid, mouldy, 'off' flavour, texture, metallic.

Data was collected using a structured data sheet (Table 6.11). The attributes tested and scores obtained are presented for traditional Kalamata-style olives. Perfect olives should have a clean skin, a shape consistent with the variety and any misshapen, damaged or small fruit should be removed at the beginning or after processing. Olives with greater colour intensity are more appealing than pale olives. Olives that have not completely reached the black-ripe stage after processing by fermentation in brine can take on different shades of colour. For example, turning colour *Verdale* variety olives take on a pink colour, turning colour *Jumbo Kalamata* variety turn pink to brown and partially black-ripe *Kalamata* variety olives range from chocolate brown to dark brown.

Few processed table olives have a sweet taste, indicating the absence of sugars. Sugars in the flesh are mostly consumed or lost during processing. A characteristic of all olives is that they have some degree of bitterness, albeit weak, in olives treated with lye. Sometimes sugar is added to sweeten the olives and in this case the olives would be scored high for sweetness.

For appearance and colour, scores of 10 would reflect perfect skin appearance and colour intensity and homogeneity. Scores for aroma and taste reflect intensity and very high values would indicate imbalance, for example too salty or too bitter. Such negative features can be scored under others for negative attributes. Some tasters experience a sensation of sweetness with strong sodium chloride brines whereas others report sweetness in the absence of bitterness.

Table 6.11. Organoleptic evaluation protocol for table olives with typical scores for black Kalamata-style olives

Positive attributes (high scores are desirable)

Score		0	1	2	3	4	5	6	7	8	9	10
Appearance	Skin									X		
	Shape										X	
	Size								X			
Colour	Yellow to green											
	Light brown to dark brown									X		
	Intensity									X		
	Homogeneity							X				
Aroma	Strength										X	
	Acid/winey/vinegary						X					
Taste	Acid/winey/vinegary						X					
	Salty								X			
	Bitter			X								
	Sweet	X										
Texture	Firmness									X		
	Crispness								X			
Flesh:Stone	Pit – flesh detachment										X	
Other												

Negative defects (low scores are desirable)

		0	1	2	3	4	5	6	7	8	9	10
Appearance	Spots			X								
	Blistering		X									
	Sloughing		X									
	Shrivelling		X									
	Holes		X									
	Shape			X								
	Size		X									
Aroma	Rancid		X									
	Mouldy		X									
Taste	Rancid		X									
	Mouldy		X									
	'Off' flavour		X									
Texture	Soft				X							
	Woody		X									
	Fibrous		X									
Other												

Following evaluation by expert tasters, an organoleptic panel or a consumer group, median defect scores can be used to categorise the organoleptic quality of the olives. The highest numerical value of an individual negative attribute is taken as the score for a particular sample even if there were no other defects. If, after evaluation of the Kalamata-style olives in the above example, the maximum defect score was 1, these olives would be classified as 'Extra'. In general, the higher the defects score for a particular table olive product, the lower the quality (Table 6.12).

Table 6.12. Organoleptic grading of table olives based on negative defects (after Marsilio 2002)

Grading	Median score
Extra	<1.5
First	>1.5 <3.5
Second	>3.5 <5.5
Third	>5.5 <7.5
Unsuitable for consumption	>7.5

Samples of table olives with negative scores above 7.5 are considered inedible. With respect to negative attributes, higher quality olives have the lowest scores. Olives with scores of less than two would be considered excellent for a particular attribute. Major problems can occur with respect to skin and texture, for example soft olives. Others can show degrees of shrivelling, probably due to processing or storing in brines with a salt concentration that is too high. 'Off' flavours can also be of a chemical nature rather than mouldiness or rancidity. These include solvent-like aromas and tastes due to ethyl acetate formed during processing (i.e. acetic acid + ethanol), soapy tastes due to lye treatments, and chemical taste, for example olives treated with iron salts. Lactic tastes are probably due to the addition of lactic acid to brines during pH adjustment. Overuse of vinegar gives an unpleasant tart taste. Metallic defect tastes can be present when olives are packed in metal cans or in jars with metal lids.

Organoleptic assessment of olives in Australia. We have tested a large number of table olives (Australian, imported and processed by the authors). The results are shown in Tables 6.13–6.15. As a guide, desirable scores are suggested. The results for positive attributes indicated that the appearance and colour of the olives were generally acceptable. Some individual table olive samples were pale or had darkened through interaction with air, whereas others were marked and lacked lustre.

On average, aromas were weak to moderate unless substantial amounts of herbs and spices had been included. Olive samples with very weak aromas are generally those repacked in brine after processing; these lack the aromas of the fermentation brine. Although, on average, processed olives were firm on squeezing, they generally lacked crispness unless the raw olives had been picked at the appropriate maturation state.

The flesh came off easily from many of the samples, but this is generally a function of variety. For example with *Manzanilla* and *Kalamata* olives the flesh attachment was very weak, whereas with *UC13A6, Verdale* and *Jumbo Kalamata* it was stronger. The flesh of black table olives generally came off more freely than that of green table olives. Of course this will depend on the variety, maturation state and processing method. Included in the 174 samples were 77 samples that had been packed for sale sourced from Australian processors and randomly purchased from retail outlets. It was interesting to note that all parameters followed a similar pattern except that the salt attribute gave a lower score. The data presented in Tables 6.13–6.15 are baseline values and more extensive studies are required.

Quality and safety | 267

Table 6.13. Positive attributes for table olive samples evaluated by the authors
Median scores for each attribute are presented.
*, if sugar is added after processing then values will be higher.

Attribute	Parameter	Realistic median scores for all olives (n = 174)	Median scores for olives in consumer packs (n = 77)	Desirable scores
Appearance	Skin appearance	7.4	7.1	10
	Shape	7.80	7.5	10
	Size	7.50	7.4	10
Colour	Intensity	6.6	6.8	10
	Homogeneity	6.6	6.6	10
Aroma	Strength	4.5	4.1	5
	Acid/winey/vinegary	3.0	3.0	3
Taste	Acid/winey/vinegary	3.0	3.0	3
	Salty	7.0	5.6	6
	Bitter	2.6	1.6	2–3
	Sweet	0.1	0.1	0.1*
Texture	Firmness	7.2	6.5	10
	Crispness	5.8	4.9	7
Flesh:Stone	Stone flesh detachment	6.6	6.9	7–8

Table 6.14. Negative attributes for all olives tested (174 samples). Numbers of samples/score for each defect score are presented.

Attribute	Parameter	0	1	2	3	4	5	6	7	8	9	10
Appearance	Spots		9	11	11	4	18	1	9			
	Blistering		1		1	1	2				1	
	Sloughing		1		2	2	1	1	1			
	Shrivelling		4	10	11	2	6	5	3		1	7
	Holes		2	2	1	1		1	3			
	Shape		34	20	12	1	9	6	5	2	3	4
	Size		16	20	16	4	13	3	11	1		
Aroma	Rancid		4	3		2						
	Mouldy		2	2								
Taste	Rancid		6	8	3	1						
	Mouldy		1	2	2	1						
	'Off flavour'		13	5	4	4	1	1	5	2	3	2
Texture	Soft		29	18	24	17	16	9	6	8	7	
	Woody		5	3	2	2						
	Fibrous		13	8	3					1		

Table 6.15. Negative attributes for olives in consumer packs (77 samples)
Numbers of samples/score for each defect score are presented.

Attribute	Parameter	1	2	3	4	5	6	7	8	9	10
Appearance	Spots	3	5	9	1	7	1	3			
	Blistering	1		1	1						
	Sloughing	1		1	2	1	1				
	Shrivelling	2	5	4	1	4	4	2			
	Holes	1	1	1	1		1				
	Shape	12	10	7		2	3	1	1		
	Size	10	7	6	2	3	2	5	1		
Aroma	Rancid	1	1								
	Mouldy	2									
Taste	Rancid	2	1	1						1	
	Mouldy	1	1	1							
	'Off flavour'	9	1	1							
Texture	Soft	10	14	14	9	9	5	1	2	3	
	Woody	1		1							
	Fibrous	1		2							

In summary, organoleptic attributes of table olives are listed in Table 6.16, along with a description of what tasters should be looking for when taste testing table olives. There are also comments on the attributes indicating what may have caused the positive or negative attributes. The organoleptic testing of table olives is in its developmental stage and still needs refining. More organoleptic testing needs to be undertaken on table olives to give processors and consumers a degree of confidence in table olive products. As mentioned earlier in the chapter, physical, chemical and microbiological testing can guarantee the safety of the product but the organoleptic testing will give table olive products a profile in the commercial marketplace.

Table 6.16. Summary of organoleptic attributes of table olives

Parameter		Attributes	Comment
Appearance	Sight Colour (intensity and homogeneity, size, shape, firmness, glossiness)	Positive Intact skin, uniform fruit size and shape Colour relevant to variety, ripeness and method Negative Abnormal colour, non-uniform fruit size and shape, skin spots, insect punctures, sloughing (relevant to method), gas pockets	Use quality olives Ripening stage Type Processing technology Keeping qualities
Aroma	Olfactory	Positive Olfactory intensity of the olives: primary fermentation, esters, alcohols, aldehydes, ketones and hydrocarbons Acids: Acetic, Lactic	Fermentation type Homofermentative: lactic acid Heterofermentative: lactic and acetic acids and ethanol + carbon dioxide (gas) Metabolites Olive variety

Parameter		Attributes	Comment
		Negative	
		Off odours: sensation of disagreeable or anomalous odour	Enzymatic reactions
			Process conditions: control salt and pH levels
		• putrid	
		• butyric	
		• *Zapateria* (bad leather)	
		Mouldy (e.g. mouldy foods)	Ensure anaerobic conditions
		Rancid aroma of aged fats	Turn over products as quickly as possible
Taste	Acid	*Positive*	
		Acid taste typical of fermented foods	Apparent to consumer
		Olive fruit metabolites – lactic, malic, citric and succinic	Ensure fruit is at correct maturation
		Fermentation: lactic, acetic	Control any additions of food acids to adjust pH
		Corrective measures: lactic, acetic, citric	
		Negative	
		Too vinegary	
	Salt	*Positive*	
		Typical of sodium chloride in salty foods	
		Negative	
		Too salty	Reduce salt levels and maintain pH at correct levels
	Sweet	*Positive*	
		Sensation completely devoid of bitter notes	Ensure primary processing is complete so that olives are debittered.
		Due to sugars, alcohol and glycerol	
		Due also to absence of bitters	
		Weak salt solutions	
	Bitter	*Positive*	
		Bitter taste similar to caffeine or quinine	Stimulates appetite and a characteristic sought after by consumers
		Phenolics such as oleuropein and related compounds	
		Negative	
		Potassium salts – bitter	Ensure processing end point is monitored
	Other	*Positive*	
		Hot/Spicy: spiced olives	Specific to olive processing method
		Negative	
		Californian black: ferrous gluconate, metallic taste	
		Off flavours of unpleasant or spoiled food: butyric acid taste, sour or sickly milk taste, solvent taste (ethyl acetate)	Use acetic acid to adjust pH of brines rather than lactic acid
	Balance	Aroma and taste combinations	Occurs if one taste character overpowers others e.g. too acid, too salty

Parameter		Attributes	Comment
Touch Finger Mouth	Firmness	*Positive* Force needed to press olive fruit between thumb and index finger and to bite the olive fruit with incisors Skin should be fine (not tough) *Negative* Sensation of thick hard skin Soft olives Grainy, woody, fibrous flesh	Processing technology Measure of keeping qualities Consistency • olive variety • ripeness • processing method • storage conditions Reduced with lye Consistency tests: compression test penetration test
Mouth texture	Crisp	*Positive* Greater force required to crunch the fruit with the back molars *Negative* Soft olives with no texture	Use olives at correct maturation Control fermentation conditions
Other	Oily	*Positive* Smooth feel when eating *Negative* Unpleasant if oil is rancid	Use quality olive oil
	Woody	*Positive* Flesh is smooth, non-granular and not too chewy *Negative* Granular woody feel	Use olives at correct maturation Use alternative variety
	Flesh and stone detachment	*Positive* Tendency of olive flesh to detach from stone easily *Negative* Flesh hard to remove from stone	Use freestone varieties Ensure processing is complete

Proposed IOOC organoleptic assessment of table olives

The International Olive Oil Council are working towards a standardised methodology for evaluating the organoleptic qualities of table olives. Towards this direction a number of attributes have been identified and these are presented in Table 6.17.

Table 6.17. Proposed vocabulary for table olives (after IOOC)

Attribute	Definition	Influencing factors	
		Processing factors	Other factors
Positive attributes			
Odour (aroma)	Sensation perceived nasally or retronasally, typical of the trade preparation that olives have undergone.	Affected by processing	Fermentation type
Taste	Gustatory sensation typical of the trade preparation that olives have undergone.	Affected by processing	Variety and maturation state
Negative attributes			
Abnormal fermentation (malodour)	Olfactory sensation (smell), perceived directly or retronasally, typical of putrid, butyric, *Zapateria* fermentations. It is reminiscent of the odour of decomposing organic matter, or rancid butter.	Processing problems	Poor quality water

Attribute	Definition	Influencing factors	
		Processing factors	Other factors
Musty (malodour)	Olfactory sensation (smell), characteristic of olives attacked by moulds and yeasts due to prolonged storage in damp conditions before processing.	Oxidative moulds in fermentation tanks	Olive storage before processing
Rancid (malodour)	Olfactory sensation (smell), characteristic of olives that have undergone oxidation (oil in olives oxidises).	Post-production	Oxidation of added olive oil
Winey-Vinegary (malodour)	Olfactory sensation (smell), characteristic of olives that have undergone fermentation leading to the formation of acetic acid and over-production of alcohol and ethyl acetate (nail polish odour).	Over-active yeast fermentation	Wine or wine vinegar added post-processing
Pasteurised (malodour)	Olfactory sensation (smell), characteristic of olives that have undergone excessive heating during thermic processes.	Poor control of pasteurisation	Overheating and/or prolonged heating
Cheese (malodour)	Olfactory sensation (smell), reminiscent of stale cheese due to abnormal breakdown of proteins.	Can occur during processing	Can occur after processing during storage
Other attributes			
Salty (taste)	Basic taste produced by aqueous solutions of sodium chloride.	Over-salting during processing	Salt levels too high in packing solutions
Bitter (taste)	Basic taste produced by oleuropein which is reminiscent of weak aqueous solutions of bitter substances such as quinine or caffeine.	Incomplete debittering during fermentation	• Genetic variation in tasting bitter substances • Herbs and spices
Acid/sour (taste)	Basic taste produced by aqueous solutions of most acid substances including tartaric and citric acids.	Acids produced during fermentation or added to adjust brine acidity and pH	May be due to additives including vinegar, citric acid, lemon slices
Metallic (taste)	Sensation related to an excessive iron presence due to processing treatments.	Mainly affects lye treated olives that have been oxidised and stabilised with iron	Olives processed in rusty vats or packed in contact with metal, e.g. lids, cans
Soapy (taste)	Sensation reminiscent of soap and lye.	Affects mainly lye treated olives	Equipment and containers washed with soap or detergent and not rinsed properly
Kinaesthetic attributes			
Hardness/firmness	Textural attribute related to the force required to achieve a given deformation of a product. In the mouth, it is perceived by compressing the product between the teeth (solids) or between the tongue and palate (semi-solids).	Can be affected by processing	Variety and maturation state
Crispness	Textural attribute related to the force necessary to break a product into crumbs (small pieces) between the teeth. It is assessed by compressing the fruit between the back molars.	Can be affected by processing	Variety and maturation state
Skin firmness	Resistance of the olive skin to rupture when placed between the incisors.	Affected by lye treatment	Variety and maturation state
Flesh/stone adherence	Resistance of the olive flesh to detach from the stone.	Can be affected by processing	Variety and maturation state
Fibrousness	Attribute related to the orientation of particles in a product. It is evaluated by perceiving the fibres between the tongue and palate when chewing the olives.	Can be affected by processing	Variety and maturation state

The proposed methodology in assessing table olives is to sample the olives from a glass similar to that used for organoleptic evaluation of olive oil. Elaboration of the method follows: the sample size equals the number of olives that cover the bottom of a testing glass; the olives are covered with the packing brine in the testing glass where appropriate; and testing is undertaken at a room temperature of 20–22°C. The intensity of each attribute is scored by a taster along a line and the responses measured and determined as a score of 10.

Organoleptic evaluation of olive pastes and tapenades. A system needs to be developed to evaluate olive pastes and tapenades. The authors propose that the above terminology for table olives (Table 6.17) could be used, except for the kinaesthetic attributes, together with additional attributes such as firmness of paste (whether runny/dry), texture of paste (coarse or smooth) and separation (in which oil separates from other ingredients) (Table 6.18). The development of a systematic method for the organoleptic evaluation of olive pastes and tapenades needs further discussion and refinement.

Table 6.18. Vocabulary for the organoleptic evaluation of olive pastes and tapenades based on the proposed IOOC nomenclature for table olives

Attribute	Definition	Influencing factors
Positive attribute		
Odour (aroma)	Sensation perceived nasally or retronasally, typical of the type of paste/tapenade.	Affected by • method used to process olives • other ingredients: anchovy, herbs and spices
Taste	Gustatory sensation typical of the type of olive paste/tapenade.	Affected by • method used to process olives • variety and maturation state of olives • other ingredients: anchovy, herbs and spices
Negative attributes		
Abnormal fermentation (malodour)	Olfactory sensation (smell), perceived directly or retronasally, typical of putrid, butyric, *Zapateria* fermentations. It is reminiscent of the odour of decomposing organic matter, or rancid butter.	• defective table olives used • rancid oil used • poor quality embellishments used
Musty (malodour)	Olfactory sensation (smell), characteristic of olives attacked by moulds and yeasts due to prolonged storage in damp conditions before processing.	• defective processed olives used • moulds develop in paste during storage
Rancid (malodour)	Olfactory sensation (smell), characteristic of olives that have undergone oxidation (oil in olives oxidises).	• defective table olives used • rancid olive oil used • oxidation of added olive oil • oxidation of paste/tapenade during storage
Winey-vinegary (malodour)	Olfactory sensation (smell), characteristic of olives that have undergone fermentation leading to the formation of acetic acid and over-production of alcohol and ethyl acetate (nail polish odour).	• defective table olives due to over-active yeast fermentation used • wine or wine vinegar included in the paste/tapenade

Pasteurised (malodour)	Olfactory sensation (smell), characteristic of olives that have undergone excessive heating during thermic processes.	• defective table olives used or overheated during pasteurisation or sterilisation • paste/tapenade overheated during pasteurisation
Cheese (malodour)	Olfactory sensation (smell), reminiscent of stale cheese due to abnormal breakdown of proteins.	• defective table olives used • can occur in the paste/tapenade during storage
Other attributes		
Salty (taste)	Basic taste produced by aqueous solutions of sodium chloride.	• oversalted table olives used • additions contribute to salt levels: anchovy, salted capers
Bitter (taste)	Basic taste produced by oleuropein which is reminiscent of weak aqueous solutions of bitter substances such as quinine or caffeine.	• bitter table olives used • genetic variation in tasting bitter substances
Acid/sour (taste)	Basic taste produced by aqueous solutions of most acid substances including lactic, acetic, tartaric and citric acids.	• acidic table olives used • additions contribute to acid/sour: vinegar, citric acid, lemon slices
Metallic (taste)	Sensation related to an excessive iron presence.	• Californian/Spanish-style black table olives stabilised with iron salts used • defective table olives used – olives processed in rusty vats or packed in contact with metal, e.g. lids, cans
Soapy (taste)	Sensation reminiscent of soap and lye.	• lye treated table olives used • equipment and containers washed with soap or detergent and not rinsed properly
Firmness	Sensation between fingers/sensation in the mouth.	• runny paste: too much oil • dry paste: insufficient oil
Texture	Sensation between fingers/sensation in the mouth.	Coarse paste • affected by type of table olive used, variety of olive and its maturation state • insufficient time in food processor
Separation	Visual examination of paste indicates oil separating from other ingredients	• too much oil used • ineffective mixing of ingredients

Table olive spoilage and deterioration

The principal causes of spoilage in olives are the growth of microorganisms (such as bacteria, yeasts and moulds), enzyme action, or oxidation either through direct action or indirectly through microorganisms.

Microorganisms can be introduced during processing through unhygienic procedures or by the proliferation of undesirable microorganisms if processing procedures are poorly controlled. Monitoring brine pH, salt and microorganism levels is important throughout processing, especially in the early stages, to reduce the risk of harm to consumers and to avoid spoilage. The chief types of deterioration in table olives are summarised in Table 6.19.

Growth of microorganisms and the levels of olive spoilage are influenced by factors such as: temperature, moisture, oxygen levels, available nutrients, degree of contamination and presence or absence of growth inhibitors.

Levels of spoilage organisms can increase with temperature, so during processing the fermentation brines should be maintained between 15°C and 25°C. Low temperatures can impair fermentation, increasing the processing time. Correctly processed heat-dried, sun-dried or salt-dried olives are less prone to microbial spoilage because of their lowered water content. The higher the initial levels of a contaminating microorganism, the greater the risk of spoilage. Ensuring hygienic practices throughout processing and packaging can markedly reduce levels of contaminating microorganisms.

Naturally occurring growth inhibitors in the olive, such as polyphenols, provide some degree of protection against microorganisms; however, with processing methods involving prolonged periods of soaking in water, this advantage can be lost. Such inhibitors can also affect the activity of fermentative microorganisms. Many spoilage microorganisms are also inhibited by high salt and low pH levels, hence the need for careful control of these two parameters during processing and storage.

Oxidative moulds growing on the surface of fermentation brines release metabolites that can taint the olives. Also, exposing processed green olives in brine to air changes their green colour to a dull grey-green colour. Dried olives can develop a rancid taste because of the oxidation of fats in the flesh and any added olive oil.

Addition of herbs, spices and marinades can introduce unwanted microorganisms, including *Clostridium botulinum*, particularly if fresh herbs are used. Additives used should be sourced from reliable suppliers and treated by irradiation or fumigation to destroy microorganisms. More detailed information on the types of table olive deterioration is presented in Table 6.19.

Further explanations of these and other defects follow:

Soft olives.   The reasons for olive softening include: the olives are over-ripe; picked olives overheat in the orchard; enzyme action in olive flesh due to prolonged storage before processing; and the actions of microorganisms, such as moulds, yeasts and some types of bacteria.

Possible moulds involved in the softening of olives, reported by Greek researchers, includes *Penicillium*, *Aspergillus*, *Geotrichum* and even *Verticillium*. Bacteria implicated in the softening of olives, including *Bacillus*, *Aerobacter* and *Escherichia* spp., generally break down pectin, a structural carbohydrate in olive flesh (Garrido Fernández et al. 1997).

As microorganisms are the most likely cause of softening, this can be reduced by carefully controlling pH and salt levels of the fermentation, storage or packing brines. A rapid drop in brine pH by adding food grade acid at the beginning of fermentation and maintaining anaerobic conditions prevents this type of spoilage. Starter cultures can also be used to quickly decrease pH as well as prevent the growth of microorganisms.

Sloughed skin.   Sloughed skin can result from the formation of blisters between the skin and flesh of the olives. Such blisters can be associated with fish-eye gas pockets. Blisters may also form in green olives when they are being treated with a lye solution that is too strong. Choosing the most appropriate lye concentrations can prevent skin sloughing during lye treatment of olives. The right lye concentration can be determined by testing samples of the olives with a range of lye concentrations.

Table 6.19. Chief types of deterioration associated with table olives (after Marsilio 1993).

Deterioration	Signs and symptoms	Cause	Prevention
Soft olives	Flesh loses firmness due to excessive pectin loss in cell walls	Yeasts and moulds that breakdown pectin in the olive flesh; irrational processing technology, e.g. excess water soaking	Good processing techniques Avoid too many water changes during processing
	Naturally black-ripe olives	Overripe olives	Correct timing of olive harvest
Sloughed skin	Skin lifts or is shed	Coliforms: bacteria that break down cellulose in the fruit; inappropriate processing technology Strong lye solutions can also cause sloughing during debittering	Brine acidification; brine adjustment so that NaCl ≥8% w/v; pasteurise olives Use correct strength of lye solution
Cloudy brine	Brine looks streaked and mucilaginous	Uncertain cause, possibly due to microorganisms or substances leaching into the brine	Replace brine; acidify; NaCl ≥8% w/v
White skin spots	White to cream spots appear on the skin around fruit stomata	Some species of yeast. In Spain with *Gordal* variety, spots are due to *Lactobacillus* colonies.	Control of yeast development by adjusting brine and pH levels
Gas pockets	Formation of gas pockets, fish eyes, fissures and cavities in the flesh	Coliforms: spore forming bacilli; yeasts break down pectin in the olive flesh Can occur early in processing, but with directly brined turning colour or naturally black olives can occur any time depending on conditions.	Ensure strict anaerobic conditions; scrupulous standards of hygiene; brine acidification with acetic or lactic acid; brine adjustment so that NaCl ≥8% w/v Starter cultures can be used
Putrid and butyric fermentations	Occur early in processing. Malodours of varying intensity, e.g. faecal, rancidity	Bacteria and moulds	Good processing and storage techniques; scrupulous standards of hygiene; brine adjustment so that NaCl ≥8% w/v A starter culture can be used
'Zapateria'	Occurs late in processing. Smell of old leather		
Galazoma or cyanosis (bluish)	Unpleasant odours; black olives turn bluish-ashen colour	Presence of ferrous (iron) salt; low salt concentration; high pH; exposure to air; excessive orchard irrigation	NaCl ≥8%; right irrigation; good processing techniques; ensure anaerobic conditions; fruit pasteurisation

Skin marks. Skin marks such as bruising are likely to be caused by rough handling of the olives during harvesting, transporting and processing. These types of skin marks are less likely to be caused by microorganisms.

Darkening. All olives darken in colour during or after processing when allowed to stand in air. This is advantageous in the case of black olives. Processed black-ripe olives are deliberately exposed to air making them darker. However, with green olives, and in particular green Spanish-style olives, darkening reduces their visual appeal because they turn grey-green in colour. Ways of overcoming this problem include undertaking all operations as quickly as possible, especially during sorting and grading. Antioxidants can reduce this problem in packed olives.

Galazoma or cyanosis. Cyanosis is characterised by black olives turning a dark blue colour in contrast to the expected brown to black when exposed to air. Cyanosis is often associated with unpleasant odours. Obviously both problems impact on the organoleptic qualities of the olives, reducing their consumer appeal and commercial value. Strict control of salt concentrations in the brines (above 8% w/v salt) and pH (less than 4.5) reduces the risk of cyanosis. Other factors may also be involved. Greek researchers have suggested adherence to hygiene and using quality water are important in the prevention of cyanosis. Olives processed in brine and stored in salt concentrations of greater than 8% w/v do not develop cyanosis. Furthermore, pasteurising olives prevents cyanosis, suggesting microbial involvement. Cyanosis is less of a problem in small olives from unirrigated trees compared to large olives, and mature olives are more resistant than immature fruit (Garrido Fernández et al. 1997).

Green stains. Patches of green staining due to copper sprays are observed when green-ripe olives are placed into lye when processing Spanish-style green olives. At the time of delivery growers should advise processors if copper sprays, for example copper oxychloride, have been used in the orchard. Best practice is to use copper treatments after harvest and pruning and ensure that the olives are well washed before processing.

Brine surface yeasts and moulds. Yeasts and moulds that can utilise oxygen from air above the brine surface can grow in the upper levels of the fermentation tank/container, lowering the brine acidity and so raising the pH. As high brine pH levels promote olive deterioration, this situation must be avoided. The problem can be prevented in fermentation tanks with well-sealing lids and minimal volumes of air between the lids and the brine surfaces to produce anaerobic conditions (Plate 31). In packaged olives, formation of yeast films on the surface of brines after containers are opened can be prevented by adding a preservative such as potassium sorbate.

White skin spots. White skin spots are small pinhead-sized white spots that are easily visible from the surface of green olives after processing. They are due to microorganisms such as yeasts forming colonies under the skin around the small stomatal structures that the fruit uses for respiration. Yeast spots reduce consumer appeal and commercial value of the olives. Similar spots can be caused by overgrowth of *Lactobacilli*, including *Lactobacillus pentosus* (formerly *Lactobacillus plantarum*). Low sugar levels in the brine can be a reason why microorganisms appear on the olives. Adding fermentable sugar can reduce the problem.

Gas pockets. Gas pockets, also known as fish-eyes, are pockets or blisters in the olive flesh filled with gas (carbon dioxide/hydrogen) produced by microorganisms. Gas pockets can form in both green and black olives during processing. Culprit gas producing organisms are Gram-negative bacteria, for example species of *Enterobacter*, *Citrobacter*, *Klebsiella*, *Escherichia*, and *Aeromonas* in Spanish-style green olives and yeasts in black olives (Plate 32). A related condition, 'Alambrado', is characterised by fissures under the skin. The olive skin has the appearance of having been pressed by wire.

These problems need to be prevented from occurring as there are no corrective procedures. Preventive measures include using:

- good hygienic practices;
- clean uncontaminated water and equipment;
- good quality ingredients that are protected from contamination;
- good manufacturing practice, including acidification, inoculation with starter cultures and appropriate salt concentration; and
- a carbon dioxide saturated atmosphere.

Microorganisms present at the beginning of fermentation can dictate the effectiveness of the process. The fermentation process is critical and when the appropriate controls are in place, gas pocket formation can be reduced to very low levels.

With Spanish-style green olives, lowering the pH of the brine to pH 5 or less with food grade acids (acetic, lactic or hydrochloric) or carbon dioxide, or the addition of a *Lactobacillus* inoculum, will inhibit the growth of Gram negative bacilli. Acidifying as well as inoculating with a starter culture is advantageous. Salt concentrations should be maintained at around 8% w/v sodium chloride. Other factors that will promote gas pocket formation include olive variety and abnormally elevated brine temperatures at the beginning of the fermentation process. Using aerobic fermentation systems will reduce the incidence of gas pockets in olives during processing (see Chapter 5). The above factors also apply to black olives.

Malodorous deterioration of processed olives

Clostridial bacteria cause malodorous fermentations and are common to putrid, butyric and *Zapateria* spoilage. Organoleptic qualities of the resulting olives, such as aroma and taste, are compromised rendering them inedible. Clostridial bacteria are relatively common in the environment and can be found, for example, in water and orchard dust. Because they can produce spores that can survive under harsh environmental conditions, such as high temperatures and humidity, they have a latent potential for spoilage. *Clostridium botulinum*, a specific member of this group, is harmful to health and life. Contaminating organisms in untreated dried herbs and spices can start to grow in poorly formulated marinades (such as those with low salt and high pH).

Using good hygienic and manufacturing practices reduces malodorous deterioration. Stagnant water, organic waste and contaminated reticulation systems and pumps are possible sources of Clostridia. In plants where olive oil is also produced, olive oil processing and its associated wastes must be kept separate from the table olive processing section to prevent cross contamination. Once a contaminated tank is identified it must be quarantined from the rest of the processing operations. Several types of malodorous spoilage have been defined.

Putrid spoilage. Putrid spoilage results in the product having the faecal aroma of decomposing organic matter. Paying attention to hygiene, practising scrupulous sanitation and using potable water are very important. Poor quality water and poorly maintained equipment can lead to 'putrid' fermentation. It is important that all equipment is cleaned and sanitised thoroughly, perhaps using steam cleaning before the start of the next processing season.

Butyric spoilage. If salt and pH levels are poorly controlled, butyric acid accumulates in the brine through the action of *Clostridium butyricum* and olives become tainted, having the aroma and taste of rancid butter. Butyric fermentations are uncommon when olives are prepared by traditional methods where brines have high salt and high polyphenol levels and fermentation temperatures are low. Adjusting brines with acid to around pH 4 is also inhibitory.

Note: Putrid and butyric spoilage occurs early in table olive processing. A raised brine pH during processing is a signal that spoilage may occur. To prevent these problems from occurring, the pH is lowered initially by adding a food acid with or without inoculation with lactic acid bacteria. With putrid or butyric spoilage it is difficult or impossible to recover the table olives for consumption.

Zapateria. Zapateria spoilage results in the product having the aroma of old leather, but it is not as malodorous as butyric fermentation. Researchers have identified chemicals with unpleasant odours (such as propionic, *n*-butyric, *n*-valleric, *n*-caproic and cyclohexanecarboxylic acids, and putrescine and cadaverine) produced during the microbial fermentation of protein. This problem can occur at the end of fermentation due to an overgrowth of *Clostridia* and *Propionibacteria* (Garrido Fernández et al. 1997). These types of organisms proliferate in poorly controlled fermentations, particularly at lower salt concentrations and when there is an increase in environmental temperatures. These organisms consume the food acids (lactic and acetic acids) produced during normal fermentation resulting in loss of brine acid and an increase in brine pH. Ensuring that the brine levels are around 6% w/v salt and the brine pH is 4.5 or less (preferably between pH 3.8–4.0) during fermentation controls this problem. pH values greater than 4.5 will support *Zapateria* spoilage so it is necessary to add food grade acid to reach the required level of acidity. A further precaution is to raise the sodium chloride levels after fermentation to above 8.5% w/v salt, especially during periods of hot weather.

Taking the following actions can prevent malodorous problems:

- ensure good hygienic practices by operators and workers;
- use potable water;
- use food quality ingredients protected from contamination;
- use clean uncontaminated equipment;
- remove dregs from the bottom of the tank;
- use good manufacturing practice (acidification, inoculation with starter cultures and appropriate salt concentration);
- monitor and adjust the brine pH; and
- monitor sodium chloride levels in the brine.

As dregs at the bottom of the fermentation tanks can harbour *Clostridial* organisms, tanks should be cleaned and sanitised between batches. If deterioration is suspected, then partial correction can be attempted by removing a portion of the brine, followed by the addition of fresh brine with enough salt and acid to stop the problem.

Processing considerations. The fermentation process is critical and when the appropriate controls are in place gas pocket formation can be reduced to very low levels. Microorganisms present at the beginning of fermentation can dictate the effectiveness of the process. Poor control at the beginning of table olive processing can lead to proliferation of Gram negative organisms that have the potential to induce food poisoning if the olives are consumed. Growth of microorganisms can increase with temperature, so during processing the fermentation brines should be maintained between 15°C and 25°C. Low temperatures can impair fermentation resulting in longer process times, whereas high temperatures (>30°C) can lead to the growth of anomalous or harmful microorganisms.

Packaging and labelling olive products

Table olives as foodstuff need to be labelled according to Australian and New Zealand food standards, and if exported, labelled according to international standards and other country specific requirements. Relevant organisations responsible for establishing labelling requirements for table olives are: Food Standards Australia New Zealand (FSANZ), the Codex Alimentarius Commission and the International Olive Oil Council (IOOC).

FSANZ labelling requirements can be summarised as follows:

Nutrition labelling
Percentage labelling
Name and description of food
Food recall information
Information for allergy sufferers
Date marking
Ingredient list
Truthful labelling
Food additives
Legibility requirements
Storage requirements
Country of origin

As labelling requirements can vary from country to country, those wishing to export Australian olives need to become familiar with individual country requirements.

If table olive products have been processed and packed correctly they have a long shelf-life of one to two years. However, they are best consumed within 12 months of production. The organoleptic qualities of table olives deteriorate on storage, particularly when edible oils, herbs, spices and marinades are included. Packing solution specifications are given in Chapter 5. As salt levels in olives can be a problem for people with cardiovascular and renal disease, table olives in low salt brines are preferred and such products require pasteurisation or even sterilisation.

A typical jar or can of olives sold in Australia has the information shown in Table 6.20.

Table 6.20. Information required on a typical label for table olives

Brand	Feature
Name/description of product	Green olives in brine with chilli and garlic
Olive variety (optional)	*Manzanilla*
Size range of olives (optional)	121/140
Source of product	Product of Australia
Gross weight (volume)	375 ml
Drained weight	200 grams
Percentage ingredients	Olives, water, salt, vinegar, olive oil, chilli, garlic
Nutritional panel	See Table 6.21
Number of serves/container	18 (serves of 6 olives)
Serving size/number of olives	Edible portion – 20 grams/6 olives
Contact details	Postal address, phone no, facsimile no, email address
Storage/consumer information	Store in a cool place and refrigerate after opening
Use by date/best before date	Two years
Batch identification	Number/bar code
Other	Tamper-proof container

Name and description.   Packaged table olives must be labelled with an accurate name and description. Labels must not include misleading information. The consumer often finds it very hard to know which olive variety they are eating. In many cases, particularly with green olives, the variety of olive is not included on the label.

It is worth noting that some table olives are labelled according to the style of production. Spanish-style olives refer to olives debittered with sodium hydroxide then subjected to fermentation in salt brine. *Kalamata* olives should be only of that variety and not look-alikes. Kalamata-style olives, carrying this nomenclature, can be produced from other olive varieties such as *Hojiblanca, Leccino* and *Barnea*. Naturally black-ripe olives (Greek-style) are fermented in brine. Californian/Spanish-style black olives are those treated with sodium hydroxide, darkened by oxidation then packed in salt brine.

The Codex Alimentarius (1987)/IOOC (2004) Table Olive Standards have comprehensively defined the different types of table olive preparations, albeit some having only subtle differences. Information on these, detailed in Chapter 4, can be used as a guide when describing a specific table olive product.

Source of product.   In Australia, packaged table olive products from any source are required to have the country where the food was made or processed on the label. This could mean stating the country where the olives were packed for retail sale. If some of the ingredients do not originate from that country, the label should state 'made from imported or local and imported ingredients'.

In Australia, 'Product of Australia' means it must be made in Australia from Australian ingredients. If there are significant imported ingredients then the label should state 'Made in Australia'. Packaged imported olives that have not undergone significant transformation should be labelled 'packaged in Australia'. These definitions are summarised below.

Product of Australia.   Australian grown olives, all Australian ingredients, olives processed and packed in Australia.

Made in Australia.   Australian grown olives, processed in Australia but containing imported herbs, spices and other additives.

Packaged in Australia from imported ingredients.   Imported olives packed in the original imported brine.

Packaged in Australia from local and imported ingredients.   Imported table olives packed in Australian marinade made from local and imported materials.

Product of Greece.   Table olives processed and packaged in Greece.

Percentage labelling.   Packaged foods are required to carry labels that show the percentage of the key or characterising ingredients or components in the food product. For table olives this includes the following: olives, water, vinegar, salt, olive oil, herbs and spices.

Ingredients must be listed from the greatest to the smallest quantities by ingoing weight, including added water. If small amounts of multi-component ingredients are under 5%, the composite ingredient can be listed, for example herbs and spices. Depending on the shape and capacity of jars and tins, processed olives occupy around 50–60% of the volume. As an example, specific information on the label could be: olives (60%), water (20%), salt (5%), olive oil (5%), vinegar (2%) and herbs (0.5%).

Nutritional labelling. All manufactured foods sold in Australia and New Zealand, including processed table olives, are required to include on the label a nutrition information panel so that consumers can make informed decisions before purchase. Of particular importance for those with cardiovascular disease, diabetes and high blood pressure would be the fat and salt content of table olives as well as the levels of saturated fats. Because of the relatively low levels of sugars in most processed table olive products this is less of a problem. Sugar levels are high in table olive products when sugars are added after processing or the products are dried olives that have not previously undergone fermentation, for example sun-dried or oven-dried where the sugars concentrate.

Olives sold loose or packed at the point of sale require no nutritional labelling, but it is sometimes included.

In Australia and New Zealand, Food Standards Australia New Zealand (FSANZ) directs the requirements for a nutritional label. Only limited information is available on the FSANZ food composition calculator from which to develop a nutritional label for table olives. Therefore, chemical analyses of a representative sample of the table olive product need to be undertaken by a registered chemical laboratory. The main analyses are protein, carbohydrate, total sugar, fat (total), fatty acids (as saturated, monounsaturated and polyunsaturated), energy and sodium levels. The amount is determined by grams per serving size and g/100 g. Table 6.21 shows the main structure of a nutritional label. As the edible portion of the olive is the flesh, Table 6.22 provides information on the number of olives for a 20 g serve. Two worked examples from data derived from actual analysis by the authors are given in Table 6.23 (naturally black-ripe *Kalamata* olives) and Table 6.24 (green-ripe *Manzanilla* olives processed by spontaneous fermentation without prior treatment with lye). A typical nutrition panel for tapenade is presented in Table 6.25.

Quantity per serve. Each serve of processed table olives is nominally 20 g of the edible flesh portion. The approximate number of olives per serve can be calculated from the average weight of the olives and the average flesh weight (Table 6.22). The average flesh weight is related to the average Flesh:Stone ratio of the olive.

This and other labelling information can be obtained from the FSANZ website: http://www.foodstandards.gov.au/thecode/foodstandardscode.cfm.

Table 6.21. Schematic of the nutritional information required for labelling table olive products

Servings per package: (Insert number of servings)		
Serving size: in grams	Quantity per serve	Quantity per 100 g
Energy	kJ	kJ
Protein	g	g
Fat, total	g	g
– Saturated	g	g
– Monounsaturated	g	g
– Polyunsaturated	g	g
Carbohydrate, total	g	g
Sugars	g	g
Sodium	g	g
Insert any other nutrient or biologically active substance (or other units as appropriate) that are to be declared.	g, mg, µg	g, mg, µg

Table 6.22. Number of olives/20 g serve based on average olive weight and average olive flesh weight

Olive weight grams	Flesh:Stone ratio (F:S)				
	For (F:S) 3:1 No. of olives	For (F:S) 4:1 No. of olives	For (F:S) 5:1 No. of olives	For (F:S) 6:1 No. of olives	For (F:S) 8:1 No. of olives
3	9	8	8	8	7
4	7	6	8	6	6
5	5	5	5	5	5
7	4	4	3	3	3
10	3	3	2	2	2

Table 6.23. Nutritional label for naturally black-ripe *Kalamata* olives processed by the *Kalamata* method
Servings per jar, 15; serving size: 20 g.

	Average quantity per 20 g serving	Average quantity per 100 g
Energy	223 kJ	1120 kJ
Protein	0.4 g	2.1 g
Oil		
Total	5.1 g	25.5 g
– Saturated	0.6 g	3 g
– Monounsaturated	3.8 g	19 g
– Polyunsaturated	0.7 g	3.5 g
Carbohydrate		
– Total	1.6 g	8.0 g
– Sugars	0.02 g	0.10 g
Sodium	0.38 g	1.9 g
Potassium	0.06 g	0.29 g

Table 6.24. Nutritional label for natural green-ripe *Manzanilla* olives processed by spontaneous fermentation in brine
Servings per jar, 15; serving size, 20 g.

	Average quantity per 20 g serving	Average quantity per 100 g
Energy	164 kJ	821 kJ
Protein	0.2 g	1.2 g
Oil		
Total	3.7 g	18.4 g
– Saturated	0.7 g	3.4 g
– Monounsaturated	2.8 g	14 g
– Polyunsaturated	0.2 g	1 g
Carbohydrate		
– Total	1.4 g	7.1 g
– Sugars	0.06 g	0.31 g
Sodium	0.36 g	1.8 g
Potassium	0.04 g	0.21 g

Information for allergy sufferers and food intolerance. Labels should include information for any ingredient or component that can cause severe adverse reactions in susceptible individuals. Such ingredients or components include stuffing materials (nuts, fish or cheeses), preservatives or other additives. Allergens must be listed on the label, however small the amount. As it is dangerous for people taking medication of the monoamine oxidase inhibitor group to also consume fermented foods, olive products should be labelled appropriately if fermentation is used during processing. The powerful vasodilator histamine, a chemical that can cause vascular collapse in humans, has also been detected in processed table olives, particularly those with *Zapateria* defect, but at insignificant levels. As commercial processing/bottling enterprises use their equipment for many different types of food products appropriate warnings should be included on labels. As an example,

'This tapenade has been made using equipment also used for milk products and tree nuts'.

Table 6.25. Nutritional label for tapenade made with Spanish/Californian-style olives (black sterilised olives stabilised with ferrous gluconate)

Servings per jar, 20; serving size, 5 g.

	Average quantity per 5 g serving	*Average quantity per 100 g*
Energy	41 kJ	820 kJ
Protein	0.1 g	2 g
Oil		
Total	0.9 g	18 g
– Saturated	0.1 g	2 g
Carbohydrate		
Total	0.3 g	6 g
– Sugars	0.09 g	0.18 g
Sodium	75 mg	1.5 g

Date marking. Foods with a shelf-life of less than two years must include a best before date. Olives prepared in brine according to proper procedures have a relatively long shelf-life. Pasteurisation ensures microbiological safety; however, quality and nutritional value can deteriorate over time.

Legibility requirements of labels. Labels must be clear and legible in a type that can be seen easily against the background. Small black print is difficult to read against dark backgrounds such as olive green. Legal warning statements must be at least 3 mm in height except on very small packages, for example tapenade in small jars. In practice, much of the information on labels is difficult to find and read.

Storage requirements. Information on the safe storage of olives should be included on the label. Correctly processed and packaged table olives can be safely stored at room temperature. Once containers are opened it is advisable to store the olives under cool conditions to prevent microbiological contamination. In warm climates, typical of most of Australia, this means refrigeration (2–4°C). Once containers are opened, olives should be consumed within two to three weeks. Storage of processed table olives below 0°C changes their structure and texture. Fresh processed olives purchased without brine should also be refrigerated. Olive pastes and tapenades are pasteurised for long-term storage. The label should indicate that after the container is opened, it should be refrigerated and the contents consumed within one week.

Useful terms include: 'Refrigerate after opening', and, 'Consume contents within one week of opening'.

Food recall information. For consumer safety all packaged foods need to be labelled so that if problems arise products can be recalled effectively and efficiently. In Australia, labels must have the name and business name of the manufacturer or importer.

Safety issues can occur due to incorrect labelling of the container as well as physical, chemical and microbiological problems. Physical problems include foreign bodies such as glass particles and any environmental contaminants. Chemical problems can occur if contaminated ingredients or those not meeting food standards are used as well as the inadvertent introduction of incorrect chemicals during processing.

Although there is a high margin of safety with table olives when prepared correctly, over the years a number of reports on adverse health effects resulting from eating table olives have emerged, particularly in the international arena. Such problems are mostly self-limiting, and often ignored by the sufferer; however, more serious problems include food poisoning due to staphylococcal and botulinum bacteria.

Food additives. A number of natural and synthetic additives are often included with table olives. These are used as manufacturing aids or to ensure the preservation and quality of the olives. All food additives used with olives must have a specific use, must have been assessed and approved by FSANZ and must be used in the lowest possible quantity for their purpose. Food additives must be listed in the ingredient list by name or number. A full list of additives and numbers is available on the FSANZ website. Table 6.26 includes some of the commonly used food additives and processing aids for table olives. The artificial colouring agents Caramel (150d, prepared by the ammonium sulfate process) is often added to the covering solution of black table olives. The resulting colour is pale brown and not the rich burgundy colour of the fermentation brines after processing naturally black-ripe olives by spontaneous fermentation.

Table 6.26. Food additives and processing aids for olives
(After Codex Alimentarius (1987)/IOOC (2004) Table Olive Standards.)

Component	Maximum level: g/kg (expressed as mass/mass of flesh)	Comment
Sorbic acid and its sodium and potassium salts expressed as sorbic acid Sorbic acid INS No. 200 Sodium sorbate INS No. 201 Potassium sorbate INS No. 202	0.5 g/kg as sorbic acid (500 ppm)	Preservative Solid Water solubility: Potassium sorbate (58.5 g/100 ml)
Benzoic acid and its sodium and potassium salts expressed as benzoic acid Benzoic acid INS No. 210 Sodium benzoate INS No. 211 Potassium benzoate INS No. 212	1 g/kg as benzoic acid (1000 ppm)	Preservative Solid Water solubility: Sodium benzoate (52 g/100 ml) Potassium benzoate (60 g/100 ml)
Ascorbic acid INS No. 300	Limited by GMP	Antioxidant Solid Water solubility: 6 g/100 l
Acetic acid INS No. 260	Limited by GMP	Acidity and pH regulator Glacial acetic acid (% acetic acid = 99%) Liquid: miscible with aqueous solutions
Lactic acid INS No. 270	15 g/kg	Acidity and pH regulator Solid or 50% aqueous solution Miscible with aqueous solutions

Component	Maximum level: g/kg (expressed as mass/ mass of flesh)	Comment
Citric acid INS No. 330	15 g/kg	Acidity and pH regulator Solid Water solubility: 19.2 g/100 ml
L(+)Tartaric acid INS No. 334	15 g/kg	Acidity and pH regulator Solid Water solubility: 15 g/100 ml
Spices and extracts and aromatic herbs Natural flavours	Limited by GMP	Flavouring agents
Monosodium glutamate INS No. 621	5 g/kg	Flavour enhancer for olives stuffed with anchovies Solid Freely soluble in water
Ferrous gluconate INS No. 579	0.15 g/kg (as total Fe in the fruit)	Maintains black colour in alkaline treated olives darkened by oxidation Solid Water solubility: 12.8 g/100 ml
Ferrous lactate INS No. 585	0.15 g/kg (as total Fe in the fruit)	Maintains black colour in alkaline treated olives darkened by oxidation Solid Water solubility: 2.1 g/100 ml
Calcium lactate INS No. 327	1.5 g/kg as calcium in the final product	Used as firming agent Solid Water solubility: 3.18 g/100 ml
Calcium citrate INS No. 333	1.5 g/kg as calcium in the final product	Used as firming agent Solid Water solubility: 0.85 g/100 ml
Calcium chloride INS No. 509	1.5 g/kg as calcium in the final product	Used as firming agent Solid Water solubility: 74.5 g/100 ml

Sorbic acid. Sorbic acid and its salts have antifungal activity against yeasts and moulds, and are effective in retarding the growth of many food spoilage organisms. The commonly used form of sorbic acid is potassium sorbate. The latter is a white crystalline powder that has a greater solubility in water than the parent compound. Concentrated solutions of 50% w/v can be prepared by weighing 50 g of potassium sorbate and adding water to give a final volume of 100 ml. This solution can be mixed with olive pastes and tapenades, or aliquots can be added to brines or packing solutions. Sorbates can be used at pH values up to 6.5.

Specifically with table olives, sorbates (0.05–0.10% w/v) inhibit the formation of oxidative yeast at the surface of brines for several months after containers are opened. Adding mixtures of sorbic (0.05%) and benzoic (0.1%) acids to covering solutions of unpasteurised table olives with added herbs and spices gives the products a shelf-life of up to three months. As some aflatoxin producing moulds have been found to grow in olive pastes, such products should have an added preservative and be pasteurised.

Benzoic acid. Benzoic acid, an antimicrobial preservative against yeasts and moulds, is a white granular powder with a sweet astringent taste. Salts of benzoic acid are more soluble than the parent compound and the preferred forms for industrial use. Sodium benzoate has an optimum pH range of between pH 2.5–4.0. For preservation of table olives amounts of 0.1% w/w or less are used.

Note: Sodium metabisulphite, which has antimicrobial and antioxidant actions, is a permitted preservative for inclusion in commercially manufactured foods. It has the disadvantage that it breaks down during heating and may cause adverse reactions in susceptible consumers. It is therefore best avoided in preserving table olive products.

Ascorbic acid. Ascorbic acid (Vitamin C) can be added to the covering brines of processed green table olives to prevent discolouration by oxidation of brines and olives.

Acetic acid. Acetic acid has antimicrobial activity because of its pH effect.

Lactic acid. Lactic acid has antimicrobial activity because of its pH effect. Lactic acid is a liquid. When diluted in brines it is odourless or has a slight, not unpleasant, odour.

Tartaric acid. Tartaric acid has antimicrobial activity because of its pH effect. Tartaric acid is a colourless crystalline compound or a white powder with a strongly acid taste.

Citric acid. Citric acid has antimicrobial activity because of its pH effect. Citric acid is a colourless crystalline compound or a white powder with a strongly acid taste reminiscent of citrus fruit. It is used as a synergist to enhance the effectiveness of antioxidants such as ascorbic acid. As citric acid has good water solubility, concentrated aqueous solutions (15% w/v = 15 g/100 ml) can be prepared that are useful for adding to brines, particularly for pH and acidity adjustment.

Herbs and spices. Various herbs and spices contain essential oils and antioxidants and specific antimicrobial inhibitors. This effect is mostly additional to other means of inhibiting microorganisms, such as through the use of salt and acid. Garlic contains allicin that has antibacterial and antifungal activity. Cinnamon, cloves, oregano, thyme and rosemary have antimicrobial effects.

Monosodium glutamate. Monosodium glutamate is added to a variety of foodstuffs as a flavour enhancer. This compound has been implicated in adverse health reactions in humans.

Iron salts. Iron salts are used in table olive processing to stabilise the colour of treated olives darkened by oxidation.

Calcium salts. Calcium chloride is used as a firming agent in the pimento used in stuffed olives. Calcium salts have also been tried as firming agents for soft olives, but effects are unclear and inconsistent.

Food-borne disease and table olives

Consumers expect processed table olives and table olive products to be safe and suitable for consumption. The objective of table olive processors should be to achieve zero risk for food-borne illness and injury.

This can only be achieved if processors follow practices that ensure olives selected for processing are: produced by Good Agricultural Practices (GAP); processed by the principles of Good Manufacturing Practice (GMP) – that is, following product specifications, controlled methods and the use of potable water and food grade ingredients; and produced at premises with equipment and by personnel that can meet Good Hygienic Practices (GHP).

Under most circumstances contaminated or spoilt foodstuffs can produce discomfort and/or temporary debilitation but at worst can be fatal. Consequences for individual sufferers are an inability to work or attend school, as well as loss of jobs and earnings. Wider implications include a reduction in consumer confidence affecting trade, and with serious cases of food poisoning the implications to food processing enterprises can be catastrophic. When food spoilage occurs during or after processing, it is not only wasteful but can reduce commercial viability.

Outbreaks of food-borne illness are not uncommon. Based on the incidence of food recalls reported to Food Standards Australia New Zealand (FSANZ) (Table 6.27), the presence of physical, chemical or microbiological contaminants in foods during processing and packaging operations is real. From 1990–2004 there were 668 food recalls. Over one-third of these were due to microbiological problems and together with physical contamination made up more than 50% of recalls. Careful control during processing is, therefore, essential. Risks to consumer health can be minimised or eliminated by using documented product specifications and testing procedures.

Table 6.27. Causes and distribution of 668 food recalls reported to FSANZ for the years 1990–2004

Reason for recall	Proportion (%) and number
Microbiological	34.9 (233)
Physical: foreign matter	21.6 (144)
Chemical	10.0 (67)
Labelling	17.4 (116)
Processing faults/product deterioration	12.0 (80)
Other	4.1 (28)
Total number of recalls	100 (668)

Food safety and table olives. Safety in the table olive industry starts in the olive grove and ends with the consumer. Although much emphasis is placed on safety during processing, problems can occur at any point in the chain. Control of microbiological quality is of major importance in the prevention of food-borne illness and is very relevant to table olive production.

Table olives are a food, hence their manufacture, distribution, storage and consumption are governed by food, health and safety regulations. At the national level, Food Standards Australia New Zealand (FSANZ) is responsible for food safety. At the state level, government health departments and local government agencies share food health and safety responsibilities. There have been very few health-related problems or recalls associated with Australian processed or imported olives.

Safety issues. Safety issues are a high priority for table olive processors and packers because processing requires a combination of preservation methods such as microbiological fermentation, salting, acidification, heat, pasteurisation, sterilisation and refrigeration that, if not effectively controlled, can lead to spoilage and possibly food poisoning. Final products should be subject to physical, chemical, microbiological and organoleptic evaluation against available international standards. Potential hazards are summarised in Table 6.28. Overall food hazards are detailed later in this section.

Table 6.28. Potential hazards relevant to table olive processing

Hazards	Examples
Physical	• Particulate matter in brine • Glass fragments, foreign objects, hair, insects • Stones in pitted olives
Chemical	• Environmental contaminants in olives: heavy metals, agricultural and industrial agents • Contaminated raw materials: water, salt, sodium hydroxide, aromatics • Dubious quality and poorly treated barrels and processing tanks • Use of non–food grade cleaning and sanitising agents
Microbiological	• Bacteria: e.g. Coliforms, Protozoa – Cryptosporidia • Food poisoning: e.g. *Salmonella*, Clostridia • Toxins: e.g. Staphylococcal, Clostridial, fungal

Physical contaminants: foreign matter

Foreign matter found in processed foods consists predominantly of metal, glass or plastic objects. Less common contaminants include rubber, animal/insect matter and plant material.

Entry of foreign matter into foods may be inadvertent, accidental or deliberate. Particulate matter in water or impurities in water-soluble inputs such as salt or caustic soda can lead to cloudy solutions. Such situations should be avoided by using potable water and quality chemicals. Other sources of foreign matter could result from: poorly cleaned barrels, tanks or containers; hair, loose items, jewellery and watches worn by operatives falling into solutions; vermin (bird or rodent droppings, insects); inadequate checking of new jars and bottles prior to packing with olives.

A specific problem with table olives is the potential for olive stones or stone fragments to be inadvertently packed with depitted, stuffed olives or olive pastes. Information about this hazard should be added to labels of such products to warn consumers. For example, the following can be included on labels for table olives, 'This product contains pits'; for destoned olives, 'This product may contain stones'. Packaged olive products have been recalled in Australia when found to be contaminated with glass fragments.

Chemical contaminants

Chemical contaminants can cause immediate toxic effects such as food poisoning or allergic reactions or have long-term carcinogenic or mutagenic effects. There are a number of chemicals used in a table olive processing facility that are potential contaminants if used incorrectly, ranging from cleaning agents and sanitisers to pesticides used in the facility to control vermin. Careful storage of chemicals and controlling processing procedures will reduce the inadvertent or accidental contamination of olives. All calculations of quantities of inputs should be checked and signed off before actual additions are made. The importance of sourcing quality herbs and spices has been recently highlighted overseas by the discovery that the colour of some chilli preparations in processed foods had been enhanced with a known carcinogen (Sudan 1) that can permanently alter DNA. Furthermore, there have been reports on undeclared sulphites in products such as sun-dried tomato pesto and sun-dried tomato and olive spreads.

Like foreign matter, entry of chemicals may be inadvertent, accidental or deliberate. Poor quality groundwater or dam water may be contaminated with heavy metals and farm chemicals and should not be used for processing table olives without testing. Using second-hand barrels or containers of dubious past history should also be avoided. Even if second-hand barrels are used they should be steam-cleaned and sanitised before use. Deliberate chemical contamination raises a biosecurity problem.

Chemical residues. Information is available on the maximum permissible amounts of chemical residues and toxic metals in foods. Chemical levels in table olives will vary according to their initial concentration in the raw olives, the processing method and the quality of water or chemicals (salt, caustic soda, food acids) used. Furthermore, prolonged soaking of olives in water or debittering with caustic soda tends to reduce the amounts of water-soluble components in the flesh. Some farm chemicals are broken down during caustic soda treatments, thus reducing their quantity in the olive flesh. More specific information on chemical levels in table olives needs to be developed.

Allergens and food additives. Allergen management is a major focus in food production at all levels of the food chain. The problem increases where manufacturers, processors and packagers of foods use shared equipment. Allergen related recalls, involving undeclared allergens, is a significant problem. Food allergies are abnormal responses of the immune system to food proteins. In some consumers the response may be life threatening. The incidence of food allergies is thought to be 1–2% in the general population rising to 5–8% in children.

Great care must be taken to exclude unnecessary allergens and food additives, prevent cross contamination of raw materials during processing and ensure operatives understand the reasons for specific controls.

Reactions to foods and food additives are not uncommon and represent a significant number of recalls. Examples of food derived allergens include peanuts, tree nuts, seafood, fish, milk, gluten, eggs and soybeans and these must be declared on the label. Allergic reactions range from skin rashes, swelling of mucous membranes, asthma and life-threatening anaphylaxis. Relevant to the olive industry is the use of materials such as anchovies, nuts and cheeses to stuff pitted olives, or include as olive salad ingredients. Most recalls of this type are due to failure to declare on the label a warning that an allergen is present in the food product.

Food additives must be identified, usually by a number, and included in the ingredient list. Caution must be used when adding preservatives to olive products. Olive pastes containing sulfur dioxide (sodium metabisulphite) as a preservative have been subject to recalls because its inclusion has not been declared on the label.

Microbial contaminants

Bacterial contaminants. There have been relatively few reports of microbial problems with table olives in Australia. The main types of bacterial food poisoning are presented in Table 6.29 and the causes and sources in Table 6.30. Problems can occur either by direct infection, the production of toxins or a combination of both. Affected persons experience symptoms such as fever, abdominal pain, diarrhoea or vomiting. Common sources are animal faeces, soil and dust. The risk of infection can be eliminated or reduced by paying strict attention to hygiene and cleanliness, using quality inputs and applying effective controls during processing. Testing end products will confirm their level of safety.

Table 6.29. Summary of food poisoning bacteria relevant to table olives

Bacterial organism	Infection	Toxin	Source
Salmonella	yes		animal faeces human carriers
Escherichia coli	yes	yes	water, animal faeces
Staphylococcus aureus		yes (heat stable)	human skin, nasal passages
Clostridium perfringens	yes	yes	animal faeces soil, dust, air, water
Clostridium botulinum		yes, fatal (destroyed by heat)	soil, animals, honey (spores – heat resistant)
Bacillus cereus	yes	yes	soil, dust
Listeria monocytogenes	yes	yes	animals, insects, soil, raw milk

The use of high quality olives having low levels of contamination (that is, olives are washed and stored to prevent contamination and infection from dust, insects, rodents and other animals), and processed under hygienic conditions, reduces the risk of microbial contamination.

The three most common microbial contaminants associated with food recalls in Australia are: *Listeria monocytogenes*, *Salmonella* and *Escherichia coli*. Other microbial contaminants associated with food poisoning include Enterobacteriaceae, *Staphylococcus aureus*, *Clostridium perfringens*, *Bacillus cereus* and Hepatitis A virus.

Microorganisms are introduced during processing through unhygienic procedures or by proliferation of undesirable microorganisms if processing procedures are poorly controlled. Monitoring brine pH, salt and microorganism levels is important throughout processing, especially in the early stages, to reduce the risk of harm to consumers.

The growth of microorganisms is influenced by factors such as: temperature, moisture (water activity), environmental acidity (measured by pH and acid levels), oxygen levels, available nutrients, degree of contamination and the presence or absence of growth inhibitors.

Microbial contaminants can be introduced into table olive operations from: raw olives coming from the orchard, inputs such as contaminated water, fresh or non-irradiated herbs and spices, contaminated equipment or infected operatives handling olives.

Table 6.30. Causes/sources of food poisoning relevant to table olive production

Causes/sources	Comment
Health of operatives	Low levels of personal hygiene Undetected carriers of food-borne infective agents Recent acute illness Bad food handling skills
Unclean/unhygienic plant and equipment	Poorly designed processing plant
Inadequate control of processing environment	Poor control of air quality Insect, bird and rodent pests
Poor quality inputs	Water Salt Herbs and spices
Poor control of processing	Temperature Brine acidity and pH Brine salt levels

As olives are exposed for prolonged periods in an orchard environment, microbial contamination is common. Spray washing olives with potable water prior to primary processing reduces the hazard.

With olive processing, water is required for all washing procedures as well as preparing brine and lye solutions. As water can contain a number of contaminants (physical, chemical or microbiological), all water used for table olive processing must meet the microbiological standards of drinking water (potable water).

Scheme water (water from state or municipal authorities) is generally of potable water quality and so can be used as is. However, this water source should be tested at the point of entry into the processing plant and thereafter should be tested annually. Groundwater and tank rainwater can contain a number of microbial contaminants, often of faecal and/or plant origins, so water from these sources should be tested and treated before use.

At-risk groups. Food-borne pathogens present a hazard to particular groups of consumers such as: the elderly, newborns and those with a weakened resistance to infection, such as those with HIV, leukaemia or people taking immunosuppressive medication.

Enterobacteriaceae. This is a family of bacteria that are found in the human or animal intestinal tract, including human pathogens such as *Salmonella* and *Shigella*. *Enterobacteriaceae* are useful indicators of hygiene and post-processing contamination of heat processed foods. Their presence in high numbers in ready to eat foods indicates an unacceptable level of contamination has occurred.

Escherichia coli. While most *E. coli* are normal residents in the small intestine of humans where they aid digestion and produce vitamin K, there are some pathogenic strains, such as *E. coli* O157:H7, that produce a deadly toxin that can cause severe illness in humans and animals. Sources include uncooked food, especially meat, raw milk and unpasteurised juice, and contaminated water. Incubation in humans is usually three to four days (but can be 1–10 days) after consuming contaminated food or water. Symptoms include nausea, severe abdominal cramps and diarrhoea (with or without blood) lasting five to eight days. In at-risk groups the pathogen can cause kidney damage that can lead to death. The presence of *E. coli* in foods is undesirable because it indicates poor hygienic conditions leading to contamination or that heat treatments have been inadequate.

Salmonella. This is a group of bacteria that can cause diarrhoeal illness in humans. Most types of *Salmonella* live in the intestinal tracts of animals and birds and are transmitted to humans by contaminated foods of animal origin. Symptoms can cause death among at-risk groups. Incubation in humans is 12–72 hours after eating contaminated food, with symptoms such as diarrhoea, fever and abdominal cramps lasting four to seven days. The presence of this type of organism is indicative of poor food preparation and handling or cross contamination. Investigation of the health status of operatives should be considered to establish if any are suffering of salmonellosis or are asymptomatic carriers of the organism.

Shigella. This bacterium, carried only by humans, causes diarrhoeal illness. Poor hygiene, especially improper hand washing, allows shigella to be passed from person to person via food. Once the bacterium is in food, it multiplies rapidly at room temperature. Sources of shigella include raw animal products, shellfish and unclean water. Incubation in humans is one to seven days after eating contaminated food, with symptoms lasting five to seven days and including diarrhoea, fever, abdominal cramps, vomiting and bloody stools.

Coagulase Positive Staphylococcus aureus. Contamination with Coagulase Positive Staphylococci is largely due to human contact. Staphylococci are carried on the skin in infected cuts or pimples and in the nasal passages and throat of humans. It produces a toxin that causes vomiting in as little as 30 minutes after ingestion. In contaminated food, it multiplies rapidly at room temperature. Incubation in humans is usually rapid, occurring within 30 minutes to eight hours after eating contaminated food, with symptoms including nausea, abdominal cramps, vomiting and diarrhoea lasting 24–48 hours. Contamination can be minimised through good food handling processes and limiting the growth of the bacteria by adequate temperature controls.

Clostridium botulinum. This bacterium, found in soil and at the bottom of lakes and oceans, produces a toxin that causes botulism, a disease characterised by muscle paralysis. Proper heat processing (sterilisation) destroys *Clostridium botulinum* in canned food. Freezer temperatures inhibit its growth in frozen food, low moisture controls its growth in dried food, and high oxygen levels control its growth in fresh foods. Reported sources of *Clostridium botulinum* include home-canned and prepared food, vacuum packed and tightly wrapped food, meat products, seafood and herbal cooking oils. Incubation is usually 4–36 hours after consuming contaminated food or water. Symptoms include a dry mouth, double vision followed by nausea, vomiting and diarrhoea. Constipation, weakness, muscle paralysis and breathing problems may develop later. It can take one week to a full year to recover.

During the early part of the 20th century, *Clostridium botulinum* toxin poisoning was associated with olives canned in low acid, low salt brines under anaerobic conditions. Following serious cases of botulism poisoning after olive consumption in which 35 people died after consuming improperly canned black olives, methods were developed so that the offending organism and other harmful organisms were eliminated through temperature control (especially heat treatments), and the use of salt and food acids. Even so, a recent report from Italy attributed processed green olives as the possible cause of botulism in 16 people (Cawthorne *et al.* 2005). Although none of the olives consumed by those affected were available for testing, olives prepared at the same time had a pH of 6.2 – well above the level of pH 4.6 normally required to prevent the growth of *Clostridium botulinum*. Also, as these olives had been processed in salt water for 35 days then placed in water, it is likely that the salt levels would not have prevented the growth of *Clostridium botulinum*.

Olives at pH greater than 4.6 are considered to be low acid, and if packed in sealed containers in the absence of oxygen and not refrigerated, can host *Clostridium botulinum*. This organism, generally in the spore (dormant) form, germinates into the vegetative (active) form with cells producing toxins as they multiply.

Recalls for olives that are suspected of containing *Clostridium botulinum* are not uncommon at the international level. To illustrate this, two such table olive recalls made in 2005 follow. (Identifying information has been deleted.)

Recall 1.
The Food Agency and Company xxx are warning the public not to consume xxx brand of natural green olives described below because the product may be contaminated with *Clostridium botulinum*. Toxins produced by this bacterium may cause botulism, a life-threatening illness.

The affected olives, a product of zzz, are sold in one litre containers bearing the batch codes aaa. The affected product is known to have been distributed nationally.

There have been no reported illnesses associated with the consumption of this product.

Food contaminated with *Clostridium botulinum* toxin may not look or smell spoiled. Consumption of food contaminated with the toxin may cause nausea, vomiting, fatigue, dizziness, headache, double vision, dry throat, respiratory failure and paralysis. In severe cases of illness, people may die.

The importer is voluntarily recalling the affected product from the marketplace and the Food Agency is monitoring the effectiveness of the recall.

For more information, consumers and industry can call the Food Agency [contact details].

Recall 2.

Company xxx is recalling 200 jars of black olive tapenade, because they have the potential to be contaminated with *Clostridium botulinum*, a bacterium that can produce a toxin that can cause life-threatening illness or death. Consumers are warned not to use the product even if it does not look or smell spoiled.

Botulism, a potentially fatal form of food poisoning, can cause the following symptoms: general weakness, dizziness, double vision and trouble with speaking or swallowing. Difficulty in breathing, weaknesses of muscles, abdominal distension and constipation may also be common symptoms. People experiencing these problems should seek immediate medical attention.

The recall was initiated after it was discovered that the product had an unstable pH, an indication of inadequate processing. Subsequent investigation indicates the problem was caused by a temporary breakdown in the co-packer's production and packaging process. The product was produced by Co-packer Company ooo.

Consumers who have the black olive tapenade should discard the product, or return the product to the place of purchase for a refund. Consumers with questions can also contact the company.

Clostridium perfringens. This is a food-borne pathogen that persists as heat stable spores. If foods are only moderately cooked and allowed to remain at room temperature, spores germinate and produce a harmful toxin. Sources of this organism include meat products. Incubation is usually 8–12 hours after consuming contaminated food, with symptoms including abdominal pain, diarrhoea, and sometimes nausea and vomiting lasting for 24 hours or less. Symptoms can be more serious in at-risk people. The detection of high levels of *Clostridium perfringens* is indicative of a food-handling problem and poses a health risk; it should be investigated.

Listeria monocytogenes. *Listeria monocytogenes* is a food-borne pathogen that produces a toxin and is capable of causing the food-borne illness Listeriosis. Listeriosis has a mortality rate of about 30%. *Listeria* species are commonly found in many environments including soil, dirt and water, and can be carried by both domestic and wild animals, hence it is a contaminant of raw foods. Fresh vegetables, herbs and soft cheeses may be contaminated with *Listeria monocytogenes*. *Listeria monocytogenes*, a non-spore forming Gram positive rod, can grow with or without oxygen, at temperatures between 3°C and 45°C. It can survive freezing, tolerates a wide pH range (4.6–9.5) and can grow in up to 20% sodium chloride. In affected food this organism has the ability to multiply slowly under refrigeration and may eventually develop a population large enough to be of danger to susceptible, at-risk consumers. This organism is killed by pasteurisation. Listeriosis can cause high fever, severe headache, neck stiffness and nausea. Infected pregnant women may experience only a mild flu-like illness; however, infections during pregnancy can lead to premature delivery.

Detection of *Listeria monocytogenes* in foods indicates inadequate preparation or contamination post-preparation. This organism is widespread and can be found in a wide range of foods. Recently, Italian researchers have shown that *Listeria monocytogenes* present on green olives can survive in the fermented product despite a brine pH <4.5, titratable acidity 0.20–0.45% (as lactic acid) and sodium chloride levels between 6.2% and 7.5%. They recommend that an appropriate heat treatment be applied to ensure a reduction in *Listeria monocytogenes* (Caggia *et al.* 2004).

Bacillus cereus and other Bacillus species. The detection of high levels of *Bacillus cereus* is an indicator of poor food handling and can result in food-borne illness. Other *Bacillii*, for example *Bacillus subtilis* and *Bacillus licheniformis*, have also been associated with food-borne illness.

Exclusion of harmful organisms during olive processing can be managed by balancing the pH and salt concentration of brines and the temperature during fermentation and storage. Most harmful organisms are unable to grow or survive at salt concentrations greater than 6–7% together with pH levels of 4.3 or less. Outside these limits, harmful and/or spoilage organisms can proliferate.

Aflatoxins. Olives, olive pastes and tapenades have the potential to support aflatoxin-producing fungi such as *Aspergillus flavus* and *Aspergillus parasiticus*. International research has shown that olives are a poor substrate for *Aspergillus* growth and that oleuropein significantly slows down the onset of fungal growth and development, but the risk cannot be excluded (Garrido Fernández *et al.* 1997). Aflatoxins are a recognised group of chemical compounds released by moulds often associated with peanuts, tree nuts and other foods. Although outbreaks of aflatoxicosis are uncommon, aflatoxins have the potential to cause liver necrosis, cirrhosis and cancer as shown by animal studies. Salt-dried olives, naturally black olives in brine and lye treated naturally black olives ('façon Grecque') are prone to this type of spoilage and even though aflatoxins are not detected, when moulds are present during processing a health risk exists. It is essential that the water activity of the olives is low enough to prevent this type of spoilage. This can be achieved by ensuring: a low moisture and/or high salt content in dried olives; olives are sufficiently salted; and olives are completely immersed in brine during and after processing.

Final table olive products

In summary, processed table olives should retain some level of bitterness and fruitiness. The final salt and pH levels of packed olives depend on the style, variety and ripeness of the fruit as long as safety requirements are met. In all cases, processed table olives should be determined as microbiologically safe by an accredited laboratory before being offered for sale. Olives should be packed at the appropriate salt, pH and acid levels according to the style, and pasteurised or sterilised as required. Once packed, representative samples of the olives must be taken and the following parameters assessed: brine pH and free acidity, brine salt levels, and brine and fruit microbiology.

According to the Codex Alimentarius (1987)/IOOC (2004) Table Olive Standards, packing brines for olives are solutions of food grade salts dissolved in potable water with or without the addition of all or some of the following: vinegar, olive oil, sugars, edible material as an accompaniment or stuffing, herbs, spices or their natural extracts, and authorised additives.

Pasteurising packaged olives at 70–80°C provides some protection against microorganisms, but not all. Only preventive measures will stop the introduction of serious infective agents. Pasteurisation does prevent growth of fermentative bacteria and yeasts in packed olives, resulting in stable products.

Bibliography and additional reading

Agar, I. T., Hess-Pierce, B., Sourour, M. M., and Kader, A. A. (1998). Quality of fruit and oil of black-ripe olives is influenced by cultivar and storage period. *Journal of Agricultural and Food Chemistry 46(9)*, 3415–3421.

Agar, I. T., Hess-Pierce, B., Sourour, M. M., and Kader, A. A. (1999). Identification of optimum preprocessing storage conditions to maintain quality of black-ripe 'Manzanillo' olives. *Postharvest Biology and Technology 15(1)*, 53–64.

Allaby, M. (1998). 'Oxford Dictionary of Plant Sciences.' (Oxford University Press, Oxford.)

Aloni, N. (2000). Irrigation and sustainable nutrition – modern olive groves need supplementary nutrients. *The Olive Press, Spring Edition*, 15–16.

Amelio, M., and De Mauro, E. (2000). Naturally fermented black olives of Taggiasca variety (*Olea europaea* L.). *Grasas Y Aceites 51(6)*, 429–438.

Angerosa, F., Lanza, B., d'Alessandro, N., Marsilio, V., and Cumitini, S. (1999). Olive oil off-odour compounds produced by *Aspergillus* and *Penicillium*. In 'Proceedings of the 3rd International Symposium on Olive Growing'. (Eds I. T. Metzidakis and D. G. Voyiatzis.) *Acta Horticulturae, 474(2)*, 695–699.

Anon. (2003). Table olives and the world market. *Olivae 99*, 45–47.

Anon. (2004). Report: Olives in a nutshell. *Choice, July edition*, 24–27.

Anon. (2005). Reviving olives in Robinvale. *Australian and New Zealand Olivegrower and Processor 42* (Mar.–Apr.), 31.

Anon. (2006). Food microbiology and risk assessment conferences. *Food Australia 58(1–2)*, 32–40.

Antognozzi, E., and Proietti, P. (1995). Effects of CPPU (cytokinin) on table olive trees (*Ascolana Tenera*) under nonirrigated and irrigated conditions. In 'Quality of Fruit and Vegetables – Pre and post-harvest factors and technology'. (Eds D. Gerasopoulos, C. H. Olympios and H. Passam.) *Acta Horticulturae 379*, 159–166.

AOAC. (2000). Official methods of analysis of the AOAC International. (Ed. W. Horwitz.) 17th edn. Vols 1 and 2. (Association of Official Analytical Chemists Incorporated, Arlington, USA.)

Asehraou, A., Peres, C., Brito, D., Faid, M., and Serhrouchni, M. (2000). Characterization of yeast strains isolated from bloaters of fermented green table olives during storage. *Grasas y Aceites 51(4)*, 225–229.

Ateyyeh, A. F., Stosser, R., and Qrunfleh, M. (2000). Reproductive biology of the olive (*Olea europaea* L.) cultivar 'Nabali Baladi.' *Journal of Applied Botany 74(5–6)*, 255–270.

Australian New Zealand Food Authority. (2001). Request to include herbs, spices, herbal infusions, peanuts, cashew nuts, almonds and pistachio nuts. Final Assessment Report, 19 September 2001. Standards A17 and 1. 5. 3 – Irradiation of Foods in the Foods Standard Code. ANZFA, Australia.

Australian Pesticides and Veterinary Medicines Authority. (2005). Interim permits for chemicals used in olives. *Australian and New Zealand Olivegrower and Processor 41*, (Jan.–Feb.), 60–61.

Avidan, B., Ogrodovitch, A., and Lavee, S. (1999). A reliable and rapid shaking extraction system for determination of the oil content in olive fruit. In 'Proceedings of the 3rd International Symposium on Olive Growing'. (Eds I. T. Metzidakis and D. G. Voyiatzis.) *Acta Horticulturae 474(2)*, 653–658.

Baker, G. (2005). Black scale briefing. *Australian and New Zealand Olivegrower and Processor 41* (Jan.–Feb.), 9–10.

Baker, G. (2005). Managing major olive pests. *Australian and New Zealand Olivegrower and Processor 44* (Jul.–Aug.), 38–41.

Baker, G. (2006). Survey of black scale parasites in South Australia. *Australian and New Zealand Olivegrower and Processor 47* (Jan.–Feb.), 34–43.

Barattà, B., Caruso, T., Crescimanno, P. L., and Inglese, P. (1990). Using urea as thinning agent in olive: the influence of concentration and time of application. In 'Proceedings of the International Symposium on Olive Growing'. (Eds L. Rallo, J. M. Caballero and R. Fernández-Escobar.) *Acta Horticulturae 286*, 163–166.

Barattà, B., Caruso, T., Di Marco, L., and Inglese, P. (1985). Effects of irrigation on characteristics of olives in 'Nocellara del Belice' variety. *Fruttocoltura 3(4)*, 61–66.

Barranco, D. and Kreuger, W. W. (1990). Timing of NAA application in olive thinning. In 'Proceedings of the International Symposium on Olive Growing'. (Eds L. Rallo, J. M. Caballero and R. Fernández-Escobar.) *Acta Horticulturae 286*, 167–169.

Barranco, D., Fernández-Escobar, R., and Rallo, L. (2004). 'El Cultivo del Olivo'. 5th edn. (Ediciones Mundi-Presna: Madrid, Spain.)

Bastoni, L., Bianco, A., Piccioni, F., and Uccella, N. (2001). Biophenolic profile in olives by nuclear magnetic resonance. *Food Chemistry 73*, 145–151.

Battock M., and Azam-ali, S. (1998). Fermented Fruits and Vegetables: a global perspective. Food and Agriculture Organization Agricultural Services Bulletin No. 134, United Nations, Rome.

Baxter, P. (1997). 'The Complete Guide to Growing Fruit in Australia.' 5th edn. (Pan Macmillan: Sydney, Australia.)

Bellini, E., Giordani, E., and Parlati, M. V. (2004). The new 'Arno', 'Tevere' and 'Basento' crossbred cultivars. *Olivae 102*, 42–46.

Beltran-Heredia, J., Torregrosa, J., Dominguez, J. R., and Garcia, J. (2000). Aerobic biological treatment of black table olive washing wastewaters: effect of an ozonation stage. *Process Biochemistry 35(10)*, 1183–1190.

Beltran-Heredia, J., Torregrosa, J., Dominguez, J. R., and Garcia, J. (2000). Ozonation of black-table-olive industrial wastewaters: effect of an aerobic biological pre-treatment. *Journal of Chemical Technology and Biotechnology 75(7)*, 561–568.

Beltran, F. J., Garcia-Araya, J. F., Frades, J., Alvarez, P., and Gimeno, O. (1999). Effects of single and combined ozonation with hydrogen peroxide or UV radiation on the chemical degradation and biodegradability of debittering table olive industrial wastewaters. *Water Research 33(3)*, 723–732.

Benitez, M. L., Pedrajas, V. M., del Campillo, M. C., and Torrent, J. (2002). Iron chlorosis in olive in relation to soil properties. *Nutrient Cycling in Agroecosystems 62(1)*, 47–52.

Ben-Tal, Y., and Wodner, M. (1994). Chemical loosening of olive pedicels for mechanical harvesting. In 'Proceedings of the 2nd International Symposium on Olive Growing'. (Eds S. Lavee and I. Klein.) *Acta Horticulturae 356*, 297–301.

Bertrand, E. (2002). The beneficial cardiovascular effects of the Mediterranean diet. *Olivae 90*, 29–31.

Bianco, A., and Uccella, N. (2000). Biophenolic components of olives. *Food Research International 33(6)*, 475–485.

Bianco, A., Buiarelli, F., Cartoni, G., Coccioli, F., Muzzalupo, I., Polidori, A., and Uccella, N. (2001). Analysis by HPLC-MS/MS of biophenolic components in olives and oils. *Analytical Letters 34(6)*, 1033–1051.

Bicknell, D., Baxter, A., and Denham, R. (1998). Site preparation for successful revegetation for agricultural regions with less than 600 mm rainfall. Farmnote No. 37/98, Department of Agriculture, Western Australia.

Bignami, C., Natali, S., Menna, C., and Peruzzi, G. (1994). Growth and phenology of some olive cultivars in central Italy. In 'Proceedings of the 2nd International Olive Growing Symposium'. (Eds S. Lavee and I. Klein.) *Acta Horticulturae 356*, 106–109.

Bitonti, M. B., Chiappetta, A., Innocenti, A. M., and Uccella, N. (2002). Biomolecular characterisation and histological distribution of biophenols in green mature fruit of *Olea europea* Cassanese cv. In 'Proceedings of the 4th International Symposium on Olive Growing'. (Eds C. Vitagliano and G. P. Martelli.) *Acta Horticulturae 586(2)*, 515–519.

Blekas, G., Vassilakis, C., Harizanis, C., Tsimidou, M., and Boskou, D. G. (2002). Biophenols in table olives. *Journal of Agricultural and Food Chemistry 50*, 3688–3692.

Bogani, P., Cavalieri, D., Petruccelli, R., Polsinelli, L., and Roselli, G. (1994). Identification of olive tree cultivars by using random amplified polymorphic DNA. In 'Proceedings of the 2nd International Olive Growing Symposium'. (Eds S. Lavee and I. Klein.) *Acta Horticulturae 356*, 98–101.

Bongi, G., and Palliotti, A. (1994). Olive. In 'Handbook of Environmental Physiology of Fruit Crops. Vol. 1'. (Eds B. Schaffer and P. C. Andersen.) pp. 165–187. (CRC Press, Inc.: Florida, USA.)

Borja Padilla, R., Férnandez, A. G., and Barrantes, M. M. D. (1992). Kinetic study of anaerobic digestion process of wastewaters from ripe olive processing. *Grasas y Aceites 43(6)*, 317–328.

Borzillo, A., Iannotta, N., and Uccella, N. (2000). Oinotria table olives: quality evaluation during ripening and processing by biomolecular components. *European Food Research and Technology 212(1)*, 113–121.

Botha, J., Poole, M., Taylor, D., and Hardie, D. (2003). Olive lace bug *Froggattia olivinia* Frogatt [Hemiptera: Tingidae]. Farmnote No. 82/2003, Department of Agriculture, Western Australia.

Bouranis, D. L., Zakynthinos, G., Kapetanos, C., Chorianopoulou, S. N., Kitsaki, C., and Drossopoulos, J. B. (2001). Dynamics of nitrogen and phosphorus partition in four olive tree cultivars during bud differentiation. *Journal of Plant Nutrition 24(10)*, 1535–1550.

Brenes, M., Garcia, P., Duràn, M. C., and Garrido, A. (1993). Concentration of phenolic compounds change in storage brines of ripe olives. *Journal of Food Science 58(2)*, 347–350.

Brenes, M., Garcia, P., Romero, C., and Garrido, A. (2000). Treatment of green table olive waste waters by an activated-sludge process. *Journal of Chemical Technology and Biotechnology 75(6)*, 459–463.

Brenes, M., Romero, C., Garcia, P., and Garrido, A. (1995). Effect of pH on the colour formed by Fe-phenolic complexes in ripe olives. *Journal of the Science of Food and Agriculture 67(1)*, 35–41.

Brenes, M., Romero, C., Garcia, P., and Garrido, A. (2004). Absorption of sorbic and benzoic acids in the flesh of table olives. *European Food Research and Technology 219(1)*, 75–79.

Brenes, M., Rejano, L., Garcia, P., Sánchez, A. H., and Garrido, A. (1995). Biochemical changes in phenolic compounds during Spanish-style green olive processing. *Journal of Agricultural and Food Chemistry 43(10)*, 2702–2706.

Broughton, S., de Lima, F., and Woods, B. (2004). Control of Mediterranean fruit fly (Medfly) in backyards. Gardennote No. 24, Department of Agriculture, Western Australia.

Burnik-Tiefengraber, T., Weis, K. G., Webster, B. D., Martin, G. C., and Yamada, H. (1994). Phosphorus effects on olive leaf abscission. *Journal of the American Society for Horticultural Science 119(4)*, 765–769.

Caballero, J. M., del Río, C., and Eguren, J. (1990). Further agronomical information about a world collection of olive cultivars. In 'Proceedings of the International Symposium on Olive Growing'. (Eds L. Rallo, J. M. Caballero and R. Fernández-Escobar.) *Acta Horticulturae 286*, 45–48.

Cabras, P., Angioni, A., Garau, V. L., Melis, M., Pirisi, F. M., Karim, M., and Minelli, E. V. (1997). Persistence of insecticide residues in olives and olive oil. *Journal of Agricultural and Food Chemistry 45*, 2244–2247.

Caggia, C., Randazzo, C. I., di Salvo, M., Romeo, F., and Giudici, P. (2004). Occurrence of *Listeria* monocytogenes in green table olives. *Journal of Food Protection 67*, 2189–2194.

Campestre, C., Marsilio, V., Lanza, B., Iezzi, C., and Bianchi, G. (2002). Phenolic compounds and organic acids change in black oxidized table olives. In 'Proceedings of the 4th International Symposium on Olive Growing'. (Eds C. Vitagliano and G. P. Martelli.) *Acta Horticulturae 586(2)*, 575–578.

Canadian Food Inspection Agency Food Safety Directorate. (2005). Update–Allergy

Report–Undeclared sulphites in sun-dried tomato pesto and sun-dried tomato and olive spreads. June 29.

Castillo, P., Vovlas, N., Nico, A. I., and Jiménez-Diáz, R. M. (1999). Infection of olive trees by *Heterodera mediterranea* in orchards in southern Spain. *Plant Disease 83(8)*, 710–713.

Catulo, L., Leitão, F., Oliveira, M. M., Peres, C., Silva, S., Gomes, M. L., Peito, M. A., Fernandes, I., and Gordo, F. (2002). Table olive fermentation of *Galega* Portuguese variety. Microbiological, physico-chemical and sensorial aspects. In 'Proceedings of the 4th International Symposium on Olive Growing'. (Eds C. Vitagliano and G. P. Martelli.) *Acta Horticulturae 586(2)*, 611–615.

Çavusoglu, A., Özahçi, E., Caran, D., and Oktar, A. (2000). Researches on the natural development of fruit maturity in olives. In 'Proceedings of the International Olive Growing Symposium'. (Eds L. Rallo, J. M. Caballero, and R. Fernández-Escobar.) *Acta Horticulturae 286*, 429–432.

Cawthorne, A., Celentano, L. P., D'Ancona, F., Bella, A., Massari, M., Anniballi, F., Fencia, L., Aureli, P., and Salmaso, S. (2005). Botulism and preserved green olives. *Emerging Infectious Diseases 11(5)*, 1088. Available at http://www.cdc.gov/ncidod/EID/vol11no05/04–1088.htm [Verified 9 August 2006].

Chaney, D. E., Drinkwater, L. E., and Pettygrove, G. S. (1992). Organic soil amendments and fertilizers. Publication 21505. University of California, Division of Agriculture and Natural Resources.

Chen, S. C., Lin, C.-A., Fu, A.-H., and Chuo, Y. W. (2003). Inhibition of microbial growth in ready-to-eat food stored at ambient temperature by modified atmosphere packaging. *Packaging Technology and Science 16*, 239–247.

Chidgey, M. (2001). Frost damage control. *The Olive Press*, Autumn Edition, 23–25.

Ciarfardini, G., and Zullo, B. A. (1998). Assay of microbial enzymes in opaque samples. *Journal of Microbiological Methods 34(1)*, 73–79.

Ciarfardini, G., and Zullo, B. A. (2001). ß-glucosidase activity in *Leuconostoc mesenteroides* associated with fermentation of "Coratina" cultivar olives. *Italian Journal of Food Science 13(1)*, 41–51.

Ciafardini, G., Marsilio, V., Lanza, B., and Pozzi, N. (1994). Hydrolysis of oleuropein by *Lactobacillus plantarum* strains associated with olive fermentation. *Applied and Environmental Microbiology 60(11)*, 4142–4147.

Cimato, A., Cantini, C., and Sani, G. (1990). Climate-phenology relationships on olive CV Frantoio. In 'Proceedings of the International Olive Growing Symposium'. (Eds L. Rallo, J. M. Caballero, and R. Fernández-Escobar.) *Acta Horticulturae 286*, 171–174.

Cimato, A., Cantini, C., and Sillari, B. (1990). A method of pruning for the recovery of olive productivity. In 'Proceedings of the International Olive Growing Symposium'. (Eds L. Rallo, J. M. Caballero, and R. Fernández-Escobar.) *Acta Horticulturae 286*, 251–254.

City of South Perth. (2002). Guidelines for the establishment of a food premises. City of South Perth, Western Australia.

Civantos, L. (1998). 'The Olive Tree, the Oil, the Olive.' (International Olive Oil Council: Madrid, Spain.)

Claros, M. G., Crespillo, R., Aguilar, M. L., and Canovas, F. M. (2000). DNA fingerprinting and classification of geographically related genotypes of olive-tree (*Olea europaea* L.). *Euphytica 116(2)*, 131–142.

Codex Alimentarius Commission. (1987). Codex Standard for Table Olives, Codex Stan. 66–1981. (Rev. 1–1987.)

Coimbra, M. A., Delgadillo, I., Waldron, K. W., and Selvendran, R. R. (1996). Isolation and analysis of cell wall polymers from olive pulp. In 'Modern Methods of Plant Analysis, Plant Cell Wall Analysis Vol. 17'. (Eds H. F. Linskens and J. F. Jackson.) pp. 19–43. (Springer-Verlag: Berlin-Heidelberg.)

Coimbra, M. A., Rigby, N. M., Selvendran, R. R., and Waldron, K. W. (1995). Investigation of the occurrence of xylan-xyloglucan complexes in the cell walls of olive pulp (*Olea europaea*). *Carbohydrate Polymers 27(4)*, 277–284.

Coimbra, M. A., Rodrigues, M. R., Garcia, E., and Catulo, L. D. (1999). Effect of lactic acid fermentation on cell wall polysaccharides of green table olives. In 'Proceedings of the 3rd International Symposium on Olive Growing'. (Eds I. T. Metzidakis and D. G. Voyiatzis.) *Acta Horticulturae 474(2)*, 595–599.

Coimbra, M. A., Waldron, K. W., Delgadillo, I., and Selvendran, R. R. (1996). Effect of processing on cell wall polysaccharides of green table olives. *Journal of Agricultural and Food Chemistry 44(88)*, 2394–2401.

Conacher, H. B. S. (1975). Gas-liquid chromatographic determination of docosenoic acid in fats and oils. Collaborative Study. *Journal of the Association of Official Analytical Chemists 58*, 488–91.

Conlan, D., Beckingham, C., Robbins, M., Mailer, R., and Ayton, J. (2004). Irrigating oil olives with limited water supplies. *Australian and New Zealand Olivegrower and Processor 39* (Sep.–Oct.), 13–19.

Cordeiro, J., da Silva, T. B., Delgado, A., Pereira, S., Brito, D., and Peres, C. (2002). Antimicrobial activity of *Lactobacillus plantarum* strains isolated from traditional lactic acid fermentation of Portuguese table olives. In 'Proceedings of the 4th International Symposium on Olive Growing'. (Eds C. Vitagliano and G. P. Martelli.) *Acta Horticulturae 586(2)*, 633–636.

Cortesi, N., Rovellini, P., and Fiorino, P. (1999). The role of drupe different anatomic parts on chemical olive oil composition. In 'Proceedings of the 3rd International Symposium on Olive Growing'. (Eds I. T. Metzidakis and D. G. Voyiatzis.) *Acta Horticulturae 474(2)*, 643–647.

Cross, N. (1998). Irrigation of olives. *Australian Olive Grower 9* (Jul.–Nov.), 17–18.

Cuevas, J., and Polito, V. S. (1997). Compatibility relationships in Manzanillo olive. *Hortscience 32(6)*, 1056–1058.

Cuevas, J., Rallo, L., and Rapoport, H. F. (1994). Crop load effects on floral quality in olive. *Scientia Horticulturae 59(2)*, 123–130.

Cuevas, J., Rallo, L., and Rapoport, H. F. (1994). Initial fruit set at high temperature in olive, *Olea Europaea* L. *Journal of Horticultural Science 69(4)*, 665–672.

Cunha, S. C., Ferreira, I. M., Fernandes, J. O., Faria, M. A., Beatriz, M., Oliveira, P. P., and Ferreira, M. A. (2001). Determination of lactic, acetic, succinic, and citric acids in table olives by HPLC/UV. *Journal of Liquid Chromatography and Related Technologies 24(7)*, 1029–1038.

d'Andria, R., Morelli, G., Patumi, M., and Fontanazza, G. (2002). Irrigation regime affects yield and oil quality of olive trees. In 'Proceedings of the 4th International Symposium on Olive Growing'. (Eds C. Vitagliano and G. P. Martelli.) *Acta Horticulturae 586(1)*, 273–276.

D'hallewin, G., Mulas, M., and Schirra, M. (1990). Characteristic of eleven table-olive clones from Nera cultivar. In 'Proceedings of the International Olive Growing Symposium'. (Eds L. Rallo, J. M. Caballero, and R. Fernández-Escobar.) *Acta Horticulturae 286*, 49–52.

da Silva, T. B., Pereira, S., Oliveira, M. M., Peres, C., and Rocheta, M. (2002). Homology studies between *Lactobacillus plantarum* strains isolated from Portuguese table olives and producing regions. In 'Proceedings of the 4th International Symposium on Olive Growing'. (Eds C. Vitagliano and G. P. Martelli.) *Acta Horticulturae 586(2)*, 653–656.

De Castro, A., and Brenes, M. (2001). Fermentation of washing waters of Spanish-style green olive processing. *Process Biochemistry 36(8–9)*, 797–802.

De Gregorio, A., Dugo, G., Arena, N., and Patumi, M. (2000). Lipoxygenase activities in ripening olive fruit tissue. *Journal of Food Biochemistry 24(5)*, 417–426.

Deidda, P., Dettori, S., Filigheddu, M. R., and Canu, A. (1990) Water stress and physiological parameters in young table olive trees. In 'Proceedings of the International Olive Growing Symposium'. (Eds L. Rallo, J. M. Caballero, and R. Fernández-Escobar.) *Acta Horticulturae 286*, 255–258.

del Río, C., and Caballero, J. M. (1994). Preliminary agronomical characterization of 131 cultivars introduced in the olive germplasm bank of Cordoba in March 1987. In '2nd International Symposium on Olive Growing'. (Eds S. Lavee and I. Klein.) *Acta Horticulturae 356*, 110–115.

Delatorre, J. E., Moya, E. R., Bota, E., and Sancho, J. (1993). Physical, chemical and microbiological studies of the fermentation of Arbequina green olives. *Grasas y Aceites 44(4–5)*, 274–278.

Delgado, A., Brito, D., Fevereiro, P., Peres, C., and Marques, J. F. (2001). Antimicrobial activity of *L. plantarum*, isolated from a traditional lactic acid fermentation of table olives. *Le Lait 81(1–2)*, 203–215.

Di Marco, L., Giovannini, D., Mara, F. P., and Viglianisi, G. (1990). Reproductive and vegetative behaviour of four table-olive cultivars. In 'Proceedings of the International Olive Growing Symposium'. (Eds L. Rallo, J. M. Caballero, and R. Fernández-Escobar.) *Acta Horticulturae 286*, 187–190.

Doorenbos, J., and Kassam, A. H. (1979). Yield response to water. FAO irrigation and drainage paper (on Olive) No. 33, pp. 105–108. Food and Agricultural Organisation of the United Nations, Rome, Italy.

Durán, R. M. (1990). Relationship between the composition and ripening of olive and quality of oil. In 'Proceedings of the International Olive Growing Symposium'. (Eds L. Rallo, J. M. Caballero, and R. Fernández-Escobar.) *Acta Horticulturae 286*, 441–451.

Durán Quintana, M. C., Garcia, P. G., and Fernández, A. G. (1999). Establishment of conditions for green table olive fermentation at low temperatures. *International Journal of Food Microbiology 51(2–3)*, 133–143.

Durán, M. C., Garcia, P., Brenes, M., and Garrido, A. (1994). Induced lactic acid fermentation during the preservation stage of ripe olives from Hojiblanca cultivar. *Journal of Applied Bacteriology 76(4)*, 377–382.

Durán, M. C., Garcia, P., Brenes, M., and Garrido, A. (1994). *Lactobacillus plantarum* survival in aerobic, directly brined olives. *Journal of Food Science 59(6)*, 1197–1201.

Durán Quintana, M. C., Barranco, C. R., Garcia, P. G., Balbuena, M. B., and Férnandez, A. G. (1997). Lactic acid bacteria in table olive fermentations. *Grasas y Aceites 48(5)*, 297–311.

Elliott, R. (1993). 'Pruning: A Practical Guide.' (Lothian Publishing Co.: Port Melbourne, Victoria.)

El-Makhzangy, A., and Abdel-Rhman, A. (1999). Physico-chemical properties of Azizi green pickled olives as affected by alkali process. *Nahrung-Food 43(5)*, 320–324.

Eltem, R. (1996). Growth and aflatoxin B1 production on olives and olive paste by moulds isolated from 'Turkish-style' natural black olives in brine. *International Journal of Food Microbiology 32(1–2)*, 217–223.

Fabbri, A., and Benelli, C. (2000). Flower bud induction and differentiation in olive. *Journal of Horticultural Science and Biotechnology 75(2)*, 131–141.

Fabbri, A., Hormaza, J. I., and Polito, V. S. (1995). Random amplified polymorphic DNA analysis of olive (*Olea europaea* L.) cultivars. *Journal of the American Society for Horticultural Science 120(3)*, 538–542.

Fabbri, A., Bartolini, G., Lambardi, M., and Kailis, S. (2004). 'Olive Propagation Manual.' (CSIRO Publishing: Collingwood, Victoria.)

Faci, J. M., Berenguer, M. J., Espada, J. L., and Gracia, S. (2002). Effect of variable water irrigation supply in olive (*Olea europaea* L.) cv. Arbequina in Aragon (Spain). I. Fruit and Oil Production. In 'Proceedings of the 4th International Symposium on Olive Growing'. (Eds C. Vitagliano and G. P. Martelli.) *Acta Horticulturae 586(1)*, 341–344.

Faci, J. M., Berenguer, M. J., Espada, J. L., and Gracia, S. (2002). Effect of variable water irrigation supply in olive (*Olea europaea* L.) cv. Arbequina in Aragon (Spain). II. Extra Virgin Oil Quality Parameters. In 'Proceedings of the 4th International Symposium on Olive Growing'. (Eds C. Vitagliano and G. P. Martelli.) *Acta Horticulturae 586(2)*, 649–652.

Failla, O., Tura, D., and Bassi, D. (2002). Genotype-environment-year interaction on oil antioxidants in an olive district of northern Italy. In 'Proceedings of the 4th International Symposium on Olive Growing'. (Eds C. Vitagliano and G. P. Martelli.) *Acta Horticulturae 586(1)*, 171–174.

Famiani, F., Proietti, P., Farinelli, D., and Tombesi, A. (2002). Oil quality in relation to olive ripening. In 'Proceedings of the 4th International Symposium on Olive Growing'. (Eds C. Vitagliano and G. P. Martelli.) *Acta Horticulturae 586(2)*, 671–674.

Farinelli, D., Boco, M., and Tombesi, A. (2002.) Intensity and growth period of the fruit components of olive varieties. In 'Proceedings of the 4th International Symposium on Olive Growing'. (Eds C. Vitagliano and G. P. Martelli.) *Acta Horticulturae 586(2)*, 607–610.

Ferguson, L., Sibbett, S., and Freeman, M. (1990). 'Manzanillo' olive harvest timing. In 'International Symposium on Olive Growing'. (Eds L. Rallo, J. M. Caballero and R. Fernández-Escobar.) *Acta Horticulturae 286*, 433–436.

Fernández, A. G., and Barranco, C. R. (1999). Quality of table olives. *Grasas y Aceites 50(3)*, 225–230.

Fernández, A. G., Brenes, M. B., Garcia, P. G., and Durán Quintana, M. C. (1996). Preservation in brine of green or turning colour olives. *Grasas y Aceites 47(3)*, 197–206.

Fernández, A. G., García, P. G., Lopéz, A. L., and Lopéz, F. N. A. (2004). Nutritional characteristics of olive oil and table olive. In 'TDC Olive Encyclopaedia'. European Community Priority 5 on Food Quality and Safety (Contract number Food-CT–2004–505524 Specific Targeted Project).

Fernández, A. G., García, P. G., Lopéz, A. L., and Lopéz, F. N. A. (2004). Processing technology in olive oil and table olive. In 'TDC Olive Encyclopaedia'. European Community Priority 5 on Food Quality and Safety (Contract number Food-CT–2004–505524 Specific Targeted Project).

Fernández, J. E., Moreno, F., Girón, I. F., and Blazquez, O. M. (1997). Stomatal control of water use in olive tree leaves. *Plant and Soil 190(2)*, 179–192.

Fernández, J. E., Palomo, M. J., Diaz-Espejo, A., Clothier, B. E., Green, S. R., Giron, I. F., and Moreno, F. (2001). Heat-pulse measurements of sap flow in olives for automating irrigation: tests, root flow and diagnostics of water stress. *Agricultural Water Management 51(2)*, 99–123.

Fernández Escobar, R., Barranco, D., and Benlloch, M. (1993). Overcoming iron chlorosis in olive and peach trees using a low-pressure trunk-injection method. *Horticultural Science 28(3)*, 192–194.

Ferrara, E., Papa, G., and Lamparelli, F. (2002). Evaluation of the olive germplasm in the Apulia Region: biological and technological characteristics. In 'Proceedings of the 4th International Symposium on Olive Growing'. (Eds C. Vitagliano and G. P. Martelli.) *Acta Horticulturae 586(1)*, 159–162.

Fiorino, P. (1990). Orchard management. In 'Proceedings of the International Olive Growing Symposium'. (Eds L. Rallo, J. M. Caballero, and R. Ferández-Escobar.) *Acta Horticulturae 286*, 231–245.

Firestone, D., and Horwitz W. (1979). IUPAC gas chromatographic method for determination of fatty acid composition. Collaborative study. *Journal of the Association of Official Analytical Chemists 62(4)*, 709–21.

Fontanazza, G., Patumi, M., Solinas, M., and Serraiocco, A. (1994). Influence of cultivars on the composition and quality of olive oil. In 'Proceedings of the 2nd International Olive Growing Symposium'. (Eds S. Lavee and I. Klein.) *Acta Horticulturae 356*, 358–361.

Food Standards Agency. (2002). 'The Composition of Foods: 6th summary edition.' (Royal Society of Chemistry.)

Food Standards Agency, Scotland. (2004). Contamination of Pran™ pickles with Sudan 1 dye (second update). March 8.

Food Standards Australia New Zealand. http://www.foodstandards.gov.au/foodstandardscode/ [Verified 9 August 2006].

Food Standards Australia New Zealand. (2004). Food industry recall protocol – a guide to writing a food recall plan and conducting a food recall. 5th edn. ANZFA, Australia.

Food Standards Australia New Zealand. (2005). Final assessment report – proposal P292 – country of origin labelling of food. Food Standards Australia New Zealand, Canberra, Australia.

Forsythe, S. J. (2000). 'The microbiology of safe food.' (Blackwell Science: Oxford, UK.)

Gálvez, M. C., Barroso, C. G., and Pérez-Bustamante, J. A. (1995). Influence of the origin of wine vinegars in their low molecular weight phenolic content. In 'Viticulture and Enology'. (Eds F. Pérez-Camacho and M. Medina.) Acta Horticulturae 388, 269–272.

Gambella, F., Piga, A., Agabbio, M., Vacca, V., and D'hallewin, G. (2000). Effect of different pre-treatments on drying of green table olives (Ascolana tenera var.). Grasas y Aceites 51(3), 173–176.

Gandul-Rojas, B., and Mínguez-Mosquera, M. I. (1996). Chlorophyllase activity in olive fruits and its relationship with the loss of chlorophyll pigments in the fruits and oils. Journal of the Science of Food and Agriculture 72(3), 291–294.

Gandul-Rojas, B., Cepero, M. R., and Mínguez-Mosquera, M. I. (1999). Chlorophyll and carotenoid patterns in olive fruits, Olea europaea cv. Arbequina. Journal of Agricultural and Food Chemistry 47(6), 2207–2212.

Ganz, T. R., Kailis, S. G., and Abbott, L. K. (2002). Mycorrhizal colonization and its effects on growth, phosphorous uptake and tissue phenolic content in the European olive (Olea europaea L.) Advances in Horticultural Science 16, 109–116.

Garcia, J. M., Seller, S., and Pèrez-Camino, C. (1996). Influence of fruit ripening on olive oil quality. Journal of Agriculture and Food Chemistry 44, 3516–3520.

Garcia, J. M., Yousfi, K., Mateos, R., Olmo, M., and Cert, A. (2001). Reduction of oil bitterness by heating of olive (Olea europaea) fruits. Journal of Agricultural and Food Chemistry 49, 4231–4235.

García, J. M., Gutiérrez, F., Castellano, J. M., Perdiguero, S., Morilla, A., and Albi, M. A. (1996). Influence of storage temperature on fruit ripening and olive oil quality. Journal of Agricultural and Food Chemistry 44, 264–267.

Garcia, P., Brenes, M., Romero, C., and Garrido, A. (1995). Respiration and physicochemical changes in harvested olive fruits. Journal of Horticultural Science 70(6), 925–933.

Garcia-Garcia, P., Brenes-Balbuena, M., Hornero-Mendez, D., Garcia-Borrego, A., and Garrido-Fernández, A. (2000). Content of biogenic amines in table olives. Journal of Food Protection 63(1), 111–116.

Gardiman, M., Tonutti, P., Pizzale, L., Conte, L., and Carazzolo, A. (1999). The effect of hypoxic and CO_2-enriched atmospheres on olive ripening and oil quality. In 'Proceedings of the 3rd International Symposium on Olive Growing'. (Eds I. T. Metzidakis and D. G. Voyiatzis.) Acta Horticulturae 474(2), 525–528.

Garrido, A., Garcia, P., Brenes, M., and Romero, C. (1995). Iron content and colour of ripe olives. Nahrung 39(1), 67–76.

Garrido Fernández, A., Férnandez Diez, M. J., and Adams, M. R. (1997). 'Table Olives – Production and Processing.' (Chapman and Hall: London.)

Georget, D. M. R., Smith, A. C., and Waldron, K. W. (2001). Effect of ripening on the mechanical properties of Portuguese and Spanish varieties of olive (Olea europaea L.). Journal of the Science of Food and Agriculture 81(4), 448–454.

Georget, D. M. R., Smith, A. C., Waldron, K. W., and Rejano, L. (2003). Effect of 'Californian' process on the texture of Hojiblanca olive (Olea europaea L.) harvested at different ripening stages. Journal of the Science of Food and Agriculture 83(6), 574–579.

Giammanco, M., Tripoli., E., Tabacchi, G., Di Majo, D., Giammanco, S., and Guardia, M. (2005). The phenolic compounds of olive oil: structure, biological activity and beneficial effects on human health. *Nutrition Research Reviews 18(1)*, 98–112.

Goldhammer, D. A. (1999). Regulated deficit irrigation for California canning olives. In 'Proceedings of the 3rd International Symposium on Olive Growing'. (Eds I. T. Metzidakis and D. G. Voyiatzis.) *Acta Horticulturae 474(2)*, 369–372.

Goldhammer, D. A., Dunai, J., and Ferguson L. (1993). Water use requirements of Manzanillo olives and responses to sustained deficit irrigation. In 'Irrigation of Horticultural Crops'. (Ed. J. López-Gálvez.) *Acta Horticulturae 335*, 365–372.

Goldhammer, D. A., Dunai, J., and Ferguson, L. (1994). Irrigation requirements of olive trees and responses to sustained deficit irrigation. In 'Proceedings of the 2nd International Olive Growing Symposium'. (Eds S. Lavee and I. Klein.). *Acta Horticulturae 356*, 172–175.

González, M. J. F., Garcia, P. G., Fernández, A. G., and Durán Quintana, M. C. (1993). Microflora of the aerobic preservation of directly brined green olives from Hojiblanca cultivar. *Journal of Applied Bacteriology 75(3)*, 226–233.

Goren, R., Huberman, M., and Martin, G. C. (1998). Phosphorus-induced leaf abscission in detached shoots of olive and citrus. *Journal of the American Society for Horticultural Science 123(4)*, 545–549.

Grati-kammoun, N., Khlif, M., Rekik, H., Hamdi, M. T. (1999). Evolution of oil characteristics during olive maturation (Chemlali variety). In 'Proceedings of the 3rd International Symposium on Olive Growing'. (Eds I. T. Metzidakis and D. G. Voyiatzis.) *Acta Horticulturae 474(2)*, 701–704.

Gregoriou, C. (1996). Assessment of variation of landraces of olive tree in Cyprus. *Euphytica 87(3)*, 173–176.

Gucci, R., and Cantini, C. (2000). 'Pruning and Training Systems for Modern Olive Growing.' (CSIRO Publishing: Collingwood, Australia.)

Guerin, J., Collins, G., Jones, G., Wu, S., Mekuria, G., Burr, M., and Sedgley, M. (2002). Current Australian olive research: olive biotechnology for the renascent Australian industry. In 'Proceedings of the National Olive Industry Convention Conference'. pp. 36–40. Olives South Australia, Adelaide, Australia.

Guillén, R., Fernández Bolaños, J., and Heredia, A. (1993). Component changes in olive (Hojiblanca var.) during ripening. *Grasas y Aceites 44(3)*, 201–203.

Hartmann, H. T. (1953). Effect of winter chilling on fruitfulness and vegetative growth in olive. *Proceedings of the American Society of Horticultural Science 62*, 184–190.

Hartmann, H. T., and Porlingus, L. (1957). Effect of different amounts of winter chilling on the fruitfulness of several olive varieties. *Botanical Gazette*, 102–104.

Hassapidou, M. N., Balatsouras, G. D., and Manoukas, A. G. (1994). Effect of processing upon the tocopherol and tocotrienol composition of table olives. *Food Chemistry 50(2)*, 111–114.

Heredia, A., Ruiz Gutiérrez, V., Felizon, B., Guillén, R., Jiménez, A., and Férnandez Bolaños, J. (1993). Apparent digestibility of dietary fibre and other components in table olives. *Nahrung 37(3)*, 226–233.

Hobman, F. (1995). Economic study into irrigated olive growing and oil processing in southern Australia. RIRDC Research Paper, 95/5. Rural Industries Research and Development Corporation, Canberra, Australia.

Holzapfel, W. H. (2002). Appropriate starter culture technologies for small-scale fermentation in developing countries. *International Journal of Food Microbiology 75*, 197–212.

Hornero Méndez, D., and Garrido Fernández, A. (1994). Biogenic amines in table olives – analysis by high performance liquid chromatography. *Analyst 119(9)*, 2037–2041.

Hornero Méndez, D., Gallardo-Guerrero, L., Jaren-Galan, M., and Mínguez-Mosquera, M. I. (2002). Differences in the activity of superoxide dismutase, Polyphenol oxidase and Cu-Zn content in the fruits of Gordal and Manzanilla olive varieties. *Zeitschrift für Naturforschung 57(1–2)*, 113–120.

Hughes, B. (2004). Replacing nutrients. *Australian and New Zealand Olivegrower and Processor 39* (Sep.–Oct.), 12.

Hunt, N., and Gilkes, B. (1992). 'Farm Monitoring Handbook.' (University of Western Australia: Crawley, Western Australia.)

Improtechnology Limited. (2004). Olive tree cultivation. In 'TDC Olive Encyclopaedia'. European Community Priority 5 on Food Quality and Safety (Contract number Food-CT–2004–505524 Specific Targeted Project).

Improtechnology Limited. (2004). Promotion of nutrition in table olives and olive oil in SMEs. In 'TDC Olive Encyclopaedia'. European Community Priority 5 on Food Quality and Safety (Contract number Food-CT–2004–505524 Specific Targeted Project).

Indrissi, I. J., Rahmani, M., and Souizi, A. A. (2004). Industrial-scale biological debittering of table olives. *Olivae 101*, 34–37.

Inglese, P., Barone, E., and Gullo, G. (1996). The effect of complementary irrigation on fruit growth, ripening pattern and oil characteristics of olive (*Olea europaea* L.) cv Carolea. *Journal of Horticultural Science 71(2)*, 257–263.

Inglese, P., Gullo, G., and Pace, L. S. (1999). Summer drought effects on fruit growth, ripening and accumulation and composition of 'Carolea' olive oil. In 'Proceedings of the 3rd International Symposium on Olive Growing'. (Eds I. T. Metzidakis and D. G. Voyiatzis.) *Acta Horticulturae 474(1)*, 269–273.

Inglese, P., Gullo, G., and Pace, L. S. (2002). Fruit growth and olive oil quality in relation to foliar nutrition and time of application. In 'Proceedings of the 4th International Symposium on Olive Growing'. (Eds C. Vitagliano and G. P. Martelli.) *Acta Horticulturae 586(2)*, 507–509.

Inglese, P., Gullo, G., Pace, L. S., and Ronzello, G. (1999). Fruit growth, oil accumulation and ripening of the olive cultivar 'Carolea' in relation to fruit density. In 'Proceedings of the 3rd International ISHS Symposium on Olive Growing'. (Eds I. T. Metzidakis and D. G. Voyiatzis.) *Acta Horticulturae 474(1)*, 265–268.

Instituto de la Grasa y sus Derivados (Seville) (1990). 'Table Olive Processing. (International Olive Oil Council: Madrid, Spain.)

International Olive Oil Council. (1996). 'World Olive Encyclopaedia.' (International Olive Oil Council: Madrid, Spain.)

International Olive Oil Council. (2000). 'World Catalogue of Olive Varieties.' (International Olive Oil Council: Madrid, Spain.)

International Olive Oil Council. (2004). 'Trade standards applying to table olives.' COI/OT/NC no. 1. (International Olive Oil Council: Madrid, Spain.)

Ippolito, A., and Nigro, F. (2002). Shrivelling of olive fruits associated with water stress. In 'Proceedings of the 4th International Symposium on Olive Growing'. (Eds C. Vitagliano and G. P. Martelli.) *Acta Horticulturae 586(2)*, 745–747.

Istituto Sperimentale per la Elaiotecnica (Pescara, Italy). (2004). Sensory analysis and its application to olive and virgin olive oil. In 'TDC Olive Encyclopaedia'. European Community Priority 5 on Food Quality and Safety (Contract number Food-CT–2004–505524 Specific Targeted Project).

Jiménez, A., Guillén, R., Fernández Bolaños, J., and Heredia, A. (1994). Cell wall composition of olives. *Journal of Food Science 59(6)*, 1192.

Jiménez, A., Guillén, R., Sánchez, C., Férnandez Bolaños, J., and Heredia, A. (1995). Changes in texture and cell wall polysaccharides of olive fruit during Spanish green olive processing. *Journal of Agricultural and Food Chemistry 43(8)*, 2240–2246.

Jiménez, A., Rodriguez, R., Fernández-Caro, I., Guillén, R., Fernández-Bolaños, J., and Heredia, A. (2000). Dietary fibre content of table olives processed under different European styles: study of physico-chemical characteristics. *Journal of the Science of Food and Agriculture 80(13)*, 1903–1908.

Jordão, P. V., and Lietão, F. (1990). The olive's mineral composition and some parameters of quality in fifty olive cultivars grown in Portugal. In 'Proceedings of the International Olive Growing Symposium'. (Eds L. Rallo, J. M. Caballero, and R. Fernández-Escobar.) *Acta Horticulturae 286*, 461–464.

Kachouri, M., M'Sallem, M., Zarrouk, M., and Cherif, A. (1995). Comparative study in four olive varieties. In 'Plant Lipid Metabolism'. (Eds J. C. Kader and P. Mazliak.) pp. 567–569. (Kluwer Academic Publishers: Netherlands.)

Kader, A. A., Nanos, G. D., and Kerbel, E. L. (1989). Responses of 'Manzanillo' olives to controlled atmospheres storage. In 'Proceedings of the 5th International Controlled Atmosphere Research Conference, Wenatchee, WA, USA'. (Ed. J. K. Fellman.) 2, 119–125.

Kader, A. A., Nanos, G. D., and Kerbel, E. L. (1990). Storage potential of fresh 'Manzanillo' olives. *California Agriculture 44(3)*, 23–24.

Kailis, S. G., and Considine, J. A. (2002). The olive *Olea europaea* L. in Australia: 2000 onwards. *Advances in Horticultural Science 16*, 299–306.

Kailis, S. G., and Harris, D. (2001) Growing olives in Australia for table olive processing – part 1. *Australian Olive Grower 23*, 4–7.

Kailis, S., and Harris, D. (2001). Growing olives in Australia for table olive processing – part 2. *Australian Olive Grower 24*, 9–12.

Kailis, S. G., and Harris, D. (2004). Establish protocols and guidelines for table olive processing in Australia, RIRDC Publication No. 04/136, RIRDC Project No. UWA59A. Rural Industries Research and Development Corporation, Canberra, Australia.

Kailis, S. G., and Harris, D. (2004). Table olives. In 'The New Crop Industries Handbook'. (Eds S. Salvin, M. Bourke and T. Byrne.) RIRDC Publication No. 04/125, 321–330. Rural Industries Research and Development Corporation, Canberra, Australia.

Kailis, S. G., and Sweeney, S. (2000). Olives in Australia. In 'Proceedings of the 4th Olive International Symposium on Olive Growing'. (Eds C. Vitagliano and G. P. Martelli.) *Acta Horticulturae 586(1)*, 385–388.

Kailis, S., Harris, D., and Smyth, J. (2001). Spanish-style green table olives: fruit quality and processing method. *The Olive Press* 15–19.

Karaoulanis, G. D., and Bamnidou, A. (1995). Colour changes in different processing conditions of green olives of Chalkidiki variety. *Grasas y Aceites 46(3)*, 153–159.

Kaynas, N., Sutçu, A. R., and Fidan, A. E. (2002). Olive variety trial in Marmara region. In 'Proceedings of the 4th International Symposium on Olive Growing'. (Eds C. Vitagliano and G. P. Martelli.) *Acta Horticulturae 586(1)*, 187–189.

Keys, A. (Ed.) (1980). 'Seven Countries: multivariate analysis of death and coronary heart diseases.' (Harvard University Press: Cambridge, Massachusetts.)

Khlif, M., and Trigui, A. (1990). Olive cultivar investigations. Preliminary results. In 'Proceedings of the International Olive Growing Symposium'. (Eds L. Rallo, J. M. Caballero, and R. Fernández-Escobar.) *Acta Horticulturae 286*, 65–68.

Kiritsakis, A., Nanos, G. D., Polymenopoulos, Z., Thomai, T., and Sfakiotakis, E. M. (1998). Effect of fruit storage conditions on olive oil quality. *Journal of the American Oil Chemists Society 75(6)*, 721–724.

Klein, I., Ben-Tal, Y., Lavee, S., De Malach, Y., and David, I. (1994). Saline irrigation of cv. Manzanillo and Uovo di Piccione trees. In 'Proceedings of the 2nd International Symposium on Olive Growing'. (Eds S. Lavee and I. Klein.) *Acta Horticulturae 356*, 176–180.

Knabel, S. J., Fatemi, P., Patton, J., Laborde, L. F., Annous, B., and Spears, G. M. (2003). On farm contamination of horticultural products in the USA and strategies for decontamination. *Food Australia 52*, 580–586.

Kopsidas, G. C. (1994). Wastewaster from the table olive industry. *Water Research 28(1)*, 201–205.

Kopsidas, G. C. (1995). Multiobjective optimisation of table olive preparation systems. *European Journal of Operational Research 85(2)*, 383–398.

Krueger, W. H., Heath, Z., and Mulqueeney, B. (2002). Effect of spray solution concentration, active ingredient, additives and sequential treatments of napthalene acetic acid for chemical thinning of Manzanillo table olives (*Olea europea*). In 'Proceedings of the 4th International Symposium on Olive Growing'. (Eds C. Vitagliano and G. P. Martelli.) *Acta Horticulturae 586(1)*, 267–271.

Lavee, S. (1986). Olive. In 'CRC Handbook of Fruit Set and Development'. (Ed. S. P. Monselise.) pp. 261–278. (CRC Press Inc.: Florida, USA.)

Lavee, S., Rallo, L., Rapoport, H. F., and Troncoso, A. (1996). The floral biology of the olive – I. The effect of flower number, type and distribution on fruit set. *Scientific Horticulture 66(3–4)*, 149–158.

Lavee, S., Rallo, L., Rapoport, H. F., and Troncoso, A. (1999). The floral biology of the olive – II. The effect of inflorescence load and distribution per shoot on fruit set and load. *Scientific Horticulture 82(3–4)*, 181–192.

Lavee, S., Avidan, B., Meni, Y., Haskal, A., and Wodner, M. (2005). Three new semi-dwarf table olive varieties. *Olivae 102*, 33–41.

Lavermicocca, P., Gobbetti, M., Corsetti, A., and Caputo, L. (1998). Characterisation of lactic acid bacteria isolated from olive phylloplane and table olive brines. *Italian Journal of Food Science 10(1)*, 27–39.

Lavermicocca, P., Valerio, F., Lonigro, S. L., Baruzzi, F., Morea, M., and Gobbetti, M. (2002). Olive fermentations using lactic acid bacteria isolated from olive phylloplane and olive brines. In 'Proceedings of the 4th International Symposium on Olive Growing'. (Eds C. Vitagliano and G. P. Martelli.) *Acta Horticulturae 586(2)*, 621–624.

Lazovic, B., Miranovic, K., Gasic, O., and Popovic, M. (1999). Olive protein content and amino acid composition. In 'Proceedings of the 3rd International Symposium on Olive Growing'. (Eds I. T. Metzidakis and D. G. Voyiatzis.) *Acta Horticulturae 474(2)*, 465–468.

Lee, S-Y. (2004). Microbial safety of pickled fruits and vegetables and hurdle technology. *Internet Journal of Food Safety 4*, 21–32.

Leitao, F. (1990). Productivity of twenty olive (*Olea europaea* L.) cultivars. In 'Proceedings of the International Olive Growing Symposium'. (Eds L. Rallo, J. M. Caballero, and R. Fernández-Escobar.) *Acta Horticulturae 286*, 69–72.

Lopez-Villalta, M.C. (1999). 'Olive Pest and Disease Management.' International Olive Oil Council, Madrid, Spain.

Mafra, I., Lanza, B., Reis, A., Marsilio, V., Campestre, C., De Angelis, M., and Coimbra, M. A. (2001). Effect of ripening on texture, microstructure and cell wall polysaccharide composition of olive fruit (*Olea europaea*). *Physiologia Plantarum 111(4)*, 439–447.

Maldonado, M. B., and Zuritz, A. (2004). Determination of variable diffusion of sodium during debittering of green olives. *Food Process Engineering 27*, 345–358.

Mannino, P., and Pannelli, G. (1990). Fully mechanized harvesting of olive fruit, technical and agronomical preliminary evaluations. In 'International Symposium on Olive Growing'. (Eds L. Rallo, J. M. Caballero, and R. Fernández-Escobar.) *Acta Horticulturae 286*, 437–440.

Manrique, T., Rapoport, H. F., Castro, J., and Pastor, M. (1999). Mesocarp cell division and expansion in the growth of olive fruits. In 'Proceedings of the 3rd International Symposium on Olive Growing'. (Eds I. T. Metzidakis and D. G. Voyiatzis.) *Acta Horticulturae 474(1)*, 301–304.

Manzocco, L., Calligaris, S., Mastrocola, D., Nicoli, M. C., and Lerici, C. R. (2000). Review of non-enzymatic browning and antioxidant capacity in processed foods. *Trends in Food Science and Technology 11*, 340–346.

Márquez, J. A., Benlloch, M., and Rallo, L. (1990). Seasonal changes of glucose, potassium and rubidium in 'Gordal Sevillana' olive in relation to fruitfulness. In 'Proceedings of the International Symposium on Olive Growing'. (Eds L. Rallo, J. M. Caballero, and R. Fernández-Escobar.) *Acta Horticulturae 286*, 191–194.

Marsilio, V. (1993). Table olive production, processing and standards in Italy. *Olivae 49*, 6–16.

Marsilio, V. (2002). Sensory analysis of table olives. *Olivae 90*, 32–41.

Marsilio, V., and Lanza, B. (1995). Effects of lye-treatment on the nutritional and microstructural characteristics of table olives (Olea europaea L.). *Revista Espanola de Ciencia y Tecnologia de Alimentos 35(2)*, 178–190.

Marsilio, V., and Lanza, B. (1998). Characterisation of an oleuropein degrading strain of *Lactobacillus plantarum*. Combined effects of compounds present in olive fermenting brines (phenols, glucose and NaCl) on bacterial activity. *Journal of the Science of Food and Agriculture 76*, 520–524.

Marsilio, V., Campestre, C., and Lanza, B. (2001). Phenolic compounds change during California-style ripe olive processing. *Food Chemistry 74*, 55–60.

Marsilio, V., Lanza, B., and De Angelis, M. (1996). Olive cell wall components: physical and biochemical changes during processing. *Journal of the Science of Food and Agriculture 70*, 35–43.

Marsilio, V., Lanza, B., and Lombardi, D. S. (1999). Quality improvement of table olives by use of an oleuropein degrading strain of *Lactobacillus Plantarum*. In 'Proceedings of the 3rd International Symposium on Olive Growing'. (Eds I. T. Metzidakis and D. G. Voyiatzis.) *Acta Horticulturae 474(2)*, 601–604.

Marsilio, V., Lanza, B., and Pozzi, N. (1996). Progress in table olive debittering: degradation in vitro of oleuropein and its derivatives by *Lactobacillus plantarum*. *Journal of the American Oil Chemists Society 73(5)*, 593–597.

Marsilio, V., Lanza, B., Campestre, C., and De Angelis, M. (2000). Oven-dried table olives: textural properties as related to pectic composition. *Journal of the Science of Food and Agriculture 80*, 1271–1276.

Marsilio, V., Campestre, C., Lanza, B., and De Angelis, M. (2001). Sugar and polyol compositions of some European olive fruit varieties (*Olea Europaea* L.) suitable for table olive purposes. *Food Chemistry 72(4)*, 485–490.

Marsilio, V., Campestre, C., Lanza, B., De Angelis, M., and Russi, F. (2002). Sensory analysis of green table olives fermented in different saline solutions. In 'Proceedings of the 4th International Symposium on Olive Growing'. (Eds C. Vitagliano and G. P. Martelli.) *Acta Horticulturae 586(2)*, 617–620.

Marsilio, V., Seghetti, L., Iannucci, E., Russi, F., Lanza, B., and Felicioni, M. (2005). Use of lactic acid bacteria starter culture during green olive (*Olea europaea* L. Ascolana tenera) processing. *Journal of the Science of Food and Agriculture 85(7)*, 1084–1090.

Martin, G. C. (1990). Olive flower and fruit population dynamics. In 'Proceedings of the Olive Growing Symposium'. (Eds L. Rallo, J. M. Caballero, and R. Fernández-Escobar.) *Acta Horticulturae 286*, 141–154.

Martin, G. C. (1994). Mechanical olive harvest: use of fruit loosening agents. In 'Proceedings of the 2nd International Symposium on Olive Growing'. (Eds S. Lavee and I. Klein.) *Acta Horticulturae 356*, 284–291.

Martin, G. C., Connell, J. H., Freeman, M. W., Krueger, W. H., and Sibbett, G. S. (1994). Efficacy of foliar application of two naphthaleneacetic acid salts for olive fruit thinning. In 'Proceedings of the 2nd International Symposium on Olive Growing'. (Eds S. Lavee and I. Klein.) *Acta Horticulturae 356*, 302–305.

Mazomenos, B. E., Pantazi-Mazomenou, A., and Stefanou, D. (2002). Attract and kill of the olive fruit fly *Bactrocera oleae* in Greece as a part of an integrated control system. In 'Proceedings of the meeting – Pheromones and other biological techniques for insect control in orchards and vineyards'. (Eds P. Witzgall, B. Mazomenos, and M. Konstantopoulo.) International Organisation of Biological Control /wprs Bulletin 25, 137–146.

Mazzuca, S., and Uccella, N. (2002). ß-Glucosidase releasing of phytoalexin derivatives from secobiophenols as defense mechanism against pathogenic elicitors in olive drupes. In 'Proceedings of the 4th International Symposium on Olive Growing'. (Eds C. Vitagliano and G. P. Martelli.) *Acta Horticulturae 586(2)*, 529–531.

McClure, P. Personal communication. Mildura Olive Trial 1969–1978 – Personal communication, P. McClure, Sunraysia Horticultural Centre, Victoria.

McEvoy, E., Gomez, E., McCarrol, A., and Sevil, J. (1998). Potential for establishing an olive industry in Australia. RIRDC Publication No. 98/5, RIRDC Project No. AQ-210. Rural Industries Research and Development Corporation, Canberra, Australia.

McQuaker, N. R., Brown, D. F., and Fluckner, P. D. (1979). Digestion of environmental materials for analysis by inductively coupled plasma – atomic emission spectrometry. *Analytical Chemistry 51*, 1082–1084.

Mekuria, G. T., Collins, G. G., and Sedgley, M. (1999). Genetic variability between different accessions of some common commercial olive cultivars. *Journal of Horticultural Science and Biotechnology 74(3)*, 309–314.

Metzidakis, I. (2002). Effect of regeneration pruning for the recovery of olive productivity and fruit characteristics in ten olive cultivars. In 'Proceedings of the 4th International Symposium on Olive Growing'. (Eds C. Vitagliano and G. P. Martelli.) *Acta Horticulturae 586(1)*, 333–336.

Michelakis, N. (1990). Yield response of table and oil olive varieties to different water use levels under drip irrigation. In 'Proceedings of the International Symposium on Olive Growing'. (Eds L. Rallo, J. M. Caballero and R. Fernández-Escobar.) *Acta Horticulturae 286*, 271–274.

Michelakis, N. (2002). Olive Orchard Management: advances and problems. In 'Proceedings of the 4th International Symposium on Olive Growing'. (Eds C. Vitagliano and G. P. Martelli.) *Acta Horticulturae 586(1)*, 239–245.

Mínguez Mosquera, M. I., Gandul Rojas, B., and Gallar Doguerrero, L. (1993). De-esterification of chlorophylls in olives by activation of chlorophyllase. *Journal of Agricultural and Food Chemistry 41(12)*, 2254–2258.

Mínguez Mosquera, M. I., Gandul Rojas, B., and Mínguez Mosquera, J. (1994). Mechanism and kinetics of the degradation of chlorophylls during the processing of green table olives. *Journal of Agricultural and Food Chemistry 42(5)*, 1089–1095.

Mínguez Mosquera, M. I., Gallardo Guerrero, L., Hornero Méndez, D., and Garrido Férnandez, J. (1995). Involvement of copper and zinc ions in green staining of table olives of the variety Gordal. *Journal of Food Protection 58(5)*, 564–569.

Moir, C. J., Andrew-Kabilafkas, C., Arnold, G., Cox, B. M., Hocking, A. D., and Jenson, I. (2001). 'Spoilage of processed foods.' Australian Institute of Food Science and Technology Inc. (NSW Branch.) Food Microbiology Group.

Monselise, S. P., and Goldschmidt, E. E. (1982). Alternate bearing in fruit trees. (Eds D. P. Coyne, D. Durkin, and M. W. Williams.) *Horticultural Reviews 4*, 128–173. (Avi Publishing Company Inc.: Westport, USA.)

Montano, A., De Castro, A., Rejano, L., and Brenes, M. (1996). 4-Hydroxycyclohexanecarboxylic acid as a substrate for Cyclohexanecarboxylic acid production during the Zapatera spoilage of Spanish-style green table olives. *Journal of Food Protection 59(6)*, 657–662.

Montedoro, G., Baldioli, M., Selvaggini, R., Begliomini, A. L., Taticchi, A., and Servili, M. (2002). Relationships between phenolic composition of olive fruit and olive oil: the importance of the endogenous enzymes. In 'Proceedings of the 4th International Symposium on Olive Growing'. (Eds C. Vitagliano and G. P. Martelli.) *Acta Horticulturae 586(2)*, 551–556.

Montemurro, N., Benedetto, P., Lacertosa, G., Castoro, V., and Martelli, S. (2002). Quality olive oils production in Basilicata region: chemical characteristics investigations. In 'Proceedings of the 4th International Symposium on Olive Growing'. (Eds C. Vitagliano and G. P. Martelli.) *Acta Horticulturae 586(2)*, 533–536.

Monties, B. L. (1989). Lignins. In 'Methods in Plant Biochemistry Vol. 1.' (Eds P. M. Dey and J. B. Harborne.) pp. 113–157. (Academic Press: London.)

Morales-Bernardino, J. (2002). Pruning management in traditional olive orchards. *Olivae 92*, 38–43.

Moreno, F., Fernández, J. E., Clothier, B. E., and Green, S. R. (1996). Transpiration and root water uptake by olive trees. *Plant and Soil 184(1)*, 85–96.

Morphett, B. (1993). 'Pruning for Fruit.' (Botanic Gardens of Adelaide: Adelaide, South Australia.)

Motarjemi, Y. (2002). Impact of small scale fermentation technology on food safety in developing countries. *International Journal of Food Microbiology 75*, 213–229.

Motilva, M. J., Tovar, M. J., Romero, M.P., Alegre, S., and Girona, J. (2002). Evolution of oil accumulation and polyphenol content in fruits of olive tree (*Olea europaea* L.) related to different irrigation strategies. In 'Proceedings of the 4th International Symposium on Olive Growing'. (Eds C. Vitagliano and G. P. Martelli.) *Acta Horticulturae 586(1)*, 345–348.

Mulas, M. (1994). Genetic variability of histological characteristics in olive fruits. In 'Proceedings of the 2nd International Olive Growing Symposium'. (Eds S. Lavee and I. Klein.) *Acta Horticulturae 356*, 70–73.

Mulas, M. (1995). Influenza della morfologia biometria cellulare sulla resistenza meccanica dei tessuti del frutto di olivo. In 'L'Olivicultura Mediterranea: Stato prospettive della coltura e della ricerca. Istituto Sperimentale per la Olivicultura'. (Eds N. Lombardo, N. Iannotta, and C. B. Bati.) pp. 167–174. (Istituto Sperimentale per la Olivicultura: Rende, Italy.)

Mulas, M., Virdis, F., Schirra, M., and Mura, M. (1999). Fruit quality of table-olive clones selected from 'Nera' cultivar. In 'Proceedings of the 3rd International Symposium on Olive Growing'. (Eds I. T. Metzidakis and D. G. Voyiatzis.) *Acta Horticulturae 474(2)*, 605–608.

Murillo, J. M., Lopez, R., Férnandez, J. E., and Cabrera, F. (2000). Olive tree response to irrigation with wastewater from the table olive industry. *Irrigation Science 19(4)*, 175–180.

Nanos, G. D., Agtsidou, E., and Sfakiotakis, E. M. (2002). Temperature and propylene effects on ripening of green and black 'Conservolea' olives. *Horticultural Science 37(7)*, 1079–1081.

Nanos, G. D., Kiritsakis, A. K., and Sfakiotakis, E. M. (2002). Preprocessing storage conditions for green 'Conservolea' and 'Chondrolia' table olives. *Postharvest Biology and Technology 25*, 109–115.

Nanos, G. D., Thomai, T., Sfakiotakis, E. M., and Fitsios, N. (1999). Maturity indices for green olives destined to be processed as 'Spanish-style' olives. In 'Proceedings of the 3rd International ISHS Symposium on Olive Growing'. (Eds I. T. Metzidakis and D. G. Voyiatzis.) *Acta Horticulturae 474(2)*, 521–524.

Navarro, C., Fernández-Escobar, R., and Benlloch, M. (1990). Flower bud induction in

'Manzanillo.' In 'Proceedings of the International Olive Growing Symposium'. (Eds L. Rallo, J. M. Caballero, and R. Fernández-Escobar.) *Acta Horticulturae 286*, 195–198.

Navarro, L. R. (1999). Sevillian genuine Manzanilla cultivar. *Grasas y Aceites 50(1)*, 60–66.

Nicetic, O., Spooner-Hart, R., and Rae, D. (2005). Research highlights safer black scale solution. *Australian and New Zealand Olivegrower and Processor 45* (Sept.–Oct.), 22–23.

Nuberg, I., and Yunusa, I. (2003). Olive water use and yield. RIRDC Publication No. 03/048, RIRDC Project No. UA–47A. Rural Industries Research and Development Corporation, Canberra, Australia.

Nychas, G-J. E., Panagou, E. Z., Parker, M. L., Waldron, K. W., and Tassou, C. C. (2002). Microbial colonization of naturally black olives during fermentation and associated biochemical activities in the cover brine. *Letters in Applied Microbiology 34*, 173–177.

Öngen, G., Sargin, S., Tetik, D., and Köse, T. (2005). Hot air drying of green table olives. *Food Technology and Biotechnology 43(2)*, 181–187.

O'Sullivan, G. (2003). Olive variety assessment for summer rainfall regions. RIRDC Publication No. 03/021 RIRDC, Project No. OAP–1A. Rural Industries Research and Development Corporation, Canberra, Australia.

Palomo, M. J., Moreno, F., Férnandez, J. E., Diaz-Espejo, A., and Giron, I. F. (2002). Determining water consumption in olive orchards using the water balance approach. *Agricultural Water Management 55(1)*, 15–35.

Panagou, E. Z., Katsaboxakis, C. Z., and Nychas, G-J. E. (2002). Heat resistance of *Monascus ruber* ascospores isolated from thermally processed green olives of the Conservolea variety. *International Journal of Food Microbiology 76*, 11–18.

Panagou, E. Z., Tassou, C. C., and Katsaboxakis, K. Z. (2002). Microbiological, physicochemical and organoleptic changes in dry-salted olives of Thassos variety stored under different modified atmospheres at 4 and 20°C. *International Journal of Food Science and Technology 37*, 635–641.

Papoff, C. M., Agabbio, M., Vodret, A., and Farris, G. A. (1996). Influence of some biotechnological combinations on the sensory quality of "Manna" green table olives. *Industrie Alimentari 35(347)*, 375–381.

Passioura, J. B. (1996). Drought and drought tolerance. In 'Drought tolerance in higher plants: genetical, physiological and molecular biological analysis'. (Ed. E. Belhassen.) pp. 1–5. (Kluwer Academic Publishers: Dordrecht, Netherlands.)

Pastor, M., Castro, J., and Hidalgo, J. (2002). Correction of iron chlorosis in olive. *Olivae 90*, 42–45.

Patumi, M., Fontanazza, G., Baldoni, L., and Brambilla, I. (1990). Determination of some precursors of lipid biosynthesis in olive fruits during ripening. In 'Proceedings of the International Olive Growing Symposium'. (Eds L. Rallo, J. M. Caballero, and R. Fernández-Escobar.) *Acta Horticulturae 286*, 199–202.

Patumi, M., d'Andria, R., Marsilio, V., Fontanazza, G., Morelli, G., and Lanza, B. (2002). Olive and olive oil quality after intensive monocone olive growing (*Olea europaea* L., cv. Kalamata) in different irrigation regimes. *Food Chemistry 77*, 27–34.

Perica, S., Bellaloui, N., Greve, C., Hu, H., and Brown, P. H. (2001). Boron transport and soluble carbohydrate concentrations in olive. *Journal of the American Society for Horticultural Science 126(3)*, 291–296.

Perica, S., Brown, P. H., Connell, J. H., and Hu, H. N. (2002). Olive response to foliar boron application. In 'Proceedings of the 4th International Symposium on Olive Growing'. (Eds C. Vitagliano and G. P. Martelli.) *Acta Horticulturae 586(1)*, 381–383.

Piga, A., Gambella, F., Vacca, V., and Agabbio, M. (2001). Response of three Sardinian olive cultivars to Greek-style processing. *Italian Journal of Food Science 13(1)*, 29–40.

Pinney, K., and Polito, V. S. (1990). Flower initiation in 'Manzanillo'. In 'Proceedings of the International Olive Growing Symposium'. (Eds L. Rallo, J. M. Caballero, and R. Fernández-Escobar.) *Acta Horticulturae 286*, 203–205.

Portman, T., Frankish, E., and McAlpine, G. (2002). Guidelines for the management of microbial food safety in fruit packing houses. Bulletin 4567, Department of Agriculture, Western Australia.

Preziosi, P., and Tini, M. (1990). Preliminary observations of some maturity parameters of drupes on 39 Italian olive cultivars. In 'International Symposium on Olive Growing'. (Eds L. Rallo, J. M. Caballero and R. Fernández-Escobar.) *Acta Horticulturae 286*, 85–88.

Proietti, P., and Antognozzi, E. (1996). Effect of irrigation on fruit quality of table olives (*Olea europaea*), cultivar 'Ascolana tenera'. *New Zealand Journal of Crop and Horticultural Science 24(2)*, 175–181.

Proietti, P., Famiani, F., and Tombesi, A. (1999). Gas exchange in olive fruit. *Photosynthetica 36(3)*, 423–432.

Proietti, P., Palliotti, A., and Nottiani, G. (1999). Availability of assimilates and development of olive fruit. In 'Proceedings of the 3rd International Symposium on Olive Growing'. (Eds I. T. Metzidakis and D. G. Voyiatzis.) *Acta Horticulturae 474(1)*, 297–300.

Proietti, P., Tombesi, A., and Boco, M. (1994). Influence of leaf shading and defoliation on oil synthesis and growth of olive fruit. In 'Proceedings of the 2nd International Olive Growing Symposium'. (Eds S. Lavee and I. Klein.) *Acta Horticulturae 356*, 272–277.

Psilakis, N., and Psilakis, M. (2000). 'The Secret of Good Health – Olive Oil.' (N. Psilakis: Heraklion Crete, Greece.)

Rallo, L., Tōrreno, P., Vargas, A., and Alvarado, J. (1994). Dormancy and alternate bearing in olive. In 'Proceedings of the 2nd International Olive Growing Symposium'. (Eds S. Lavee and I. Klein.) *Acta Horticulturae 356*, 127–136.

Rallo, P., and Rapoport, H. F. (2001). Early growth and development of the olive fruit mesocarp. *Journal of Horticultural Science and Biotechnology 76(4)*, 408–412.

Rama, P., and Pontikis, C. A. (1990). In vitro propagation of olive (*Olea europaea sativa* L.). *Journal of Horticultural Science 65*, 347–353.

Ramli, U. S., Baker, D. S., Quant, P. A., and Harwood, J. L. (2002). Control mechanisms operating for lipid biosynthesis differ in oil-palm (*Elaies guineensis* Jacq.) and olive (*Olea europaea* L.) callus cultures. *Biochemical Journal 364*, 385–391.

Ranalli, A., Tombesi, A., Ferrante, M. L., and De Mattia, G. (1998). Respiratory rate of olive drupes during their ripening cycle and quality of oil extracted. *Journal of the Science of Food and Agriculture 77*, 359–367.

Ravetti, L. M., Matías, A. C., Patumi, M., Rocchi, P., and Fontanazza, G. (2002). Characterization of virgin olive oils from Catamarca and La Rioja, Argentina. General

Characteristics. In 'Proceedings of the 4th International Symposium on Olive Growing'. (Eds C. Vitagliano and G. P. Martelli.) *Acta Horticulturae 586(2)*, 603–606.

Rivas, F. J., Beltrán, F. J., and Gimeno, O. (2000). Joint treatment of wastewater from table olive processing and urban wastewater. Integrated ozonation – aerobic oxidation. *Chemical Engineering and Technology 23 (2)*, 177–181.

Rivas, F. J., Beltrán, F. J., Gimeno, O., and Acedo, B. (2001). Wet air oxidation of wastewater from olive oil mills. *Chemical Engineering and Technology 24(4)*, 415–421.

Rivas, F. J., Beltrán, F. J., Gimeno, O., and Alvarez, P. (2001). Chemical-biological treatment of table olive manufacturing wastewater. *Journal of Environmental Engineering 127(7)* 611–619.

Rivas, F. J., Beltrán, F. J., Gimeno, O., and Frades, J. (2001). Treatment of olive oil mill wastewater by Fenton's Reagent. *Journal of Agricultural and Food Chemistry 49(4)*, 1873–1880.

Rivas, F. J., Beltrán, F. J., Alvarez, P., Frades, J., and Gimeno, O. (2000). Joint aerobic biodegradation of wastewater from table olive manufacturing industries and urban wastewater. *Bioprocess and Biosystems Engineering 23*, 283–286.

Rivas, F. J., Gimeno, O., Portela, J. R., de la Ossa, E. M., and Beltrán, F. J. (2001). Supercritical water oxidation of olive oil mill wastewater. *Industrial and Engineering Chemistry Research 40(16)*, 3670–3674.

Rivas, J., Beltrán, F., Acedo, B., and Gimeno, O. (2000). Two-step wastewater treatment: sequential ozonation – aerobic biodegradation. *Ozone Science and Engineering 22*, 617–636.

Roca, M., and Mínguez-Mosquera, M. I. (2001). Changes in chloroplast pigments of olive varieties during fruit ripening. *Journal of Agricultural and Food Chemistry 49*, 832–839.

Romero, A., and Díaz, I. (2002). Optimal harvesting period for 'Arbequina' olive cultivar in Catalonia (Spain). In 'Proceedings of the 4th International Symposium on Olive Growing'. (Eds C. Vitagliano and G. P. Martelli.) *Acta Horticulturae 586(1)*, 393–396.

Romero, C., Brenes, M., Garcia, P., and Garrido, A. (1996). Respiration of olives stored in sterile water. *Journal of Horticultural Science 71(5)*, 739–745.

Romero, C., Brenes, M., García, P., and Garrido, A. (2002). Hydroxytyrosol 4-ß-D-Glucoside, an important phenolic compound in olive fruits and derived products. *Journal of Agricultural and Food Chemistry 50*, 3835–3839.

Romero, C., Garcia, P., Brenes, M., and Garrido, A. (1998). Use of manganese in ripe olive processing. *Zeitschrift fur Lebensmitteluntersuchung und-Forschung A 206(4)*, 297–302.

Romero, C., Garcia, P., Brenes, M., and Garrido, A. (2001). Colour improvement in ripe olive processing by manganese cations: industrial performance. *Journal of Food Engineering 48(1)*, 75–81.

Rosado, R., del Campillo, M. C., Martinez, M. A., Barron, V., and Torrent, J. (2002). Long-term effectiveness of vivianite in reducing iron chlorosis in olive trees. *Plant and Soil 241(1)*, 139–144.

Rotunno, T., Di Caterina, R., and Argenti, L. (1997). Decay of fenthion in green table olives. *Journal of Agricultural and Food Chemistry 45(10)*, 3957–3960.

Rugini, E., and Lavee, S. (1992). In 'Olive in Biotechnology in Agriculture No. 8: Biotechnology of perennial fruit crops'. (Eds F. A. Hammerschlag and R. E. Litz.) pp. 371–382. (CAB International Wallingford: Oxon, UK.)

Ryan, D., Robards, K., and Lavee, S. (1999). Changes in phenolic content of olive during maturation. *International Journal of Food Science and Technology 34(3)*, 265–274.

Rymon, D., and Lavee, S. (1990). Extension of the marketed product line of table olives by marketing a new low fat variety. In 'Proceedings of the International Olive Growing Symposium'. (Eds L. Rallo, J. M. Caballero, and R. Fernández-Escobar.) *Acta Horticulturae 286*, 481–484.

Sadeghi, H., and Talaii, A. R. (2002). Impact of environmental conditions on fatty acids combination of olive oil in an Iranian olive, cv. Zard. In 'Proceedings of the 4th International Symposium on Olive Growing'. (Eds C. Vitagliano and G. P. Martelli.) *Acta Horticulturae 586(2)*, 579–581.

Saija, A., and Uccella, N. (2001). Olive biophenols: functional effects on human wellbeing. *Trends in Food Science and Technology 11*, 357–363.

Salazar, D. M. (1990). Pomological typification of olive trees in Valencia. In 'Proceedings of the International Symposium on Olive Growing'. (Eds L. Rallo, J. M. Caballero, and R. Fernández-Escobar.) *Acta Horticulturae 286*, 101–104.

Sánchez, A. H., De Castro, A., Rejano, L., and Montano, A. (2000). A comparative study on chemical changes in olive juice and brine during green olive fermentation. *Journal of Agricultural and Food Chemistry 48*, 5975–5980.

Sánchez, J. C., Iraola, V. M., Sastre, J., Florido, F., Boluda, L., and Fernández-Caldas, E. (2002). Allergenicity and immunochemical characterisation of six varieties of *Olea europaea*. *Allergy 57(4)*, 313–318.

Sánchez Romero, C., Guillén, R., Heredia, A., Jimenez, A., and Fernández Bolãnos, J. (1998). Degradation of hemicellulosic and cellulosic polysaccharides in pickled green olives. *Journal of Food Protection 61(1)*, 89–93.

Sánchez Romero, C., Guillén, R., Heredia, A., Jimenez, A., and Fernández Bolãnos, J. (1998). Degradation of pectic polysaccharides in pickled green olives. *Journal of Food Protection 61(1)*, 78–86.

Sciancalepore, V., and De Stefano, G. (1995). Fractionation of phenolase from green table olives (Ascolana tenera var.) by immobilised copper affinity chromatography. *Grasas y Aceites 46(4–5)*, 251–254.

Servili, M., Baldioli, M., Mariotti, F., and Montedoro, G. F. (1999). Phenolic composition of olive fruit and virgin olive oil: distribution in the constitutive parts of fruit and evolution during the oil mechanical extraction process. In 'Proceedings of the 3rd International ISHS Symposium on Olive Growing'. (Eds I. T. Metzidakis and D. G. Voyiatzis.) *Acta Horticulturae 474(2)*, 609–613.

Sibbett, G.S., Ferguson, L., Coviello, J., and Lindstrand, M.L. (2005). 'Olive Production Manual. 2nd edition.' University of California Division of Agricultural and Natural Resources Publication No. 3353. California, USA.

Siles, F. J. S. (1999). New technologies in table olive processing (Spanish). *Grasas y Aceites 50(2)*, 131–140.

Sillari, B., and Cantini, C. (2003). Phased coppicing of olive: results of twenty years' experimentation. *Olivae 98*, 36–43.

Silva, S., Gomes, M. L., Vilas-Boas, L., Catulo, L., and Peres, C. (2002). Analytical methods for brines and fruits of *Olea europaea galega cv*. In 'Proceedings of the 4th

International Symposium on Olive Growing'. (Eds C. Vitagliano and G. P. Martelli.) *Acta Horticulturae 586(1)*, 129–132.

Smyth, J. (2002). Perspectives of the Australian olive industry. *Advances in Horticultural Science 16*, 280–288.

Smyth, J. D. (2005). Packaging of olive oil and table olives. *Australian and New Zealand Olivegrower and Processor 43* (May–Jun.), 11–16.

Sobel, J., Tucker, N., Sulka, A., McLaughlin, J., and Maslanka, S. (2004). Food botulism in the United States, 1990–2000. Emerging Infectious Diseases. Available at http://www.cdc.gov/ncidod/EID/vol10no9/03–0745. htm [Verified 9 August 2006.]

Solangaarachchi, S. M., and Gould, K. S. (2001). Anthocyanin pigmentation in the adventitious roots of *Metrosideros excelsa* (Myrtaceae). *New Zealand Journal of Botany 39*, 161–166.

Soler-Rivas, C., Espin, J. C., and Wichers, H. J. (2000). Oleuropein and related compounds. *Journal of the Science of Food and Agriculture 80(7)*, 1013–1023.

Souza, S., Gomes, L., Catulo, L., Leitão, F., and Vilas-Boas, L. (1999). Composition of brines from olive fermentations with specific lactic starters. In 'Proceedings of the 3rd International Symposium on Olive Growing'. (Eds I. T. Metzidakis and D. G. Voyiatzis.) *Acta Horticulturae 474(2)*, 729–733.

Spanedda, A. F., and Terrosi, A. (2002). Toxic residue patterns in olive fruit, oil, and waste water of the most common insecticides used for controlling olive fly in Central Italy. In 'Proceedings of the 4th International Symposium on Olive Growing'. (Eds C. Vitagliano and G. P. Martelli.) *Acta Horticulturae 586(2)*, 853–856.

Spennemann, D. H. R. (1999). 'Centenary of Olive Processing at Charles Sturt University.' (Charles Sturt University: Wagga Wagga, New South Wales.)

Spenneman, D. H. R., and Allen, L. R. (2000). From cultivar to weed: the spread of olives in Australia. *Olivae 82*, 44–46.

Spooner-Hart, R. (2005). Sustainable pest and disease management in Australian olive production. RIRDC Publication No. 05/080, RIRDC Project No. UWS–17A. Rural Industries Development Corporation, Canberra, Australia.

Spyropoulou, K. E., Chorianopoulos, N. G., Skandamis, P. N., and Nychas, G-J. E. (2001). Survival of *Escherichia coli* O157:H7 during the fermentation of Spanish-style green table olives (Conservolea variety) supplemented with different carbon sources. *International Journal of Food Microbiology 66*, 3–11.

Steiner, I., Fischer, M., and Washutt, J. (1993). Analysis of edible oils with special consideration of oils containing polyunsaturated fatty acids. *Fett Wissenschaft Technologie-Fat Science Technology 95(12)*, 461–472.

Strack, D. (1997). Phenolic metabolism. In 'Plant Biochemistry. Chapter 10'. pp. 387–416. (Academic Press: San Diego, USA.)

Sweeney, S. (2003). NOVA – The National Olive Variety Assessment Project. RIRDC, Publication No. 03/054, RIRDC Project No. SAR 23A. Rural Industries Research and Development Corporation, Canberra, Australia.

Sweeney, S. (2005). National Olive Variety Assessment – (NOVA) – Stage 2. RIRDC Publication No. 05/155, RIRDC Project No. SAR–47A. Rural Industries Research and Development Corporation, Canberra, Australia.

Sweeney, S., and Davies, G. (1998). The olive industry. In 'A Handbook for Farmers and Investors'. RIRDC Publication No. 98/034. (Ed. K. Hyde.) pp. 405–411. Rural Industries Research and Development Corporation, Canberra, Australia.

Talhinhas, P., Sreenivasaprasad, S., Neves-Martins, J., and Oliveira, H. (2005). Molecular and phenotypic analysis reveal association of diverse *Colletotrichum acutatum* groups and a low level of *C. gloeosporoides* with olive anthracnose. *Applied and Environmental Microbiology 71(6)*, 2987–2998.

Tava, A. (1998). Wax ester components of olive fruit (*Olea Europaea* L.). *Industrie Alimentari 37*, 28–32.

Thomson, T. (1998). Olive irrigation. *The Olive Press, Summer Edition*, 11–12.

Tombesi, A. (1994). Olive fruit growth and metabolism. In 'Proceedings of the 2nd International Symposium on Olive Growing'. (Eds S. Lavee and I. Klein.) *Acta Horticulturae 356*, 225–232.

Tombesi, A., Boco, M., Pilli, M. (1999). Influence of light exposure on olive fruit growth and composition. In 'Proceedings of the 3rd International Symposium on Olive Growing'. (Eds I. T. Metzidakis and D. G. Voyiatzis.) *Acta Horticulturae 474(1)*, 255–259.

Tombesi, A., Pilli, M., Boco, M., and Proietti, P. (1994). Evolution of olive fruit respiration, photosynthesis and oil composition during ripening. In 'Proceedings of the 2nd International Symposium on Olive Growing'. (Eds S. Lavee and I. Klein.) *Acta Horticulturae 356*, 278–283.

Tous, J. (1995). The olive in Australia. *Olivae 55*, 10–15.

Tous, J., Lloveras, J., and Romero, A. (1995). Effect of Ethephon spray treatments on mechanical harvesting and oil composition of Arbequina olives. *Journal of the American Society for Horticultural Science 120(4)*, 558–561.

Tous, J., Romero, A., and Barranco, D. (1990). Olive cultivars in Catalonia (Spain). In 'Proceedings of the International Symposium on Olive Growing'. (Eds L. Rallo, J. M. Caballero and R. Fernández-Escobar.) *Acta Horticulturae 286*, 129–132.

Tous, J., Romero, A., and Díaz, I. (1999). Fruit and oil characteristics of five Spanish olive cultivars. In 'Proceedings of the 3rd International Symposium on Olive Growing'. (Eds I. T. Metzidakis and D. G. Voyiatzis.) *Acta Horticulturae 474(2)*, 639–642.

Tous, J., Romero, A., Plana, J., and Hermoso, J. F. (2002). Behaviour of ten Mediterranean olive cultivars in the northeast of Spain. In 'Proceedings of the 4th International Symposium on Olive Growing'. (Eds C. Vitagliano and G. P. Martelli.) *Acta Horticulturae 586(1)*, 113–116.

Troncoso, A. (2000). Fertigation of the olive tree. *Australian Olive Grower 17*, 28–32.

Tukey, H. B. (1979). Structure and physiology of a fruit tree and how it grows fruit. In 'Dwarfed Fruit Trees'. pp. 57–74. (Comstock Publishing Associates, Cornell University Press: Ithica and London.)

Turantas, F., Göksungur, Y., Dinçer, H., Ünlütürk, A., Güvenç, U., and Zorlu, N. (1999). Effect of potassium sorbate and sodium benzoate on microbial population and fermentation of black olives. *Journal of the Science of Food and Agriculture 79(9)*, 1197–1202.

Tzia, C., Oreopoulou, V., Kallisperi, M., Liadakis, G. N., and Melanitis, A. (1999). Quality assurance and HACCP of olive oil. In 'Proceedings of the 3rd International

Symposium on Olive Growing'. (Eds I. T. Metzidakis and D. G. Voyiatzis.) *Acta Horticulturae 474(2)*, 667–670.

Uccella, N. (2001). Olive biophenols: novel ethnic and technological approach. *Trends in Food Science and Technology 11*, 328–339.

Uccella, N. A. (2002). The secoiridoid biophenols of *Olea europea* L. drupes and the role of their metabolites. In 'Proceedings of the 4th International Symposium on Olive Growing'. (Eds C. Vitagliano and G. P. Martelli.) *Acta Horticulturae 586(2)*, 489–492.

Uceda, M., Hermoso, M., García-Ortiz, A., Jimenez, A., and Beltrán, G. (1999). Intraspecific variation of oil contents and the characteristics of oils in olive cultivars. In 'Proceedings of the 3rd International Symposium on Olive Growing'. (Eds I. T. Metzidakis and D. G. Voyiatzis.) *Acta Horticulturae 474(2)*, 659–662.

United States Food and Drug Administration. (2005). Harry and David recalls black and Kalamata olive tapenade because of possible health risks. FDA, Nov 16 USA. Available at http://www.fda.gov/oc/po/firmrecalls/harrydavid11_05.html [Verified 9 August 2006.]

Valyasevi, R., and Rolle, R. S. (2002). An overview of small-scale food fermentation technologies in developing countries with special reference to Thailand: scope for their improvement. *International Journal of Food Microbiology 75*, 231–239.

Van Sumere, C. F. (1989). Phenols and Phenolic Acids. In 'Methods in Plant Biochemistry Vol 1'. (Eds P. M. Dey and J. B. Harborne.) pp. 29–73. (Academic Press: London.)

Vavoulidou, E., Avramides, E. J., Papadopoulos, P., Dimirkou, A., Charoulis, A., and Konstanidinou-Doltsinis, S. (2005). Copper content in agricultural soils related to cropping systems in different regions of Greece. *Communications in Soil Science and Plant Analysis 36 (4–6)*, 759–773.

Vege-Start™ Product information. (Chr. Hansen Laboratories: Victoria, Australia.)

Wahlqvist, M. L., and Kouris-Blazos, A. (2001). Health benefits of the Mediterranean diet. In 'Proceedings of the Symposium: The 3 Pillars of Mediterranean Cuisine'. (Ed. S. G. Kailis.) pp. 7–19. (New Norcia, Western Australia.), University of Western Australia, Crawley, Western Australia.

Wiesman, Z., and Lavee, S. (1994). Vegetative growth retardation, improved rooting and viability of olive cuttings in response to application of growth retardants. *Plant Growth Regulation 14(1)*, 83–90.

Wiesman, Z., and Lavee, S. (1995). Enhancement of IBA stimulatory effect on rooting of olive cultivar stem cuttings. *Scientia Horticulturae 62(3)*, 189–198.

Wilkinson, K. (2004). Compost: a versatile soil management tool. *Australian and New Zealand Olivegrower and Processor 38* (Jul.–Aug.), 29–32.

World Fertilizer Use Manual. http://www.fertilizer.org.ifa/publicat/htm/pubman/olive.htm [Verified 9 August 2006].

Zimbalatti, G., and Giametta, F. (2002). Selection and calibration of eating olives technical and economic aspects. In 'Proceedings of the 4th International Symposium on Olive Growing'. (Eds C. Vitagliano and G. P. Martelli.) *Acta Horticulturae 586(2)*, 679–682.

Index

abiotic factors 54
abnormal colour 250
abnormal texture 250
acidity, during processing 181–84, 186, 201
additives 284–86, 289
aerobic fermentation 183, 202
aflatoxins 294
allergens 282–83, 289
alternate bearing 36–37
amino acids 48
anaerobic fermentation 183–85, 199–204
annual olive events 25–26
anthocyanins 48–49
anthracnose 118–19
antipasto 234
apical buds 27–28
Arbequina 54, 80–81
Ascolana Tenera 55
Aspergillus species 179, 294
Australian olive industry 1–16
axillary buds 27–28
Azapa 55

Bacillus cereus 290, 294
Bacillus species 294
bacteria 97, 177–78
   lactic acid 175, 177–78, 180–81, 212, 261
bacterial contaminants 289–92

bark 25
Barnea 56
Barouni 56
bearing 36–37, 53, 54
biennial bearing 36–37
biotic factors 54
black olives 219–22
black scale 114, 120
black-ripe olives 123, 198–202
blemished fruit 127, 249
boron 102–103
branches 24
brine, evaluation 250, 255–56, 260, 262–63
   pH measurement 253–55
   processing 198–206
brine-cured olives 198–206
broken fruit 249
broken olives 166
bruised olives 165, 205–206
bud differentiation, reproductive 30–31
budding, onto rootstock 20
buds 27–28
bulk table olive products 188
butyric spoilage 277–78

calcium 101–102
calcium deficiency 101–102
Californian Red Scale 115

Californian-style black olives 219–22
carbohydrates 45–46, 258–59
Castelvetrano-style olives 218–19
Chalchidikis 57, 81
chemical contaminants 288–89
chemical quality criteria 250–58
chemical residues 122, 289
chemical thinning of crops 110
chilling 30
chilling hours 72
chlorosis 104
chocolate-coated dried olives 226
chopped olives 166
chromatography, high pressure liquid 251–52
cleaning of processing equipment 154–57
climatic considerations 70–73
Clinistix test strips 251
Clinitest™ tablets 251
clonal cuttings 18–19
Clostridium botulinum 188, 222, 262, 292, 293
Clostridium perfringens 263, 293
clothing 156–57
Coagulase Positive *Staphylococcus aureus* 292
colour, olive 248
colour changes 127–28, 275
Conservolea 57
consumer size products 188
consumption 2, 8, 13
contaminants 179–80, 288–92
copper 103
cracked olives 165, 205–206
critical water requirements 26
crop thinning, by chemicals 110
crop yield 69
Cucco 57–58
cultural preferences 7
cultures, starter 175–77
curculio beetle 117
cuticle 40–41
cuttings 18–19
cyanosis 275–76

darkening 275
debittering 3, 41, 47–48, 160, 206, 216
deep ripping 86
defects 249, 259–60
dehydrated olives 163
deterioration, olive 273–78

Diastix™ test strips 251
disease management 113–20
diseases, food-borne 286–94
divided olives 166
drainage 75
drained weight 187
dried olives 223–26
drought 104

embryo 34, 42
endocarp 34, 42
Enterobacteriaceae 261, 262, 291
environmental considerations 185–86
epidermis 40
equipment, in processing facilities 153–55
Escherichia coli 262, 291
ethylene 123
evaluation, of fermentation brines 260
    microbiological 167, 260–63
    of processed table olives 245–48
evapotranspiration 106–107
exocarp 40–41
exports 8, 13
extraneous material 250

fat levels 256–58
fatty acids 44, 258
feral olives 5
fermentation, 182–85, 199–204, 277, 278
    during lye treatment 211–12, 214–15
    stuck 214–15
Ferrandina-style olives 226–27
fertigation 95, 99, 100, 101
fertilisation 31–33, 34
fertiliser requirements 104–106
fertilisers 89, 91, 92–96
fertility 53
finished product specifications 169
firmness 247
flesh, 41–42
    analysis 256–58
    changes in resistance 50–51
    components in 43–50
    consistency 247–48
    microbiological analysis 262–63
Flesh:Stone ratio 39–40, 43, 247
flower drop 31
flowers 28–34

foliar sprays 31, 93, 95, 99, 100, 101
food additives 284–86, 289
food contact surfaces 155–57
food safety 287–94
food spoilage 179
food standards 244, 256, 279, 281, 287
food-borne diseases 286–94
foreign matter 288
Frantoio 58
frost prone areas 72–73
fruit, 37–40
   defects 127, 249
   flesh 41–42
   microbiological flora on 50
   ripening 35–36, 38, 53
fruit development 33–36
fruit drop 31, 32
fruit flies 113, 116–17, 120
fruit growth 33–36
fruit load 24, 35–36
fruit quality 69
fruit rot 118–19
fruit set 31–33, 34
fungicides 120–22

galazoma 275–76
gas pockets 202, 276
Good Manufacturing Practice 133–34
government requirements 70
grading, processed olives 246, 266
   raw olives 127, 133, 173–74
grafting 20
Greek-style salted olives 223
green olives, Spanish-style 209–18
green stains 276
green-ripe olives 122–23, 203, 217–18
growing regions in Australia 14–15
growth, fruit 33–36
   vegetative 26–28

halved olives 166
hand harvesting 124, 125
handling, post-harvest 126–27
   raw olives 172–73
hardness, flesh 50–51
hardwood cuttings 19
Hardy's Mammoth 64
harvesting 39, 123–26

heading cuts 112
health control 112–13
heat degree-days 72
heat-dried table olives 46, 225–26
herbicides 86
herbs 158, 229–31, 286
high pressure liquid chromatography 251–52
historical aspects of table olives 2–7
Hojiblanca 58–59
hot dry areas 73
hygiene, in processing facilities 154–57

imports 2, 8, 12–13, 192–93
inhibitors 182
insecticides 120–22
integrated pest management 120–22
international table olive trade 7–8
iron 103
irrigation 73–74, 84–85, 88–89, 106–109

Jumbo Kalamata 64, 203

Kalamata 6, 7, 59
Kalamata-style table olives 207–209

labelling olive products 279–86
lactic acid bacteria 175, 177–78, 180–81, 212, 261
lateral buds 27–28
leaf analysis 105
leaves 21–24
Leccino 60
lemons, preserved 230
life cycle of olive tree 20–21, 25–26
light, for flowering 29, 30–31
Listeria monocytogenes 293–94
loose table olives 193–94
lye processing, 209–19
   alternatives 216
   Californian/Spanish-style black olives 219–21
   fermentation during 211–12, 214–15
   without fermentation 217–18

magnesium 102
maintenance of trees 110–12
malodorous deterioration of processed olives 277–78
manganese 103

manipulation of microbial activity 179–82
manures 92–95
Manzanilla 60–61
marinades 229–31
maturation states 122–23
mechanical harvesting 124–25
mechanised pruning 111–12
medfly 116
mesocarp 35, 41–42
microbial activity, manipulation of 179–82
microbial contaminants 289–92
microbiological control 179–82, 244
microbiological evaluation 167, 260–63
microbiological flora on olive fruit 50
microorganisms 117–19, 177–79, 187, 199–201, 273–78
micropropagation 18
minced olives 166
mineral content, of processed olive flesh 259–60
   of raw olive flesh 49
mineral oil spray 122
Mission 61
moisture, 180
   content of raw olives 42
   in processed olive flesh 256–57
molybdenum 104
monitoring nutritional status 104–106
moulds 172, 179, 261, 274, 276, 285, 294
mutilated fruit 249

Nab Tamri 64
National Olive Variety Assessment 52, 54
natural history, of olive production 20–21
   of the olive tree 18–20
natural microbiological flora on olive fruit 50
natural olives 161–62
nematodes 119–20
nitrogen 95–99
nitrogen deficiency 98
nitrogen fertilisers 92, 95–96
Nocellara del Belice 61–62, 81
NOVA 52, 54
nutrients, effect in processing 181
  loss through harvesting 92
  soil 84, 90–105
nutritional labelling 281–82, 283
nutritional problems 104–105

occupational health and safety 155–56
oil accumulation 35
oil content of raw olive flesh 43–45, 53
oil cured olives 234
Olea europaea 17–66
Olea oleaster 2
oleander scale 115
oleuropein 37–39, 42, 46–48, 196, 198, 199, 200, 209, 216
Oliva di Cerignola 62
olive bud mite 115–16
olive consistency 247–48
Olive de Nimes 218
olive fruit fly 113, 116–17, 120
olive growing in Australia 14–15
olive knot 118
olive lace bug 115
olive moth 113, 120
olive oil, as packing solution 228
olive production, natural history 20–21
olive scale 114–15
olive tree 17–66
olives darkened by oxidation 164
olives turning colour 123, 161
olives with capers 166
orchards, establishment 81–89
  health 112–13
  nutrition specifics 97–102
  planning 69–76
  soil requirements 82–84
organic acids in raw olive flesh 50
organic manures 93–95
organoleptic evaluation 263–73
over-watering 107–108
oxidation, to darken olives 164
oxygen levels, in processing 180–81

packaged table olive products 194–96
packaging 187–88, 279–86
packing, black-ripe olives 201–202
  Kalamata-style table olives 208
  Spanish-style green olives 215–16
  water-cured olives 197
packing solutions 227–29
parthenocarpic fruit 33
pastes 235–37, 239, 272–73
pasteurisation 188–91
peacock spot 117–18

pest management 113–22, 136
pesticides 86, 288
pH, during processing 46, 172, 175, 176, 181–85, 188, 253–55
   soil 74, 75, 83–84, 90–92, 95–96, 102, 105, 106
phenolic substances 46–48
phenolphthalein 254
phosphate fertilisers 96
phosphorus 99–100
phosphorus deficiency 99–100
photosynthesis 22–24
physical contaminants 288
physical defects 249–50
physical quality criteria 248–49
physical testing of processed table olives 245–48
phytophthora 119
Picholine 62–63
Picholine-style table olives 218
pigments in raw olive flesh 48–49
pitted olives 166
pitting olives 234–35
planning orchards 69–76
planning processing facilities 135–36
plant maintenance 154–57
planting densities 81
planting trees 86–88
pollen 28–29, 31, 32
pollination 31–33, 53
pollinators, specific 78
post-harvest handling 126–27
post-processing operations 227–31
potassium 100–101
potassium deficiency 100–101
potassium fertilisers 96
preservation methods 186–87
preservatives 182
primary processing specifications 167
processed table olives, 191–92
   grading 246, 266
   salt levels 252–56, 260
processing, acceptance of raw olives 170
   adicity during 181–84, 186, 201
   alternatives to lye 216
   brine 198–206
   dried olives 223–26
   environmental considerations 185–86
   Ferrandina olives 226–27
   filling tanks 174–75
   general aspects 131–36, 175–77
   Kalamata-style table olives 207–209
   methods 132–33, 160–61, 167–68, 175–77, 240
   microorganisms in 177–79
   oxygen levels in 180–81
   pH during 46, 172, 175, 176, 181–85, 188, 253–55
   and raw olives 161
   salt use 158–60
   sources of inputs 158
   specifications 167–69, 227–31
   standards 131–34, 136
   temperature 181
   untreated olives in brine 203–204
   water control 180
   water requirements 134–35
   with lye 209–22
   with water 196–98
processing controls 167
processing facilities,
   cleaning of 154–57
   equipment 153–55
   hygiene in 154–57
   planning 135–36
processing protocol, generic 169–70
processing tanks 174–75
producing raw olives 67–129
product recall 157, 283–84, 292–93
production, maturation states 122–23
production planning 69
productive phases 21
productivity 53–54, 109–10
products available in Australia 191–92
propagation 18–20
protein 48, 258
Provencale olives 194
pruning 110–12
putrid spoilage 277–78

qualities of packed table olives 244
quality criteria 248–58
quality evaluation of processed olives 245–278
quartered olives 166
Queensland fruit fly 116

radiation 72, 82
rainfall 84
raw olive production 67–129

raw olives, acceptance 170
   entering processing line 172–74
   grading 127, 133, 173–74
   handling 126–27, 172–73
   moisture content 42
   oil content 43–45, 53
   organic acids 50
   pigments 48–49
   sorting 173–74
   storage 126–27, 171–72
   transporting 126
   undesirable qualities 127–28
   use in table olive production 161
   vitamins 50
   washing 173
Readily Available Water 82–83, 109
recall procedures 157, 283–84, 292–93
recipes 232–33
regulated deficit irrigation 109
reproductive development 28–31
ripening, fruit 35–36, 38, 53
ripening stages 122–23
ripping soil 86
roots 24, 54, 91
Rutherglen bug 117

safety, food 248, 286–94
   in packaging 187–88
salad olives 166
saline water irrigation 85
salmonella 291
salometer 252–53
salt levels in processed olives 252–56, 260
salt refractometer 252
salt, used in processing 158–60
salt-dried olives 223–25
salted olives 223
sanitation of processing equipment 154–57
scale insect pests 114–15, 122
sclerification 34
secondary processing specifications 168–69, 227–31
seedlings 19–20
self-rooted cuttings 18–20
Sevillana 63
shigella 291
Shot Berries 33
shrivelled fruit 249

shrivelled olives 163, 223
site requirements 74–75
size grading, of raw olives 173–74
size, processed olives 246
skin 40–41
skin blemishes 127
skin marks 275
skin spots, white 276
sliced olives 166
slit olives 165–66
sloping sites 75
sloughed skin 274
sodium chloride 158, 171, 252–56, 260
sodium hydroxide 160; *see also* lye processing
soft nose 119
soft olives 274
soil analysis 91–92, 105–106
soil management 90–92
soil nutrients 84, 90–105
soil pH 74, 75, 83–84, 90–92, 95–96, 102, 105, 106
soil requirements 82–89
soil types 74–75, 83
soil water 82–83
sorting raw olives 173–74
sources of processing inputs 158
Spanish-style black olives 219–22
Spanish-style green olives 209–18
specialty products 11, 164, 193, 195
specific pollinators 78
spices 158, 229–31, 286
split olives 165–66
spoilage 178, 273–78
standards, food 244, 256, 279, 281, 287
Staphylococcus aureus 186, 292
starter cultures 175–77
stems, plant 24
stems, post-processing 250
stone characteristics 248
stone fragments 250, 288
stoned olives 166
stones 166–67
storage of raw olives 126–27, 171–72
stress factors 112–13
stuck fermentations 214–15
stuffed olives 166, 234–35, 250
styles of table olives 160–61, 165–68
suckers 25

sugar levels in processed olive flesh 258–59
sulfur 102
sun-dried olives 46, 225–26

Taggiasca 63–64
tapenades 237–39, 272–73
temperature, and flowering 29–31
   daily 70–73
   extremely low 72–73
   in processing 181
terminal buds 27–28
testing, physical 245–48
thinning cuts 112
Throuba-style olives 223
titration measurement 254–55
trace elements 102–104
trade preparations of table olives 161–64
trade, table olive 7–11, 192–93
training young trees 89
transpiration 22
transporting raw olives 126
treated olives 162–63
tree maintenance 110–12
tree planting 86–88
tree protection 112
triacylglycerols 43
tritratable acidity 254–55
trunk 24
turning colour olives 203

UC13A6 66
UC22A11 66
UC23A13 66
UC23A9 66
UC6A7 66
undesirable qualities of raw olives 127–28
unproductive phase 20
UV light disinfection of water 135

varieties 6, 8, 14, 15, 51–64
variety selection 76–81
Vege-Start™ 176–77
vegetative growth 26–28
Verdale 66
verticillium wilt 119–20
vinegar 228–29
vitamins in raw olive flesh 50
Volhard titration 253

washing raw olives 173
waste products 185–86
wastewater 185–86
water availability 73–74, 82
water control in processing 180
Water Holding Capacity 82
water processing 196–98
water requirements, in plant growth 26, 84–85, 107
   in processing 134–35
water, UV light disinfection 135
water-curing olives 196–98
watering *see* irrigation
watering, over 107–108
watershoots 112
weed control 86
white skin spots 276
whole olives 165
wholesale trade in Australia 192–93
wild olive 2

yeasts 178–79, 276, 261
young trees 86–89

zapateria 278
zinc 103